Ethics, Machine Learning, and Python in Geospatial Analysis

Mohammad Gouse Galety
Samarkand International University of Technology, Uzbekistan

Arul Kumar Natarajan
Samarkand International University of Technology, Uzbekistan

Tesfaye Fufa Gedefa
Space Science and Geospatial Institute, Ethiopia

Tsegaye Demsis Lemma
Space Science and Geospatial Institute, Ethiopia

A volume in the Advances in Geospatial Technologies (AGT) Book Series

IGI Global
PUBLISHER of TIMELY KNOWLEDGE

Published in the United States of America by
IGI Global
Engineering Science Reference (an imprint of IGI Global)
701 E. Chocolate Avenue
Hershey PA, USA 17033
Tel: 717-533-8845
Fax: 717-533-8661
E-mail: cust@igi-global.com
Web site: http://www.igi-global.com

Library of Congress Cataloging-in-Publication Data

CIP DATA PROCESSING

Ethics, Machine Learning, and Python in Geospatial Analysis
Mohammad Gouse Galety, Arul Kumar Natarajan, Tesfaye Fufa Gedefa, Tsegaye Demsis Lemma
2024 Engineering Science Reference

ISBN: 9798369363812(hc) | ISBN: 9798369363829(sc) | eISBN: 9798369363836

This book is published in the IGI Global book series Advances in Geospatial Technologies (AGT) (ISSN: 2327-5715; eISSN: 2327-5723)

British Cataloguing in Publication Data
A Cataloguing in Publication record for this book is available from the British Library.

All work contributed to this book is new, previously-unpublished material.
The views expressed in this book are those of the authors, but not necessarily of the publisher.

For electronic access to this publication, please contact: eresources@igi-global.com.

Advances in Geospatial Technologies (AGT) Book Series

ISSN:2327-5715
EISSN:2327-5723

Editor-in-Chief: Nilanjan Dey, Techno India College of Technology, India

MISSION

The geospatial technology field is a fast-paced, high growth industry that is involved in a variety of fields, including military planning, public health, land use, environmental protection, and Google Earth mapping. With such a diverse body of applications, the research in geospatial technologies is always evolving and new theories, methodologies, tools, and applications are being developed. **Advances in Geospatial Technologies (AGT) Book Series** is a reference source and outlet for research that discusses all aspects of geographic information, including areas such as geomatics, geodesy, GIS, cartography, remote sensing, and other areas. Because geospatial technologies are so pervasive in such a variety of areas, AGT also includes books that address interdisciplinary applications of the technologies.

COVERAGE

- Environmental Monitoring
- Surveying Technologies
- Global Positioning Systems
- Geographical information systems
- Land Surveying
- Geovisualization
- Military Security and Geospatial Technologies
- Disaster Informatics
- Public Sector Use of Geospatial Technologies
- Geomatics

The Advances in Geospatial Technologies (AGT) Book Series (ISSN 2327-5715) is published by IGI Global, 701 E. Chocolate Avenue, Hershey, PA 17033-1240, USA, www.igi-global.com. This series is composed of titles available for purchase individually; each title is edited to be contextually exclusive from any other title within the series. For pricing and ordering information please visit http://www.igi-global.com/book-series/advances-geospatial-technologies/73686. Postmaster: Send all address changes to above address.

Titles in this Series

For a list of additional titles in this series, please visit:
http://www.igi-global.com/book-series/advances-geospatial-technologies/73686

Advanced Geospatial Practices in Natural Environment Resource Management
Rubeena Vohra (Bharati Vidyapeeth's College of Engineering, India) and Ashish Kumar (Bennett University, India)
Engineering Science Reference • copyright 2024 • 292pp • H/C (ISBN: 9798369313961) • US $275.00 (our price)

Emerging Trends, Techniques, and Applications in Geospatial Data Science
Loveleen Gaur (Amity University, India & Taylor's University, Malaysia & University of the South Pacific, Fiji) and P.K. Garg (Indian Institute of Technology, Roorkee, India)
Engineering Science Reference • copyright 2023 • 309pp • H/C (ISBN: 9781668473191) • US $240.00 (our price)

Addressing Environmental Challenges Through Spatial Planning
Athar Hussain (Netaji Subhas University of Technology, India) Kailash Chandra Tiwari (Delhi Technological University, India) and Alpana Gupta (Indian Institute of Technology, Delhi, India)
Engineering Science Reference • copyright 2022 • 329pp • H/C (ISBN: 9781799883319) • US $215.00 (our price)

Methods and Applications of Geospatial Technology in Sustainable Urbanism
José António Tenedório (Universidade NOVA de Lisboa, Portugal) Rossana Estanqueiro (Universidade NOVA de Lisboa, Portugal) and Cristina Delgado Henriques (Universidade de Lisboa, Portugal)
Engineering Science Reference • copyright 2021 • 684pp • H/C (ISBN: 9781799822493) • US $215.00 (our price)

Interdisciplinary Approaches to Spatial Optimization Issues
Sami Faiz (University of Tunis El Manar, Tunis, Tunisia) and Soumaya Elhosni (University of Tunis El Manar, Tunis, Tunisia)

For an entire list of titles in this series, please visit:
http://www.igi-global.com/book-series/advances-geospatial-technologies/73686

IGI Global
PUBLISHER of TIMELY KNOWLEDGE

701 East Chocolate Avenue, Hershey, PA 17033, USA
Tel: 717-533-8845 x100 • Fax: 717-533-8661
E-Mail: cust@igi-global.com • www.igi-global.com

Table of Contents

Detailed Table of Contents

Historically, statistical techniques have been a major component of geographic analysis, which is the study of geographical phenomena using geographic data. But the emergence of machine learning (ML) has completely changed the area, making it possible to analyse intricate spatial relationships and glean insightful information from enormous geographical datasets. The foundations, methods, and applications of machine learning for geographic analysis are all covered in detail in this chapter. The authors start with a brief overview of machine learning and then move on to talk about its application to geographic analysis. Subsequently, they explore particular machine learning approaches that are frequently employed in the domain, such as decision trees, support vector machines (SVMs), K-means clustering, convolutional neural networks (CNNs), geostatistics, reinforcement learning, time series analysis, and anomaly detection. The authors wrap up by outlining the possibilities of machine learning for geospatial analysis in the future and provide resources for more research.

As technology advances, the potential applications for geospatial data will only continue to grow. However, conventional techniques for evaluating geographic data frequently involve manual interpretation or rule-based strategies, which take a long time and have a limited capacity to handle big datasets. Current technology has significantly enhanced geospatial analysis by providing powerful data collection, processing, and interpretation tools. This study used machine learning to analyze geospatial data and extract insights that would be difficult or impossible to obtain using traditional methods. Literature review, various Python libraries for geospatial data, building and evaluating machine learning models for algorithms like random forest, decision tree, linear regression, and K-means clustering using freely available geospatial data were presented. Machine learning makes analyzing geospatial data more effective for deriving deep understandings and extracting insights.

Machine learning can play a critical role in geospatial analysis, providing enhanced computing efficiency, flexibility, and scalability, improved predictive capabilities, complicated problem resolution, and information extraction from big datasets. Python has emerged as the predominant language for geospatial machine learning due to its user-friendly interface, extensive library support, and versatility. This chapter has explored a diverse ecosystem of Python libraries ranging from Geopandas, Fiona, Leafmap, Geemap, PySAL, and Shapely for geospatial data manipulation to Keras Spatial, TorchGeo, Scikit-learn, and TensorFlow for deep learning applications. Complementing this, it also explored a variety of QGIS Python plugins that enhance geospatial machine learning capabilities, including smart-map, cluster analysis, PyQGIS, ClusterPoints, AI vectorizer, mapflow, deepness, and many more, offering functionalities for digital mapping, clustering, and map segmentation.

Numerous organizations regularly produce enormous volumes of geospatial data due to the widespread use of sensors and location-based services. However, traditionally collecting, storing, managing, exploring, analyzing, and visualization of geospatial data has been a complex and time-consuming task. This study proposed a big data analytics approach to collect, store, manage, explore, process, and analyze massive amounts of geospatial data. A comprehensive literature review, various Python libraries for geospatial big data, challenges in geospatial big data analytics, and big data analytics techniques such as spatial clustering, spatial regression analysis, and spatial-temporal analysis, were presented. In addition, geospatial big data analytics algorithms like K-means clustering, ordinary least squares (OLS), geographically weighted regression (GWR), Spatio-temporal clustering algorithms, Spatio-temporal regression models, and others were discussed. Finally, case studies on performing geospatial big data analytics using Pyspark were addressed.

Analysis of day-to-day weather patterns is critical and essential in daily life. Although traditional methods exist, in modern times, we have developed realistic and reliable methods to provide better insights and understanding of complex weather patterns for various surges, especially in these times of global warming. Implementing clustering and topological data analysis in this analysis has looked into a vast understanding of how regions with similar characteristics behave when weather changes occur due to heat, pressure, or wind-related phenomena. The classification model developed using Mapper analysis has produced 95.8% accuracy, concluding that weather follows a transient weather pattern due to various resources and how stagnant conditions affect transient weather patterns, causing rise in sub-clusters. Thus, fitting and interpreting newer models helps us understand weather analysis and classification.

Foreword

I am privileged to introduce this comprehensive volume, *Ethics, Machine Learning, and Python in Geospatial Analysis*, which stands at the intersection of ethics, technology, and geospatial analysis. In an era where data-driven decision-making has become paramount, this book offers invaluable insights into the ethical considerations, technical foundations, and advanced methodologies essential for navigating the complex landscape of geospatial analysis.

The chapters within this book span a diverse range of topics, each meticulously crafted to provide readers with a holistic understanding of the subject matter. From examining the ethical dimensions of GIS data privacy to mastering geospatial analysis with Python, each Chapter contributes to a nuanced field exploration.

The initial chapters (Chapter. 1, Chapter. 2, and Chapter. 3) critically examine the ethical implications of geospatial data privacy, providing readers with a framework for navigating ethical dilemmas in collaborative environments. Building upon this foundation, subsequent chapters delve into the technical intricacies of geospatial analysis using Python, offering comprehensive overviews of Python libraries, data storage and management techniques, and advanced machine learning methodologies (Chapter. 4 to Chapter. 9).

Furthermore, this book goes beyond theoretical discussions, offering practical guidance on harnessing the power of Python programming for real-world problem-solving. Chapters dedicated to machine learning applications in geospatial analysis (Chapter. 8, Chapter. 9, and Chapter. 10) equip readers with the tools and techniques necessary for enhancing spatial understanding and decision-making.

Additionally, including specialized applications, such as big data analytics and topological insights into weather dynamics (Chapter. 11 and Chapter. 12), further underscores the breadth and depth of topics covered in this volume.

For seasoned practitioners and newcomers, *Ethics, Machine Learning, and Python in Geospatial Analysis* presents a comprehensive resource, guiding readers through the dynamic realm of geospatial analysis with integrity and skill.

The authors' commitment and proficiency in assembling this significant work are worthy of praise. May this publication function as a beacon for those commencing their exploration of geospatial analysis, fostering ethical discernment and technological advancement.

E. George Dharma Prakash Raj
School of Computer Science, Engineering, and Applications, Bhrarathidasan University, Tiruchirappalli, India

Preface

In the vast realm of modern data analysis, the fusion of ethics, machine learning, and Python programming within the domain of geospatial analysis stands as a testament to the evolving landscape of technological innovation. This book, *Ethics, Machine Learning, and Python in Geospatial Analysis*, endeavors to navigate the intricate interplay between ethical considerations, advanced methodologies, and practical applications within geospatial analysis. As we embark on this intellectual journey, it is imperative to provide an overview of the subject matter, delineate its significance in the contemporary world, elucidate the intended audience, dissect the importance of each chapter submission, and culminate with a reflection on the broader implications of this work on the field of geospatial analysis.

Geospatial analysis, the process of examining and interpreting geographic data to understand patterns, relationships, and trends, has witnessed a paradigm shift in recent years with machine learning techniques and the ubiquity of Python programming. At the heart of this transformation lies the ethical imperative to safeguard individual privacy, uphold data integrity, and promote responsible data stewardship. Thus, our exploration of geospatial analysis transcends mere technical proficiency; it embodies a commitment to ethical conduct and societal well-being.

In today's interconnected world, the significance of geospatial analysis cannot be overstated. From urban planning and environmental conservation to disaster management and public health, geospatial insights inform critical decision-making processes that shape our collective future. As such, this book endeavors to equip readers with the requisite knowledge and tools to navigate the complexities of geospatial analysis ethically and effectively.

The target audience for this book is multifaceted, encompassing both novice practitioners seeking a comprehensive introduction to the field and seasoned professionals seeking to expand their skill set. Researchers, data scientists, GIS analysts, urban planners, policymakers, and educators will find value in the diverse topics covered within these pages.

Each chapter submission in this volume contributes a unique perspective to the overarching theme of ethics, machine learning, and Python in geospatial analysis.

Chapter 1: "Ethical Dimensions of GIS Data Privacy: Examining the Intersection of Ethics and Privacy" lays the groundwork for our exploration by critically examining the ethical implications of geospatial data privacy. Drawing upon real-world examples and theoretical frameworks, this chapter provides readers with a nuanced understanding of the ethical dilemmas inherent in geospatial analysis.

Chapter 2: "A Guide to Ethical Geospatial Practices: From Historical Context to Future Scenarios," builds upon the foundation laid in the preceding chapter, offering practical guidance on navigating ethical considerations in geospatial analysis. By tracing the historical evolution of ethical frameworks and envisioning future scenarios, this chapter equips readers with the tools to navigate ethical challenges confidently and ethically.

Chapter 3: "Navigating Ethical and Privacy Frontiers: Geospatial Data in Collaborative Environments," delves into the complexities of collaborative geospatial analysis, where ethical considerations intersect with practical challenges. This chapter illuminates the ethical dimensions of data sharing, collaboration, and transparency in geospatial analysis through case studies and best practices.

Chapter 4: "Geospatial Data Analysis: A Comprehensive Overview of Python Libraries and Implications," serves as a technical primer, providing readers with a comprehensive overview of Python libraries and tools essential for geospatial analysis. From data manipulation and visualization to spatial analysis and modeling, this chapter lays the groundwork for subsequent discussions on machine learning and advanced methodologies.

Chapter 5: "Python for Geospatial Data Analysis" delves deeper into the practical applications of Python programming in geospatial analysis, offering hands-on tutorials and examples. Through step-by-step demonstrations and code snippets, this chapter empowers readers to harness the power of Python for data manipulation, visualization, and spatial analysis.

Chapter 6: "Mastering Geospatial Analysis with Python: Understanding Geopandas, GDAL, Fiona, Matplotlib, Data Integration, and GIS Tools" elevates readers' technical proficiency by exploring advanced geospatial analysis techniques and tools. This chapter equips readers with the skills to tackle complex geospatial challenges, from geospatial data management and integration to spatial visualization and GIS interoperability.

Chapter 7: "Geospatial Data Storage and Management" delves into the intricacies of geospatial data storage, organization, and management. Through discussions on data formats, databases, and cloud-based solutions, this chapter provides readers with a comprehensive understanding of best practices for storing and accessing geospatial data.

Chapter 8: "Machine Learning for Geospatial Analysis: Enhancing Spatial Understanding and Decision-Making," introduces readers to the transformative potential of machine learning in geospatial analysis. By exploring supervised and unsupervised learning techniques, this chapter demonstrates how machine learning algorithms can enhance spatial understanding and inform data-driven decision-making.

Chapter 9: "From Geospatial Data to Insight: A Practical Guide to Machine Learning in Python for Real-World Problem-Solving" bridges the gap between theory and practice, offering practical guidance on implementing machine learning algorithms for real-world geospatial problems. Through case studies and hands-on exercises, this chapter empowers readers to leverage machine-learning techniques to extract actionable insights from geospatial data.

Chapter 10: "Geospatial Machine Learning Unleashing the Power of Python Programming: Libraries, Tools, Deep Learning Applications, and QGIS Plugins" explores the frontier of geospatial machine learning, delving into deep learning applications, specialized libraries, and QGIS plugins. By pushing the boundaries of traditional geospatial analysis, this chapter offers readers a glimpse into the future of the field.

Chapter 11: "Big Data Analytics for Geospatial Application using Python" tackles the challenges and opportunities presented by big data in geospatial analysis. Through discussions on data processing, scalability, and distributed computing, this chapter equips readers with the tools necessary to analyze large-scale geospatial datasets effectively.

Chapter 12: "Topological Insights into Weather Dynamics in Indian Context - An Application of Clustering and Mapper Algorithm: Topological Insights into Weather Dynamics" provides a case study on applying clustering and mapper algorithms to weather dynamics in the Indian context. This chapter highlights the interdisciplinary nature of geospatial analysis by illustrating how topological insights can enhance our understanding of complex geospatial phenomena.

In conclusion, this book represents a collaborative effort to explore the intersection of ethics, machine learning, and Python programming in geospatial analysis. By fostering a deeper understanding of ethical considerations, technical methodologies, and practical applications, this volume aims to empower readers to navigate the complexities of geospatial analysis with integrity and proficiency. We hope this book will serve as a valuable resource for researchers, practitioners, and educators alike, driving innovation and advancing the field of geospatial analysis in the years to come.

Mohammad Gouse Galety
Samarkand International University of Technology, Uzbekistan.

Arul Kumar Natarajan
Samarkand International University of Technology, Uzbekistan.

Tesfaye Fufa Gedefa
Ethiopian Space Science and Geospatial Institute (SSGI), Ethiopia.

Tsegaye Demsis Lemma
Ethiopian Space Science and Geospatial Institute (SSGI), Ethiopia.

Chapter 1
Ethical Dimensions of GIS Data Privacy:
Examining the Intersection of Ethics and Privacy

A. Vijayalakshmi
Christ University, Bengaluru, India

Sarwath Unnisa
https://orcid.org/0000-0002-2372-586X
Mount Carmel College, Bengaluru, India

Shirley Sheeba S.
Mount Carmel College, Bengaluru, India

K. Uma
Mount Carmel College, Bengaluru, India

S. S. Sunanna
Mount Carmel College, Bengaluru, India

ABSTRACT

The proliferation of mapping technologies has spurred significant attention to the ethical conduct surrounding geographic information systems (GIS). Various studies have explored these ethical considerations. Academic map and geography libraries are increasingly responsible for managing geospatial datasets, emphasizing the need for maintaining ethical standards in data archiving, cataloging, and distribution. Moral concerns about geospatial technologies encompass data accuracy, copyright, and quality assurance. Users of geospatial data must discern between ground-truth data and voluntary contributions, particularly with the rise of social media and participatory GIS. This chapter discusses the concerns concerning GIS data and the important data privacy issues while using GIS data.

DOI: 10.4018/979-8-3693-6381-2.ch001

INTRODUCTION

In the era of rapid advancements in geospatial technologies and the widespread integration of location-based services, the ethical implications of handling geographical information have become paramount. Geographic Information Systems (GIS), Global Positioning Systems (GPS), and other location-aware technologies have become omnipresent, posing a delicate balance between the benefits of enhanced services and potential threats to individual privacy. This exploration delves into the ethical considerations surrounding geospatial data practices, aiming to contribute to the ongoing discourse and provide insights into developing frameworks for the responsible use of location-based data (Antenucci et al., 1991).

Geographic Information Systems (GIS) have evolved into indispensable tools across various fields, including urban planning, environmental management, public health, and disaster response. GIS facilitates spatial data integration, analysis, and visualization, offering decision-makers a powerful means to understand complex relationships within geographic contexts. Its significance lies in providing a spatial perspective that enhances data-driven decision-making, offering insights into patterns, trends, and interconnections crucial for addressing global challenges. As the world becomes increasingly interconnected and data-driven, GIS remains pivotal in spatial analysis and informed decision-making across diverse sectors (chang, 2008).

The widespread application of GIS in both public and corporate sectors raises fundamental concerns about properly using geographical information. Geospatial data often contains sensitive details about individuals and groups, introducing unique challenges in terms of privacy protection. As GIS technology advances, the ethical considerations surrounding data privacy become more pronounced. The universal nature of GIS applications amplifies the need for a responsible and transparent approach to handling location-based data. The delicate equilibrium required to harness the potential of geospatial technologies while safeguarding the rights and privacy of individuals within a rapidly evolving digital landscape demands careful examination (De Smith, 2007).

The benefits of enhanced services through geospatial technologies must be carefully balanced with the potential threats to individual privacy. While GIS aids in efficient resource allocation, infrastructure planning, and addressing global challenges, it raises concerns about unauthorized access, surveillance, and the misuse of sensitive information. Striking a balance requires the development of robust ethical frameworks that consider the interests of both technology innovators and the individuals whose data is being processed. This balance is crucial for fostering responsible and transparent practices in the geospatial domain (Goodchild, 2009).

Developing frameworks for responsible use is essential to address the ethical challenges in geospatial data practices. These frameworks should encompass principles of informed consent, data anonymization, and stringent security measures to protect individuals' privacy. Moreover, transparency in data collection, processing, and sharing practices should be prioritized. Collaborative efforts between governments, industry stakeholders, and academia are crucial for establishing standards that ensure the ethical use of geospatial data in the modern era (Goodchild, 1996).

In conclusion, the ethical considerations surrounding geospatial data practices are integral to fostering a responsible and transparent approach. Addressing these ethical challenges as GIS and location-based technologies, advance becomes imperative for balancing innovation and privacy in our interconnected and data-driven world.

Significance of GIS Data

Geographic Information Systems (GIS) data holds immense significance in contemporary applications across diverse fields. GIS integrates spatial and attribute data, creating comprehensive maps and models that facilitate decision-making. Some of the areas are as follows:

- **Urban Planning:** GIS aids in designing efficient and sustainable cities by analyzing population density, transportation networks, land use patterns, and environmental factors.
- **Environmental Management:** GIS helps monitor environmental resources, assess pollution levels, and track deforestation, enabling effective conservation and sustainability efforts.
- **Disaster Response:** GIS facilitates disaster preparedness, response, and recovery by real-time mapping affected areas, identifying vulnerable populations, and coordinating relief efforts.
- **Public Health:** GIS monitors infectious diseases, identifies risk factors, and targets interventions for improved public health outcomes.

Ethical Dimensions in GIS

Ethical considerations in Geographic Information Systems (GIS) are of utmost importance due to the potential impact on individuals, communities, and society.

- **Privacy Concerns:** Privacy concerns in GIS arise from collecting and storing location-based data, which may include sensitive information about individuals and communities. If not handled responsibly, constantly tracking geographical information can infringe upon individuals' privacy rights. This involves the risk of unauthorized access to personal data and the potential to disclose sensitive details unintentionally. Striking a balance between the benefits of GIS data in various applications and the protection of individual privacy is essential. Ethical considerations dictate the need for transparent data collection practices, informed consent, and robust security measures to safeguard the privacy of those contributing to GIS databases.

- **Data Ownership and Control:** Determining ownership and control of GIS data is a fundamental ethical consideration. This involves establishing clear guidelines regarding who owns the collected information, how it can be used, and who controls its dissemination. Ethical data management practices require defining and respecting the rights of individuals or communities whose information is being collected. This includes intellectual property considerations, especially when individuals or organizations contribute data. Ethical guidelines ensure that GIS data is used responsibly, preventing unauthorized access and fostering a sense of trust among data contributors.
- **Potential Misuse of Information:** GIS data has the potential for misuse, raising ethical concerns about discriminatory practices, unwarranted surveillance, or malicious manipulation. The spatial information collected may inadvertently contribute to biased decision-making or reinforce existing inequalities. To address these concerns, ethical guidelines are crucial in establishing boundaries on the use of GIS data. These guidelines should emphasize non-discriminatory practices, responsible data sharing, and mechanisms to prevent the malicious exploitation of geographical information. By implementing ethical safeguards, the GIS community can mitigate the risks associated with the misuse of information and contribute to the responsible and equitable application of spatial data.
- **Data collection:** Data collection and ownership are pivotal components in information management, particularly in the context of Geographic Information Systems (GIS). Data collection involves systematically gathering information, often spatial or geographic, through various methods such as surveys, remote sensing, or sensor technologies.

DATA COLLECTION

Data Collection Methods

GIS data can be collected through various methods, each with ethical considerations.

- **Remote Sensing:** Remote sensing, employing satellite imagery and aerial photography, is a potent means of acquiring spatial data with diverse applications. While offering invaluable insights, it brings forth significant privacy concerns. The detailed nature of these images raises the potential for privacy infringement if used without proper consent or regulatory frameworks. To address these ethical challenges, it is crucial to establish clear consent mechanisms, ensuring that individuals are informed about the data collection and its purpose. Regulatory frameworks must be in place to govern the responsible use of remote sensing data, striking a balance between the benefits of spatial information and the protection of individual privacy.

- **Sensors and GPS:** Integrating sensors in devices and GPS technology has revolutionized data collection by providing real-time location-based information. However, this technological advancement brings ethical considerations, particularly concerning user privacy. Transparent data collection practices and clear privacy policies are imperative. Users should be informed about the types of data collected and their intended use, and they should have control over collecting their location information. Ethical use of sensors and GPS involves balancing functional benefits and respecting individuals' rights to privacy.
- **Surveys and Interviews:** Traditional data collection methods, such as surveys and interviews, are foundational in gathering qualitative and quantitative information. Ethical considerations are paramount in these processes. Informed consent is a key principle, ensuring participants understand the purpose of data collection and any potential risks. Protecting confidentiality is equally critical, with researchers employing measures to anonymize and secure the collected data. Adherence to ethical research protocols, including obtaining approval from review boards, is vital to maintaining the data collection process's integrity and upholding research participants' rights.

Data Ownership and Control

Determining ownership and control of GIS data is essential to ensure ethical data management and prevent unauthorized access or misuse.

- **Data Ownership**: Data ownership in GIS refers to identifying the entity or individual with the rights to a particular set of geographic information. This is a critical consideration as it determines who has the authority to use, share, and modify the data. Establishing clear ownership is essential for ethical GIS practices, guiding responsible data stewardship, and preventing unauthorized or inappropriate use. Defining data ownership helps avoid conflicts and ensures stakeholders understand their rights and responsibilities concerning the GIS datasets.

- **Data Control:** Data control involves establishing mechanisms and protocols governing how GIS data is accessed, used, and shared. This is crucial for maintaining ethical standards in data management. Implementing effective data control mechanisms ensures that access to sensitive geographic information is restricted appropriately, preventing unauthorized use or distribution. Ethical data control practices also include safeguarding against potential misuse, ensuring data integrity, and promoting responsible and transparent handling of GIS datasets.
- **Data Sharing Agreements:** In the collaborative landscape of GIS, where data is often shared between organizations, establishing clear agreements is vital for ethical data practices. These agreements outline the terms and conditions under which data is shared, specifying usage rights, confidentiality provisions, and any data modification or redistribution limitations. Ethical data-sharing agreements help build trust among organizations, promote responsible data sharing, and ensure that GIS data is utilized to align with all parties' interests and ethical standards.

Consent in GIS Data Collection

Consent in GIS data collection is a fundamental ethical principle that emphasizes obtaining permission from individuals or entities before gathering, using, or storing their geographic information (Huang et al., 2018). In the context of GIS, which often involves collecting location-based data, securing informed consent is crucial to respect the privacy rights of those contributing to the datasets. This process entails providing clear and comprehensible information about the purpose of data collection, how the data will be used, and any potential risks or implications (Ghinita, 2013).

Transparent communication is essential to ensure that individuals know how much their location information will be utilized. This involves detailing the specific projects, applications, or analyses for which the GIS data will be employed. Consent should be sought to allow individuals to make an informed decision voluntarily, without coercion or deception (Curry et al., 1997) (Van et al., 2018).

Ethical GIS practices require ongoing communication and transparency, especially in cases where the data might be used for various purposes over time (Chang et al., 2021). Regularly seeking updated consent ensures that individuals maintain control over their data and are aware of any changes in how it is utilized. By prioritizing consent in GIS data collection, practitioners uphold principles of autonomy, privacy, and ethical data stewardship (Coleman et al., 2009).

PRIVACY

Privacy Concerns in GIS

Location-related information can disclose personal information about people. We must take the utmost care to handle the sensitive information entrusted to us as GIS professionals. This section explores the various privacy concerns that arise in GIS.

- **Location-based Data Sensitivity:** Location data can unveil sensitive information, including home addresses, travel patterns, and daily schedules. It is crucial to keep this information safe from unwanted access (Onsrud et al., 1994).
- **Data Collection and Consent:** Obtaining informed consent from individuals is crucial when collecting location data. They should understand what data is being collected, for what purpose, and how it will be used.
- **Data Anonymization and De-identification:** Anonymizing or de-identifying location data is essential to protect privacy. This process involves removing or obfuscating personally identifiable information (PII) from the dataset.
- **Aggregate and Generalize Data:** Working with aggregated or generalized data, rather than granular individual data points, helps reduce the risk of identifying specific individuals.

The Challenge of Location Privacy

Preserving location privacy is a complex endeavor, especially in the age of pervasive digital tracking. This section delves into the challenges and considerations surrounding location privacy in GIS (Solymosi et al., 2023).

- **Unintended Data Sharing**: Users may be unaware that their devices or applications collect and share location data. This can happen automatically without explicit consent.
- **Third-party Access and Sharing:** When sharing location data with third parties, ensuring they have proper privacy and security measures in place is crucial. Contracts and agreements should outline data handling practices.
- **Re-identification Risks:** Even if location data is anonymized, it may still be possible to re-identify individuals by cross-referencing it with other publicly available information or datasets.
- **Balancing Utility and Privacy:** Striking the right balance between the utility of GIS and the privacy rights of individuals is a nuanced challenge.

Surveillance and Geospatial Data

Integrating surveillance with geospatial data amplifies the potential impact on individual privacy. This section explores how surveillance technologies and geospatial data intersect and the privacy implications (Elwood et al., 2011).

- **Location Tracking:** Surveillance systems often rely on geospatial data to track the movements and locations of individuals, vehicles, or assets. This can lead to pervasive tracking of individuals.
- **Monitoring of Public Spaces:** Public surveillance cameras, equipped with geospatial capabilities, are commonly used for traffic monitoring, crowd management, and law enforcement.
- **Data Integration and Fusion:** Combining surveillance data with geospatial information creates highly detailed profiles of individuals, increasing the risk of privacy breaches.

Risks to Individual Privacy

In the era of advanced technology, the risks to individual privacy are more pronounced than ever. This section addresses the specific risks and challenges individuals face in GIS.

- **Unintended Data Exposure:** Inadequate security measures can lead to data breaches or unauthorized access to GIS databases, putting an individual's privacy at risk.
- **Data Aggregation and Profiling:** Combining GIS data with other datasets can create detailed profiles of individuals, potentially infringing on their privacy rights.
- **Lack of User Awareness:** Individuals may not always be fully aware of how their location data is collected, used, or shared through GIS systems. Balancing the immense potential of GIS with the protection of individual privacy rights is a multifaceted challenge.

Ethical Considerations in GIS

Location-based data, at its core, is deeply personal. It can unveil patterns of life, habits, and routines. As GIS professionals, we are responsible for safeguarding the privacy and rights of individuals and communities. This section delves into the ethical considerations that should guide our every action.

- **Privacy and Data Protection:** Respect for individual privacy rights is a cornerstone of ethical GIS practice. This involves securing consent, protecting sensitive information, and implementing robust data security measures.
- **Transparency and Accountability:** Transparency builds trust. We have to communicate clearly about data sources, methodologies, and purposes. Accountability means taking responsibility for the consequences of our actions.
- **Avoiding Discrimination and Bias:** We must avoid discriminatory or biased data collection, analysis, and decision-making outcomes. This includes considering the potential impact on marginalized or vulnerable communities.
- **Beneficence and Non-maleficence:** Maximizing benefits while minimizing harm should guide our decisions. We must weigh the positive and negative impacts of GIS projects on individuals and communities.

Responsible Use of GIS Data

The use of GIS data goes beyond analysis; it influences policy, planning, and public opinion. This section addresses our ethical responsibilities when utilizing GIS data for decision-making.

- **Equity and Access:** Ensuring that GIS technologies and data are accessible and beneficial to all members of society is a fundamental ethical imperative. We must bridge digital divides and promote inclusivity.
- **Community Engagement and Participation:** The communities affected by GIS initiatives deserve a voice. Engaging with them leads to more inclusive and responsive decision-making.
- **Long-term Sustainability:** Considering the long-term impacts of GIS projects is crucial. From data collection to infrastructure development, we must ensure our actions are sustainable and environmentally responsible.
- **Cultural Sensitivity and Indigenous Knowledge:** Acknowledging and respecting the cultural significance of geographical features is essential. Likewise, respecting indigenous knowledge and land rights is a critical ethical consideration.

Legal and Ethical Frameworks

Legal frameworks provide the guardrails for ethical GIS practice. This section explores the laws and regulations that govern the collection, storage, and use of GIS data.

- **Compliance with Privacy Laws:** Adhering to privacy regulations, such as GDPR, HIPAA, or other regional laws, is non-negotiable. It is imperative to ensure that GIS practices align with legal requirements.
- **Regulatory Compliance:** Staying informed about evolving privacy standards and best practices in GIS is crucial. Compliance with relevant regulations helps protect individual privacy rights (Crampton, 1995).

Data Anonymization and Aggregation

Anonymization and aggregation are powerful tools for protecting individual privacy while deriving valuable insights from GIS data. This section guides these critical processes.

- **Anonymization Techniques:** Anonymizing GIS data involves removing or confusing personally identifiable information (PII) to protect privacy. Techniques such as k-anonymity, differential privacy, and noise injection can be employed.
- **Aggregation Strategies:** Aggregating data at a higher level of granularity can reduce the risk of individual identification. This involves grouping data to a level where individuals cannot be discerned (Majeed & Lee, 2020).

NAVIGATING ETHICAL FRONTIERS IN GEOGRAPHIC INFORMATION SYSTEMS (GIS)

- **Algorithmic Bias in GIS**

Algorithmic bias in GIS algorithms can inadvertently perpetuate discrimination. To address this, rigorous evaluation methods are needed to identify and rectify biases. Researchers advocate for transparency in algorithmic decision-making processes to ensure fairness and mitigate unintended consequences (Reddy et al., 2019).

- **Community Empowerment and GIS**

Community involvement in GIS ensures ethical data use. Empowering communities to participate in decision-making about data collection fosters a sense of ownership. Case studies like community-led mapping projects exemplify successful community engagement, highlighting the benefits of inclusive GIS practices (Krygier & Wood, 2016).

- **Cross-Border Data Sharing**

Cross-border data sharing in GIS requires ethical considerations due to varying privacy regulations. Establishing international agreements and frameworks is essential to navigate legal complexities. Privacy-preserving techniques and secure data transfer protocols are crucial in responsible cross-border data sharing (Crompvoets et al., 2018).

- **Ethical Considerations in Emerging GIS Technologies**

Emerging GIS technologies like AR and VR bring new ethical challenges. Researchers emphasize the need for guidelines to ensure responsible development and use. Balancing innovation with privacy safeguards requires collaboration between technologists, ethicists, and policymakers (Gordon et al., 2016).

- **Educational Initiatives on GIS Ethics**

Integrating ethics education into GIS programs is essential for preparing professionals. Educational initiatives should focus on promoting ethical awareness and decision-making. This proactive approach ensures GIS practitioners can navigate complex ethical considerations (Pei, 2021).

- **Case Studies on GIS Data Breaches**

Analyzing GIS data breaches provides valuable insights into vulnerabilities and risk mitigation. Researchers can develop best practices for preventing and responding to breaches by examining real-world cases. This proactive stance is crucial for maintaining public trust in GIS systems (Gerdan, 2018).

- **The Role of Professional Organizations**

Professional organizations play a pivotal role in setting ethical standards for GIS practitioners. Their involvement in promoting guidelines and facilitating discussions fosters a community committed to ethical GIS practices. The collaboration between professionals and organizations contributes to continually refining ethical standards (Sui, 2011).

- **Public-Private Partnerships in GIS**

Public-private partnerships in GIS projects require transparent practices to protect public interests. Ethical considerations include data ownership, privacy, and accountability. Establishing guidelines for responsible collaboration ensures that GIS projects benefit society without compromising ethical principles (Bonney, 2009).

- **International Perspectives on GIS Ethics**

Comparing GIS ethics across international contexts provides a broader understanding of ethical considerations. Examining different regulatory approaches and cultural perspectives informs the development of global standards. Collaboration between countries and international organizations is crucial for shaping ethical norms in GIS (Dueker, 2005).

- **Legal and Ethical Implications of Geofencing**

Geofencing technologies in GIS raise legal and ethical concerns about privacy and surveillance—research advocates for clear regulations and guidelines to govern geofencing practices. Understanding the implications of geofencing on individual privacy is essential for responsible implementation (Shokri et al., 2017).

FUTURE TRENDS AND CHALLENGES: EMERGING ETHICAL ISSUES

The future of Geographic Information Systems (GIS) stands at the intersection of promise and ethical complexity, ushering in a new era of challenges and considerations. As GIS technologies advance, the evolving landscape brings ethical issues that demand meticulous attention and thoughtful deliberation. One of the central concerns on this ethical frontier revolves around the ever-expanding realm of data privacy and security. The sheer volume and increasing granularity of geospatial data present pressing questions about ownership, consent, and the responsible utilization of this information. With the proliferation of location-aware technologies and the widespread integration of GIS into various aspects of daily life, the ethical imperative of safeguarding individual privacy becomes paramount. Stakeholders must define the boundaries of data ownership, ensure informed consent, and establish robust frameworks for the responsible use of geospatial information (Mittelstadt et al., 2017).

Beyond data privacy challenges, the rise of geospatial algorithms introduces an additional layer of ethical complexity. The potential for unintended consequences, such as biases and discrimination perpetuated through these algorithms, raises critical questions about fairness and equity. Addressing and rectifying algorithmic biases becomes a pivotal ethical consideration as GIS technologies become more ingrained in decision-making processes. The advent of geospatial artificial intelligence (AI) further amplifies the ethical challenges facing GIS. The opacity of AI algorithms employed in GIS applications poses a significant hurdle to understanding and accountability. Transparency and interpretability are essential to build trust among users and stakeholders. Establishing clear lines of accountability for the outcomes generated by AI algorithms in GIS becomes imperative to ensure responsible and ethical decision-making.

Interdisciplinary collaboration emerges as a key component in navigating the intricate ethical landscape of GIS. Technologists, ethicists, and policymakers must join forces to develop comprehensive frameworks that balance innovation with ethical considerations. Bridging the gap between technological advancements and ethical responsibilities requires a collective effort to anticipate and proactively address challenges. As GIS permeates diverse sectors, ranging from healthcare and urban planning to environmental monitoring, the ethical considerations must evolve in tandem. The dynamic nature of GIS applications necessitates an ongoing dialogue about potential ethical pitfalls and the development of adaptive frameworks. It is crucial to anticipate the ethical implications of GIS technologies in various contexts, paving the way for responsible and equitable deployment (Brundage et al., 2018).

In conclusion, Geographic Information Systems' future is intertwined with promise and ethical complexities. The ethical considerations surrounding data privacy, algorithmic biases, and the advent of geospatial AI underscore the need for a proactive and collaborative approach. By fostering interdisciplinary collaboration, ensuring transparency in AI algorithms, and staying attuned to the evolving ethical landscape, stakeholders in the GIS community can contribute to the responsible and ethical advancement of geospatial technologies. As GIS plays a pivotal role in shaping our interconnected world, the ethical dimensions of its evolution must remain at the forefront of discourse and action. In the dynamic realm of Geographic Information Systems (GIS), pursuing innovation is a driving force, propelling transformative capabilities that redefine how we interact with spatial data. However, this quest for progress comes with a critical ethical imperative – the delicate balancing act between innovation and privacy. As GIS technologies advance, concerns about the potential misuse and exploitation of geospatial data underscore the necessity for robust mechanisms to safeguard individual privacy.

At the heart of this delicate equilibrium lies the imperative for robust data protection measures. Data anonymization, encryption, and secure storage are pivotal components of a comprehensive strategy to shield geospatial information from unauthorized access and malicious use. By implementing these mechanisms, GIS developers can ensure that individuals' privacy is prioritized, even in the face of advancing technologies. Privacy-by-design principles emerge as a fundamental framework in GIS development, integrating ethical considerations seamlessly into the lifecycle of geospatial applications. By embedding privacy measures from the initial design stages, developers proactively address privacy concerns, mitigating potential risks and ensuring user privacy is a core consideration throughout the application's evolution (Jobin et al., 2019).

Public awareness is crucial in balancing GIS innovation and privacy preservation. Educating users about how geospatial data is collected and utilized fosters a sense of understanding and empowerment. Clear and transparent communication regarding data collection practices enables users to make informed decisions about sharing their location data. This informed consent is a cornerstone of ethical GIS practices, allowing individuals to control their geospatial information. Empowering individuals through education goes beyond mere awareness – it involves providing users with tools and mechanisms to manage and control their geospatial information actively. Implementing clear consent mechanisms, user-friendly interfaces, and easily accessible privacy settings ensures that individuals can dictate how much their location data is utilized. This user-centric approach respects individual privacy rights and contributes to a more ethical and responsible GIS ecosystem.

Collaboration emerges as a linchpin in balancing innovation and privacy in GIS. The synergy between GIS developers, privacy advocates, and regulatory bodies is vital to establishing industry standards that uphold privacy without stifling innovation. By fostering a collaborative environment, stakeholders can work towards creating guidelines that strike an equilibrium between the transformative potential of GIS technologies and the imperative to protect individual privacy. Industry standards are pivotal in setting benchmarks for responsible and ethical GIS practices. These standards, forged through collaboration, provide a roadmap for GIS developers, ensuring that their innovations align with privacy considerations. Regulatory bodies contribute by enforcing compliance with these standards, creating a regulatory framework that safeguards privacy rights while fostering a conducive environment for GIS innovation.

In conclusion, balancing GIS innovation and privacy protection requires a multifaceted approach. Robust data protection measures, integrating privacy-by-design principles, user education, and stakeholder collaborative efforts are essential to this balance. By prioritizing individual privacy rights, fostering awareness, and creating an environment encouraging responsible innovation, the GIS community can navigate the evolving landscape while ensuring that ethical considerations remain at the forefront of technological progress. In the ever-expanding realm of Geographic Information Systems (GIS), the journey through the ethical dimensions of data privacy illuminates a path fraught with challenges and opportunities. As GIS technologies evolve, the ethical considerations embedded in their application become crucial guideposts for responsible innovation. This concluding reflection encapsulates the key insights from exploring GIS ethics, emphasizing the delicate equilibrium required to navigate the intricate intersection of technological progress, privacy rights, and societal welfare.

- **Navigating the Ethical Landscape:**

The ethical considerations within GIS extend far beyond the technical intricacies of spatial data analysis. They reach into the fabric of societal values, individual rights, and the responsible stewardship of information. The chapters traversed the multifaceted terrain, uncovering layers of concerns ranging from privacy and data ownership to algorithmic fairness and the challenges ushered in by emerging technologies like artificial intelligence. In this intricate tapestry, it becomes evident that the ethical compass guiding GIS practices is essential for ensuring the technology's positive impact on society (Blatt, 2012).

- **The Tension Between Innovation and Privacy:**

At the heart of GIS lies a delicate balancing act — a tension between the inexorable march of innovation and the imperative to safeguard individual privacy. The promise of GIS lies in its transformative capabilities, offering insights that can reshape urban planning, environmental management, public health, and disaster response. However, as GIS technologies advance, concerns about geospatial data's potential misuse and exploitation cast a shadow. The need to strike a harmonious balance between progress and privacy is paramount, necessitating robust mechanisms for data protection (Kitchin & Dodge, 2014).

- **Ethical Foundations in Data Collection and Ownership:**

The ethical foundations in the chapters regarding data collection and ownership support responsible GIS practices. Whether through remote sensing, GPS, or traditional surveys and interviews, the importance of informed consent, confidentiality protection, and clear delineation of data ownership cannot be overstated. These principles are not merely procedural requirements but embody a commitment to respecting the individuals and communities contributing to the GIS datasets. Establishing data-sharing agreements further reinforces ethical norms, fostering collaborative and trustworthy relationships in the GIS community (Goodchild, 2009).

- **Privacy in the Age of GIS**

Privacy concerns, as highlighted in the exploration of location-based data sensitivity, anonymization techniques, and the challenges of location privacy, underscore the ethical tightrope walk inherent in GIS practices. If not handled responsibly, constantly tracking geospatial information can infringe individual privacy rights. The chapters have emphasized the importance of transparent data collection practices, robust security measures, and the integration of privacy-by-design principles to safeguard the sensitive details embedded in GIS datasets (Elwood & Leszczynski, 2011).

- **Interdisciplinary Collaboration and Regulatory Compliance**

A pivotal aspect that emerges from the discussions is the necessity for interdisciplinary collaboration. Ethicists, technologists, policymakers, and privacy advocates must join forces to create comprehensive frameworks that align technological innovation with ethical considerations. Regulatory compliance is crucial in setting industry standards, and prioritizing privacy without stifling innovation. The dynamic interplay between stakeholders, industry standards, and regulatory frameworks forms the bedrock of a responsible and ethical GIS ecosystem (Donovan et al., 2018).

CASE STUDY

Open data initiatives make property-level data such as crime statistics, environmental hazards, and registered sex offenders publicly available. While this data can benefit public safety efforts and inform decision-making, it can also lead to unintended consequences. The ethical issues and privacy concerns of these data are listed below:

Ethical Issues

- Geospatial Stigma: Attaching negative labels to specific locations can create a perception of inherent risk or undesirability associated with those areas. This can lead to property devaluation, decreased investment, and social exclusion for residents.
- Data Currency and Accuracy: Publicly available data may not be frequently updated, leading to an inaccurate portrayal of an area's current situation.

Privacy Concerns

- Potential for Misinterpretation: Individuals associated with past events (e.g., registered sex offenders) might face undue social stigma due to the permanent online presence of their information, hindering rehabilitation and reintegration efforts.

Mitigating Actions

- Contextualization of Data: Data should be presented alongside relevant information about the timeframe, methodology of collection, and data limitations.
- Balancing Transparency with Privacy: While public access to information is important, measures should be taken to protect the privacy of individuals.
- Data Expiry and Removal Mechanisms: Mechanisms should be established to remove outdated or inaccurate data, particularly when it can lead to persistent social stigma.

Foody's research examines how readily available geospatial data can have negative downstream effects. The study highlights the importance of considering the potential for social stigma and ensuring data accuracy and context are emphasized when making such information publicly available. This case study demonstrates the ethical considerations surrounding publicly available geospatial data and the potential for unintended consequences. It emphasizes the need for responsible data management practices to safeguard individual privacy and prevent the perpetuation of social stigma (Foody, 2006).

The provided case studies illustrate the complex interplay between the benefits of geospatial data for various applications and the inherent ethical concerns surrounding data privacy.

- Privacy Risks: Sharing location data can expose individuals to targeted advertising, stalking, and discrimination. Publicly available property data can lead to geospatial stigma and negatively impact residents of certain areas.
- Importance of User Consent and Control: Transparency regarding data collection practices and providing users with granular control over how their data is shared are crucial.
- Data Accuracy and Context: Mitigating actions such as ensuring data accuracy, providing context for public information, and implementing data expiry mechanisms are essential to minimize privacy violations and prevent misinterpretations.

Addressing these challenges requires a multifaceted approach. Social media platforms, businesses, and regulatory bodies must work together to implement robust user consent mechanisms, prioritize data minimization practices, and empower users with control over their geospatial data. Furthermore, increased public awareness about the potential risks and responsible data management practices are essential to safeguard individual privacy in the era of ubiquitous geospatial technologies.

CONCLUSION

This exploration of GIS ethics, the overarching narrative, emphasizes the symbiotic relationship between technological advancement and ethical responsibility. The future of GIS holds tremendous promise, but this promise must be tempered with a profound commitment to ethical principles. The chapters traversed have laid bare the intricacies of balancing innovation and privacy, navigating the ethical landscape in data collection and ownership, and addressing privacy concerns in the age of GIS. The path forward demands a collective dedication to fostering ethical awareness, prioritizing individual privacy rights, and perpetuating an ongoing dialogue that adapts to the evolving challenges posed by GIS technologies. The responsible use of GIS data requires a shared commitment to transparency, accountability, and equitable access. In the intricate dance between innovation and ethical considerations, GIS professionals, researchers, and policymakers must collaborate to ensure that the transformative power of GIS is harnessed for the greater good, safeguarding the rights and dignity of individuals and communities in an increasingly interconnected world.

REFERENCES

Antenucci, J. C., Brown, K., Croswell, P. L., Kevany, M. J., & Archer, H. (1991). *Geographic Information Systems: a guide to the technology* (Vol. 115). Van Nostrand Reinhold.

Blatt, A. J. (2012). Ethics and Privacy Issues in the Use of GIS. *Journal of Map & Geography Libraries*, *8*(1), 80–84. doi:10.1080/15420353.2011.627109

Bonney, R., Ballard, H., Jordan, R., McCallie, E., Phillips, T., Shirk, J., & Wilderman, C. C. (2009). *Public Participation in Scientific Research: Defining the Field and Assessing Its Potential for Informal Science Education. A CAISE Inquiry Group Report.*

Brundage, M., Avin, S., Clark, J., Toner, H., Eckersley, P., Garfinkel, B., & Amodei, D. (2018). The malicious use of artificial intelligence: Forecasting, prevention, and mitigation. *arXiv preprint arXiv:1802.07228.*

Chang, K. T. (2008). *Introduction to geographic information systems* (Vol. 4). McGraw-Hill.

Chang, V., Mou, Y., & Xu, Q. A. (2021). The ethical issues of location-based services on big data and IoT. In *Modern Industrial IoT, Big Data and Supply Chain* [Springer Singapore.]. *Proceedings of the IIoTBDSC, 2020*, 195–205.

Coleman, D., Georgiadou, Y., & Labonte, J. (2009). Volunteered geographic information: The nature and motivation of produsers. *International Journal of Spatial Data Infrastructures Research, 4*(4), 332–358.

Crampton, J. (1995). The ethics of GIS. *Cartography and Geographic Information Systems, 22*(1), 84–89. doi:10.1559/152304095782540546

Crompvoets, J., Rajabifard, A., Van Loenen, B., & Fernández, T. D. (2008). A multi-view framework to assess SDIs. *Wageningen University, RGI.*

Curry, M. R. (1997). Digital people, digital places: Rethinking privacy in a world of geographic information. *Ethics & Behavior, 7*(3), 253–263. doi:10.1207/s15327019eb0703_7

De Smith, M. J., Goodchild, M. F., & Longley, P. (2007). *Geospatial analysis: a comprehensive guide to principles, techniques and software tools.* Troubador publishing ltd.

Donovan, J., Caplan, R., Matthews, J., & Hanson, L. (2018). Algorithmic accountability. *PRiMER: Peer-Reviewed Reports in Medical Education Research.*

Dueker, K. J., & Butler, J. A. (2000). A geographic information system framework for transportation data sharing. *Transportation Research Part C, Emerging Technologies, 8*(1-6), 13–36. doi:10.1016/S0968-090X(00)00006-1

Elwood, S., & Leszczynski, A. (2011). Privacy, reconsidered: New representations, data practices, and the geoweb. *Geoforum, 42*(1), 6–15. doi:10.1016/j.geoforum.2010.08.003

Elwood, S., & Leszczynski, A. (2011). Privacy, reconsidered: New representations, data practices, and the geoweb. *Geoforum, 42*(1), 6–15. doi:10.1016/j.geoforum.2010.08.003

Foody, G. M. (2006). Privacy in the public eye: Exploring the phenomenon of geospatial stigma. *Area, 38*(3), 311–321.

Gerdan, S. (2018). GIS-based decision-support system applications in disaster management. *Yönetim ve Ekonomi Dergisi, 25*(3), 961–979. doi:10.18657/yonveek.306383

Ghinita, G. (2013). *Privacy for location-based services* (Vol. 4). Morgan & Claypool Publishers. doi:10.1007/978-3-031-02336-1

Goodchild, M. F. (2009). Geographic information systems and science: Today and tomorrow. *Annals of GIS, 15*(1), 3–9. doi:10.1080/19475680903250715

Goodchild, M. F. (2009). Geographic information systems and science: Today and tomorrow. *Annals of GIS*, *15*(1), 3–9. doi:10.1080/19475680903250715

Goodchild, M. F., Steyaert, L. T., Parks, B. O., Johnston, C., Maidment, D., Crane, M., & Glendinning, S. (Eds.). (1996). *GIS and environmental modeling: progress and research issues*.

Gordon, E., Elwood, S., & Mitchell, K. (2016). Critical spatial learning: Participatory mapping, spatial histories, and youth civic engagement. *Children's Geographies*, *14*(5), 558–572. doi:10.1080/14733285.2015.1136736

Huang, H., & Gartner, G. (2018). Current trends and challenges in location-based services. *ISPRS International Journal of Geo-Information*, *7*(6), 199. doi:10.3390/ijgi7060199

Jobin, A., Ienca, M., & Vayena, E. (2019). The global landscape of AI ethics guidelines. *Nature Machine Intelligence*, *1*(9), 389–399. doi:10.1038/s42256-019-0088-2

Kitchin, R., & Dodge, M. (2014). *Code/space: Software and everyday life*. Mit Press.

Krygier, J., & Wood, D. (2016). *Making maps: a visual guide to map design for GIS*. Guilford Publications.

Majeed, A., & Lee, S. (2020). Anonymization techniques for privacy preserving data publishing: A comprehensive survey. *IEEE Access : Practical Innovations, Open Solutions*, *9*, 8512–8545. doi:10.1109/ACCESS.2020.3045700

Mittelstadt, B. D., Allo, P., Taddeo, M., Wachter, S., & Floridi, L. (2016). The ethics of algorithms: Mapping the debate. *Big Data & Society*, *3*(2), 2053951716679679. doi:10.1177/2053951716679679

Onsrud, H. J., Johnson, J. P., & Lopez, X. (1994). Protecting personal privacy in using geographic information systems. *Photogrammetric Engineering and Remote Sensing*, *60*(9), 1083–1095.

Pei, T., Xu, J., Liu, Y., Huang, X., Zhang, L., Dong, W., Qin, C., Song, C., Gong, J., & Zhou, C. (2021). GIScience and remote sensing in natural resource and environmental research: Status quo and future perspectives. *Geography and Sustainability*, *2*(3), 207–215. doi:10.1016/j.geosus.2021.08.004

Reddy, E., Cakici, B., & Ballestero, A. (2019). Beyond mystery: Putting algorithmic accountability in context. *Big Data & Society*, *6*(1), 2053951719826856. doi:10.1177/2053951719826856

Shokri, R., Stronati, M., Song, C., & Shmatikov, V. (2017, May). Membership inference attacks against machine learning models. In 2017 IEEE symposium on security and privacy (SP) (pp. 3-18). IEEE. doi:10.1109/SP.2017.41

Solymosi, R., Buil-Gil, D., Ceccato, V., Kim, E., & Jansson, U. (2023). Privacy challenges in geodata and open data. *Area*, *55*(4), 456–464. doi:10.1111/area.12888

Sui, D., & Goodchild, M. (2011). The convergence of GIS and social media: Challenges for GIScience. *International Journal of Geographical Information Science*, *25*(11), 1737–1748. doi:10.1080/13658816.2011.604636

van der Kruk, E. S., Crompvoets, J., & van Loenen, B. (2018). Privacy issues in an urban mobility project: A case study of the CityBug app. *Applied Geography (Sevenoaks, England).*

Chapter 2
A Guide to Ethical Geospatial Practices:
From Historical Context to Future Scenarios

V. Sahithya
Coimbatore Institute of Technology, India

M. Sujithra
Coimbatore Institute of Technology, India

B. Senthilkumar
https://orcid.org/0000-0001-8117-
3314
Kumaraguru College of Technology, India

M. Shreenithi
Coimbatore Institute of Technology, India

Hari Priya
Coimbatore Institute of Technology, India

ABSTRACT

This chapter explores the ethics of geospatial data, examining how data collection methods have evolved and highlighting both benefits and past mistakes. The vast amount of geospatial data collected today creates new ethical challenges. The guide dives into core principles like privacy, security, and fairness in this context. Legal frameworks are explored to provide a complete picture. It then tackles present-day challenges like privacy concerns and bias in geospatial technologies. Looking ahead, the guide explores emerging technologies and emphasizes global collaboration for ethical practices. It presents potential future scenarios and analyzes how current actions will impact them. Regulations and ongoing ethical discussions are seen as crucial for shaping the future. Finally, the guide stresses the importance of continuous awareness, education, and ethical business practices to ensure geospatial data remains a force for good in our interconnected world.

DOI: 10.4018/979-8-3693-6381-2.ch002

INTRODUCTION

This chapter explores the significant relevance of data ethics in the geospatial domain. Addressing the ethical considerations that come with the widespread use of geospatial data across critical fields like marketing, disaster management, urban planning, and navigation is of utmost importance. This exploration is intended to thoroughly investigate these topics, providing guidance and insights tailored to your field's individuals, groups, and decision-makers. The overarching goal is to advocate for ethical data practices and heighten public awareness regarding the moral implications of geospatial data.

- ***To Scrutinize the Relationship Between Geospatial Data and Ethics:*** This exploration is not a mere surface-level examination. It aims to delve deep into the interplay between geospatial data and ethical issues, conducting a comprehensive analysis of moral dilemmas, ethical standards, and challenges that emerge during the collection, utilization, and dissemination of geospatial data. You can be confident that this exploration leaves no stone unturned in its quest for understanding.
- ***To Educate and Inform:*** The exploration strives to educate readers about geospatial data's background and current state, emphasizing the moral standards that should govern its application and the relevant laws and regulations.
- ***To Provide Guidance and Best Practices:*** Sections on best practices and ethical governance are likely included in this exploration. Its purpose is to offer direction and advice to individuals, groups, and policymakers on responsible and ethical use of geospatial data.
- ***To Foster Ethical Awareness and Discussion***: This exploration is not a one-sided presentation. It actively fosters awareness and ongoing discussion, inviting you to be a part of the conversation. By addressing current ethical issues and envisioning the future of geospatial data and ethics, it encourages you to think critically and engage in the discussion, making your voice and perspective an integral part of the exploration.

In conclusion, this exploration is a guide, an information source, an educational tool, and a catalyst for ethical awareness and discussion about geospatial data and ethics. It is an invaluable resource for anyone interested in this subject, providing a moral compass for decision-making when dealing with geospatial data.

The Significance of Geospatial Data and Ethics

The value of geospatial data and ethics is critical in many fields. It is essential to support the responsible use of location-based data, protect individual rights, and guide decision-making (Gómez et al., 2022).

- ***Making Informed Decisions:*** When used responsibly, geographic data gives individuals, governments, and organizations the ability to make educated decisions. Disaster relief, urban planning, and resource allocation offer insights into spatial relationships, environmental factors, and demographics.
- ***Privacy and Data Protection:*** Preserving privacy requires careful consideration of the ethical issues surrounding geospatial data. With the amount of data being collected and shared these days increasing, it is imperative to ensure that location data is gathered, stored, and shared in a way that respects people's right to privacy.
- ***Preservation of the Environment and Cultural Heritage:*** The environment and cultural heritage must be safeguarded, which requires the ethical use of geospatial data.
- ***Equality and Fairness:*** Addressing social and economic inequalities can be facilitated using geographic data. When analyzing and interpreting data, ethical issues aid in reducing bias and advancing equity in the distribution of resources and service accessibility.
- ***Security and Safety:*** The ethical management of geospatial data is essential to guaranteeing the security and safety of people and infrastructure. It helps with preparedness for disasters, security operations, and emergency response.
- ***Legal and Regulatory Compliance:*** When using geospatial data ethically, organizations and governments can better adhere to applicable laws and regulations. This ensures that data is collected and utilized within permitted parameters.

To sum up, the importance of geospatial data and ethics can be attributed to its capacity to support ethical and responsible decision-making, safeguard privacy, maintain the environment and culture, encourage equity, improve security, guarantee legal compliance, and foster public trust. Using geospatial data to its greatest potential while upholding social values and individual rights requires careful consideration of ethical issues.

GEOSPATIAL DATA: A HISTORICAL PERSPECTIVE

Evolution of Geospatial Data Collection

Geospatial data collection has evolved significantly, driven by societal demands and technological advancements. Initially, manual techniques and early cartography served ancient civilizations for land management and navigation. The Renaissance improved surveying and mapping accuracy, which was essential for geographic explorations and colonial expansion. The Age of Exploration witnessed accelerated data collection as explorers mapped uncharted territories using crude instruments.

In the early 20th century, aerial photography transformed mapping, benefiting industries like forestry and urban planning. The launch of Sputnik 1 in 1957 marked a turning point, introducing satellite technology and remote sensing for global geospatial data collection. The 1970s saw the advent of GPS, revolutionizing precise positioning for surveying, agriculture, and transportation. In the 21st century, geospatial big data emerged with IoT, enabling real-time data collection for applications like location-based services and smart cities. Crowdsourced data through social media platforms and the integration of machine learning and AI for data processing further expanded possibilities. However, this evolution brings ethical considerations, including data ownership, privacy, and responsible data use. In summary, the journey of geospatial data collection unfolds from manual techniques to sophisticated systems shaped by technology and guided by ethical and societal concerns (Arnould, 2020).

Historical Uses and Misuses of Geospatial Data

Geographical data has had both positive and negative uses throughout history. In addition to highlighting its more sinister uses, such as espionage and conquest, this section also reveals the historical uses of geospatial data, such as navigation, exploration, and military application. These historical applications and abuses highlight the ethical issues in geographic data.

- Navigation and Exploration
- Resource Management and Agriculture
- Military and Defence
- Conquest and Imperialism
- Espionage and Intelligence
- Infrastructure Development
- Humanitarian Efforts
- Ethical Conundrums

In conclusion, geospatial data has long been used for various objectives, from intelligence and imperialism to defense and exploration. Although its uses have advanced many industries, they have raised moral questions about data abuse, privacy, and monitoring. Navigating the current ethical issues surrounding geospatial data requires understanding this historical context.

Ethical Considerations in Early Geospatial Data Handling

Many ethical issues arose from exploration, territorial expansion, and the pursuit of geographic knowledge in the early days of geospatial data handling. This section explores the complex moral conundrums that early users of geospatial data, such as cartographers and explorers, had to deal with. From looking at these historical settings, it is clear how these early encounters have shaped and still shaped modern ethical considerations in geospatial data practices (Kedron & Trgovac, 2021).

- ***Accuracy vs. Distortion:*** Early mapmakers and explorers had to weigh the moral conundrum of balancing accuracy against the impact of political or personal biases. The integrity of geospatial data was challenged by the debate over whether to accurately represent observed data or alter it to further particular agendas.
- ***Territorial Conquest and Colonization:*** Indigenous populations were frequently harmed by the use of geographic data to aid in territorial conquest and colonization.
- ***Ethical Treatment of Indigenous Knowledge:*** The ethical treatment of geographic data and indigenous knowledge is debated due to historical reliance on local wisdom without proper credit or compensation.
- ***Geopolitical Power Dynamics:*** Ethical conundrums were raised by the consolidation of geopolitical power and influence through the use of geospatial data. Accurate maps and geographic intelligence can provide a strategic advantage and encourage aggressive territorial expansion that exacerbates geopolitical conflicts.
- ***Scientific Curiosity and Exploration:*** Early scientists and explorers grappled with ethical dilemmas, balancing intellectual curiosity with preserving delicate ecosystems and cultural heritage sites, setting the stage for contemporary discussions on responsible exploration and data collection.
- ***Legacy and Modern Ethics*:** Contemporary discussions on data privacy, intellectual property, and indigenous rights are shaped by historical ethical dilemmas faced by early geospatial data handlers, guiding modern practitioners to navigate these issues responsibly.

As a result of the conflict between exploration, territorial ambition, and the quest for geographic knowledge, early geospatial data handling was fraught with ethical dilemmas. In the rapidly developing field of geospatial data practices, these historical, moral conundrums influence discussions and ethical standards today.

THE CURRENT LANDSCAPE

Modern Geospatial Data Sources and Technologies

Geospatial data is gathered from various sources and made possible by cutting-edge technologies. An interconnected web of location data has been created by Internet of Things (IoT) sensors, mobile devices, satellites for the Global Positioning System (GPS), and remote sensing. This section examines how these contemporary sources and technologies have transformed geospatial data collection, making it possible to obtain high-precision, real-time information about our planet. It emphasizes their importance in several domains, such as emergency response and precision agriculture, and draws attention to the moral ramifications of their broad use (Al Jawarneh et al., 2020).

Applications and Impacts of Geospatial Data

Due to its abundance of location-based information, geospatial data is used in many different fields. This section examines the wide range of uses and profound effects of geospatial data, highlighting how it influences contemporary decision-making and deepens our understanding of the world.

- ***Urban Planning and Development:*** Using geographic data is essential to these processes. It aids in the optimization of public services, infrastructure, and land use in cities, creating more sustainable and effective urban environments.
- ***Environmental Monitoring:*** The conservation and monitoring of the environment depend heavily on geographic data. It helps monitor deforestation, climate patterns, ecosystem changes, and the effects of human activity on the environment.
- ***Natural Disaster Management:*** Geographic data plays a crucial role in disaster management, aiding early warning systems, evacuation planning, and post-disaster assessment to mitigate the impact of natural disasters.

- ***Logistics and Transportation:*** Geographical data is useful for the transportation industry's fleet optimization, traffic control, and route planning. It makes transportation networks safer and more effective.
- ***Public Health and Epidemiology:*** Geospatial data is useful for tracking disease outbreaks, analyzing healthcare accessibility, and identifying health disparities in public health and epidemiology. Planning for public health and epidemiological research both benefit from it.
- ***Satellite Navigation and Location-Based Services***: Location-based services, which are revolutionizing sectors, including emergency response, tourism, and navigation, are based on the Global Positioning System (GPS) and other satellite navigation systems.
- ***Business and Market Analysis:*** Business uses geospatial data for customer targeting, site selection, and market analysis. It supports businesses in making defensible choices regarding product distribution, advertising, and store locations.
- ***Defense and Intelligence:*** For intelligence gathering, mission planning, and surveillance, the defense industry uses geospatial data. It facilitates military operations and strengthens national security.
- ***Disaster Relief and Humanitarian Aid:*** Geospatial data plays a critical role in disaster relief and humanitarian aid operations by allowing organizations to assess the damage, plan recovery operations, and deliver aid to affected regions.
- ***Research and Education:*** Using geospatial data facilitates research and educational activities. It improves our world comprehension by offering useful geography, earth sciences, and social sciences resources.
- ***Security and Privacy:*** Geospatial data also affects security and privacy. It raises location tracking, data security, and personal information protection issues.

In conclusion, the uses and effects of geospatial data are extensive and varied, influencing many facets of contemporary society. It poses ethical and privacy concerns in an increasingly interconnected world, empowering decision-makers, promoting sustainability, and advancing science.

Emerging Ethical Challenges in Geospatial Data

The exponential growth of geospatial data presents ethical challenges in gathering, sharing, and use, including concerns over location privacy, consent, and data ownership. Biases in data and security risks are heightened, impacting emergency response, resource allocation, and privacy. Environmental impacts raise questions about balancing development with preservation. AI and machine learning introduce challenges aligning with human rights. Striking a balance between data protection and open data initiatives is essential, alongside addressing geopolitical concerns and ensuring accountability. Navigating these complexities requires upholding ethical standards and responsible behavior to safeguard individuals' rights and security amidst technological advancements.

Privacy Concerns in Contemporary Geospatial Data Usage

Privacy concerns in contemporary geospatial data usage have become increasingly prominent due to the proliferation of location-based services, satellite imagery, and the widespread availability of GPS-enabled devices (Ricker et al., 2015).

- Location tracking and surveillance
- Data Breach and Security Risks
- Commercial Exploitation and Profiling
- Informed Consent and Transparency
- Aggregate Data and De-Identification
- Legislative and Regulatory Frameworks
- Ethical Considerations
- Emerging Technologies and Risks

By addressing these privacy concerns and exploring potential solutions, policymakers, industry stakeholders, and individuals can work towards a more responsible and ethical approach to using geospatial data in contemporary society.

ETHICS IN GEOSPATIAL DATA: FUNDAMENTAL PRINCIPLES, LEGAL AND REGULATORY FRAMEWORK

Privacy and Anonymity

Regarding geospatial data, privacy and anonymity are the most important ethical considerations. This section explains the ethical requirements for anonymity and privacy and their importance for appropriately using location-based data (Huang et al., 2021) (Chang et al., 2021).

- ***Location Privacy and Surveillance Concerns:*** Addressing concerns about surveillance and potential misuse of precise geospatial data from smartphones and IoT devices, examining the ethical implications of tracking individuals' movements, and the need for robust privacy safeguards.

- ***Consent, Control, and Ownership Issues:*** Highlighting concerns about informed consent and users' awareness of who controls and owns their location information. They are assessing the risks of infractions stemming from limited knowledge about data ownership and utilization.
- ***Bias and Disparities in Geographic Data***: Examining how biases within geographic data, whether in mapping or algorithmic decision-making, can reinforce existing disparities. They are analyzing the ethical implications of these biases on access to opportunities, services, and resources.
- ***Reliability and Accuracy in Geographical Data:*** Exploring the ethical considerations surrounding the reliability and accuracy of geospatial data in critical sectors like emergency response, agriculture, and navigation and evaluating the potential consequences of mistakes or falsification of location data on decision-making processes.
- ***Environmental Impact and Sustainability:*** Investigating ethical questions arising from the environmental impact of geospatial data in activities like mining, deforestation, and urban development. Assessing the delicate balance between human development facilitated by geospatial data and the imperative of environmental preservation (Smith, 2020)

Security and Data Protection

Security and protection are the two most important aspects of ethical geospatial data management. Protection of sensitive location data is critical in an era where cyberattacks find geospatial data to be a valuable target. This section explores the changing risks associated with cyberattacks and data breaches and emphasizes the importance of strong encryption, strict access controls, and data protection measures.

It talks about how important it is to keep up with the most recent security procedures, such as using encrypted services and being on the lookout for phishing attempts. It also looks at how ethically organizations should approach data security holistically, incorporating threat detection and data breach response techniques. In order to safeguard geospatial data from unwanted access, malicious threats, and potential breaches, the section underlines the necessity of taking a proactive approach to data security, thereby encouraging ethical data stewardship.

Data Ownership and Sharing

An essential component of moral geospatial data practices is addressing data ownership and sharing. This section explores the intricate world of data ownership and people's rights concerning location data. It talks about the need for explicit data-sharing agreements and the obligation of data collectors to be open and honest about how their data is used. The importance of ethical data-sharing methods, including data minimization and dynamic consent settings, is highlighted. The section also recognizes how technology—including blockchain—helps to make data ownership and control easier. To practice responsible data stewardship, one must balance utilizing the advantages of geospatial data, upholding people's rights, and promoting moral data practices. The ethical issues raised by data ownership can be resolved by giving people more control over their data and promoting appropriate data sharing.

Fairness and Bias

In geospatial data practices, fairness and bias are crucial ethical considerations. This section explores the risks of bias and the moral requirements of fairness in gathering, analyzing, and using geographic data.

- Fair Data Collection
- Resolving Historical Biases
- Algorithmic Fairness
- Consent and Privacy
- Regulatory Compliance

In conclusion, the foundation of moral geospatial data practices is impartiality and fairness. Aiming for equity in the gathering, processing, and utilizing of data is essential to prevent discriminatory effects and advance inclusiveness. Ethical geospatial practitioners must recognize and address bias to guarantee the fair and impartial use of location-based data.

Environmental and Cultural Impacts

Wide-ranging effects of geographic data, both beneficial and detrimental, can be seen in the environment and culture. The ethical issues and effects of geospatial data on the environment and cultural heritage are covered in detail in this section (Kross et al., 2022).

- ***Environmental Preservation:*** By making it possible to monitor and safeguard ecosystems, natural resources, and biodiversity, geospatial data helps to preserve the environment. It supports plans for sustainable land use and conservation.

- ***Resource Management:*** Ethical geospatial data practices aid in responsibly managing resources. Making data-based decisions helps industries use resources more efficiently, produce less waste, and leave a smaller environmental impact.
- ***Reducing Environmental Damage:*** Environmental damage is the goal of ethical geospatial practices. This entails minimizing the harm that data collection causes to the environment, cutting carbon footprints, and protecting natural habitats.
- ***Data Access for Indigenous Communities:*** Giving indigenous communities access to geospatial data, resource management, and cultural preservation tools are ethical considerations.
- ***Reducing Cultural and Environmental Risks:*** **Risk** reduction is a component of ethical geospatial data use. It works to stop operations like mining and building in sensitive areas that can harm the environment or cultural heritage sites.

In conclusion, geographic data significantly affects the environment and cultural heritage. Sustainability, cultural asset protection, and environmental preservation depend on ethical data collection, analysis, and application methods. Geospatial data is an incredibly useful tool when addressing environmental and cultural issues while upholding moral standards and values.

Laws and Regulations Governing Geospatial Data

The legal and regulatory environment is intimately linked to the moral application of geospatial data. An overview of the pertinent laws that control the gathering, storing, and application of geographic data is given in this section. It talks about how crucial it is to abide by local data protection regulations and the necessity of strong legal frameworks to handle the unique problems that geospatial data presents. Ensuring that geospatial data is managed responsibly and in compliance with relevant regulations necessitates ethical and legal compliance.

Intellectual Property and Licensing

A key component of using geospatial data ethically is considering intellectual property rights. The topics of data ownership, data licensing, and intellectual property rights are examined in this section. It highlights the necessity of precise licensing contracts that specify the acceptable uses and distribution channels for geospatial data while upholding the rights of data owners and creators. Promoting responsible data management requires upholding ethical standards for data licensing and intellectual property rights.

International Agreements and Standards

International agreements and standards govern the responsible use of geospatial data on a global scale. The importance of international conventions and guidelines for geospatial data ethics is discussed in this section.

- Data Sharing and Global Collaboration
- Geospatial Data Governance
- Environmental and Cultural Preservation
- Humanitarian Aid and Disaster Response
- Harmonizing Legal and Ethical Frameworks
- Worldwide Ethical Standards.

In conclusion, international agreements and standards play a crucial role in establishing the moral guidelines for the worldwide use of geospatial data. They encourage cross-border collaboration and accountability in geospatial data ethics, responsible data sharing, privacy protection, interoperability, and the preservation of cultural and environmental assets.

ETHICAL DILEMMAS IN THE PRESENT AND GOVERNANCE BEST PRACTICES

Privacy Concerns and Geospatial Data

Geospatial data ethics places a premium on privacy concerns because location-based data can be invasive. This section covers the specific privacy issues surrounding geospatial data and the ethical issues that must be considered (Bertino et al., 2008).

- ***Location tracking:*** Tracking one's location is a major privacy concern. There are concerns about personal autonomy and invasion of privacy when someone's movements are continuously monitored.
- ***Consent and Data Collection:*** Privacy issues worsen when data is gathered without informed consent. Clear consent processes are necessary to gather and use location data to comply with ethical geospatial data practices.

- ***Data Granularity:*** Highly specific location data, like GPS coordinates, are frequently included in geospatial data sets. Greater potential privacy risks correspond with increased granularity of data.
- ***Data Sharing and Outside Parties:*** Giving away geographic data to outside parties may violate their privacy. Enforcing that third parties follow privacy standards and imposing stringent controls on data sharing are ethical considerations.
- ***Re-Identification Risk:*** Re-identification is possible even when data is anonymized. An individual's privacy may be jeopardized if seemingly anonymous data is combined with other sources of information to identify them.
- ***Ethical Data Management:*** Businesses using geographic information systems must implement ethical data management procedures. This also entails access controls, encryption, and routine data audits to preserve privacy.
- ***Minimization of Data Collection:*** The least amount of data possible is collected ethically using geospatial data. To lower privacy risks, only needed location data should be gathered.
- ***Children's Privacy:*** Extra care must be taken for children using geospatial applications. Parental consent and protecting children's location data are two ethical factors.
- ***Ethical Use of Location Data:*** Avoid using geographic data to jeopardize privacy or personal security. Instead, use it ethically. It should be against ethical guidelines to use location data for harassment or stalking.

CODE: *python*

```
import pandas as pd
# Load geospatial data (example: user location data)
# Replace 'geospatial_data.csv' with the actual file containing
your data
data = pd.read_csv('geospatial_data.csv')
# Perform anonymization of sensitive location data
# For demonstration purposes, let us drop columns containing
sensitive information
anonymized_data = data.drop(columns=['user_id', 'phone_number',
'email'])
# Save the anonymized data to a new CSV file
anonymized_data.to_csv('anonymized_geospatial_data.csv',
index=False)
print("Anonymized geospatial data saved successfully!")
```

In this example, We load geospatial data from a CSV file (replace 'geospatial_data.csv' with your file name). We perform anonymization of sensitive location data. We are dropping columns containing sensitive information like user IDs, phone numbers, and emails for demonstration purposes. Depending on your data and requirements, you may need to implement more sophisticated anonymization techniques. Finally, we save the anonymized data to a new CSV file ('anonymized_geospatial_data.csv'). In summary, location tracking, data consent, granularity, and data sharing are just a few of the many privacy issues with geospatial data. In order to appropriately address these issues, ethical geospatial data practices place a high priority on informed consent, privacy protection, and secure data management.

Bias and Discrimination in Geospatial Technologies

Because bias and discrimination in geospatial technologies have the potential to exacerbate social injustices and perpetuate inequality, they are serious ethical concerns. In using geospatial data and technologies, bias, and discrimination have ethical implications that are examined in this section (Oluoch, 2024) (Ma, 2024) (Ekeu-Wei & Blackburn, 2020).

- ***Bias in Data:*** Historical or social factors can introduce biases into geographic data, resulting in uneven representation in maps and analyses. Data biases must be found and corrected as part of ethical procedures.
- ***Algorithmic bias:*** Preexisting biases may be amplified by algorithms employed in geospatial technologies. Creating and implementing unbiased, neutral algorithms is a requirement of ethical considerations.
- ***Discriminatory Results:*** Using geospatial data and technologies can lead to discriminatory results, like zoning laws that disproportionately impact communities of color. Fairness can only be ensured through ethical decision-making.
- ***Indigenous Rights:*** The rights of indigenous communities to land and resources may be impacted by discriminatory practices. The rights and knowledge of indigenous peoples are respected in the ethical use of geospatial data.
- ***Redlining and Historical Injustices:*** Communities have been permanently impacted by historical injustices such as redlining. In order to promote equity, ethical geospatial data practices seek to recognize and address these injustices.
- ***Access to Services:*** Geospatial technologies can affect how easy it is to get public transportation, healthcare, and other necessities. Fair access is the main goal of ethical considerations.

- ***Equitable Resource Distribution:*** The distribution of resources can be ascertained using geospatial data. Ethical practices prioritize a fair distribution that does not discriminate against specific communities.

In conclusion, bias and discrimination in geospatial technologies present serious moral issues that impact environmental justice, social justice, and resource access. In order to guarantee that the use of location-based information promotes justice and equality, ethical practices in the use of geospatial data seek to recognize, address, and prevent biases.

Security Risks and Vulnerabilities

Geospatial data practices present ethical concerns regarding protecting sensitive location-based data, specifically regarding security risks and vulnerabilities. The ethical implications of geospatial technology security are examined in this section (Zhang et al., 2023) (Bera et al., 2021).

- ***Data Breach*:** People's privacy may be jeopardized when sensitive location data is exposed through data breaches in geospatial data systems. To avoid breaches, strong data security measures are the main focus of ethical considerations.
- ***Unauthorized Access*:** Unauthorized access to geographic data may lead to infringements on personal privacy, improper use of the data, and even possible harm to individuals. Preventing unwanted access is a crucial aspect of ethical behavior.
- ***Data Integrity:*** Maintaining the accuracy of geospatial data is required by ethics. Particularly in decision-making processes, data manipulation or tampering can have serious repercussions.
- ***Access Controls:*** Ensuring that only authorized individuals can access and modify geospatial data is an ethical obligation that must be carried out.
- ***Vulnerability Management:*** Vulnerability management is locating and fixing security flaws in data systems and applications. Ethical geospatial practitioners participate in this process.
- ***Data Auditing:*** To keep an eye on the security of geospatial data systems, identify possible security breaches, and guarantee data integrity, it is ethical to conduct regular data audits.
- ***Incident Response:*** Creating an incident response plan is morally required. It guarantees that procedures are in place to minimize harm and notify impacted parties during a security breach.

- ***Data Retention and Disposal:*** Both data retention and disposal are subject to ethical considerations. It is responsible for keeping data for as long as necessary and securely destroying it when it is no longer required.
- ***Ethical Hacking and Testing:*** Security testing and ethical hacking of geospatial systems are useful techniques to find and fix vulnerabilities early on.

In conclusion, there are ethical ramifications to security risks and vulnerabilities in geospatial data practices, especially regarding data breaches, unauthorized access, and privacy violations. Ethical practitioners prioritize data security, encryption, access controls, and vulnerability management to safeguard location-based data and uphold stakeholder confidence.

Cultural and Environmental Ethical Challenges

Geospatial data practices present various ethical challenges related to culture and the environment, including indigenous knowledge, cultural heritage, and environmental preservation. This section examines the moral implications of these problems in the context of geospatial technologies.

- ***Indigenous Understanding and Agreement:*** Informed consent from indigenous communities is required before using their traditional land or cultural information in geospatial applications as part of ethical geospatial data practices that respect indigenous knowledge.
- ***Preservation of Cultural Heritage*:** Preserving cultural heritage sites is a matter of ethics. Cultural sites and artifacts should be properly documented, safeguarded, and conserved through geospatial technologies.
- ***Land Rights and Ownership*:** Ownership and land rights present ethical difficulties. To support land tenure systems and acknowledge the rights of local communities and indigenous peoples, geospatial data should be used ethically.
- ***Cultural Sensitivity in Mapping*:** Cultural sensitivity is a prerequisite for ethical mapping procedures. To prevent misrepresentation or appropriation, maps should be made in collaboration with the local communities while honoring their cultural boundaries and customs.
- ***Community Engagement:*** Community engagement is prioritized in ethical geospatial practices. Participation of local communities, particularly indigenous groups, in decision-making processes concerning geospatial projects that impact their resources and land is imperative.

- ***Reducing Environmental Impact:*** Environmental impact is the goal of ethical geospatial practitioners. This involves using geospatial data to guide the assessment and mitigation of the ecological effects of human activity.
- ***Legal and Ethical Frameworks:*** Respecting legal and ethical frameworks helps address ethical dilemmas. International treaties, regional legislation, and moral standards safeguarding the environment and cultural heritage must all be followed by practitioners of geographic information science.

CODE: *Python*

```
import pandas as pd
import geopandas as gpd
import matplotlib.pyplot as plt
# Replace 'environmental_data.csv' with the actual file
containing your data
data = pd.read_csv('environmental_data.csv')
# Convert the DataFrame to a GeoDataFrame if it contains
spatial information
# For this example, let us assume the data includes latitude
and longitude columns
gdf = gpd.GeoDataFrame(data, geometry=gpd.points_from_
xy(data['longitude'], data['latitude']))
# Load a base map for visualization (e.g., world map)
world = gpd.read_file(gpd.datasets.get_path('naturalearth_
lowres'))
# Plot the environmental monitoring data on top of the base map
fig, ax = plt.subplots(figsize=(10, 6))
world.plot(ax=ax, color='lightgrey')
gdf.plot(ax=ax, marker='o', color='red', markersize=5)
plt.title('Environmental Monitoring Data')
plt.xlabel('Longitude')
plt.ylabel('Latitude')
plt.show()
```

The output of the visualization it creates can be a part of a larger workflow considering these ethical and data management principles. For instance, you might use the map to identify areas with potential privacy concerns due to high data density or to communicate the limitations of the data visually. In conclusion, geospatial data practices pose ethical challenges related to culture and the environment that call for a careful balancing act between development, conservation, cultural preservation, and indigenous rights. To ensure the ethical and sustainable use of geospatial technologies, ethical practitioners interact with local communities, respect cultural boundaries, and reduce environmental impact.

Ethical Guidelines for Geospatial Data Use

A set of guiding principles and factors for conscientious and ethical methods in the gathering, evaluating, and using location-based data is known as the "ethical guidelines for geospatial data use." Certain rules are imperative to guarantee that geospatial data is managed in a way that upholds privacy rights, fosters equity, and aids in conserving the environment and cultural heritage. Respecting individual privacy rights in the context of informed consent Individuals' right to privacy should be respected as the primary ethical precept. Geospatial practitioners should be aware of the sensitivity of location data and people's right to control their personal information. Obtain individuals' informed consent before gathering and utilizing their location data. This entails stating why data is being collected and how it will be used. People can make decisions about their data with the empowerment of informed consent (Beuthner et al., 2021) (Jafar et al., 2022).

- ***Data Minimization:*** Gather Just the Information Required: Restricting location data collection to that required for the intended purpose is an important ethical consideration. Steer clear of collecting too much or needlessly, as this raises ethical and privacy risks.
- ***Data Anonymization:*** Encrypt Data to Protect Privacy: Use effective data anonymization methods to protect privacy. The intention is to keep people's identities secret by preventing identifying specific individuals from location data.
- ***Security Measures:*** Give Data Security Top Priority: Data security is crucial to ethical geospatial practices. To prevent breaches and unwanted access to geospatial data, this entails putting encryption, access controls, and frequent security audits into place.
- ***Accountability and Transparency:*** Transparency is an essential ethical precept. Practitioners and organizations should be open and honest about gathering and using data. People can understand how their data is handled when there is transparency. Practitioners and organizations need to take responsibility for their use of geospatial data.
- ***Legal and Regulatory Compliance:*** When using geospatial data, observe all applicable laws, treaties, and moral guidelines. Maintain a responsible and moral approach by abiding by the law and upholding ethical standards.

- ***Responsible Data Management:*** When it comes to geospatial data, ensure it is securely disposed of after it is no longer needed and keep it for as long as necessary. Adopt data deletion guidelines that uphold people's right to be forgotten, encouraging ethical data handling.
- ***Active Bias Mitigation:*** To guarantee equity and equal representation in data analysis and mapping, proactively identify and mitigate biases in geospatial data, algorithms, and decision-making processes. Ethical practitioners give justice and objectivity top priority.

To sum up, the ethical guidelines for using geospatial data comprise an all-encompassing range of principles and practices that guarantee the conscientious and moral management of location-based data. These rules are essential for maintaining cultural heritage, the environment, and local communities' rights while safeguarding privacy and advancing justice.

Ensuring Privacy and Security in Geospatial Technologies

An essential component of ethical geospatial data management is ensuring security and privacy. The significance of encryption and privacy-preserving technologies in protecting location data is covered in this section. Additionally, it highlights the necessity of dynamic consent settings and granular user consent mechanisms to safeguard individuals' privacy when utilizing geospatial technologies.

- ***Privacy by Design:*** Privacy should be a top priority when developing and designing geospatial technologies. This idea, called "privacy by design," ensures that privacy is considered at every stage of technology development.
- ***Data Encryption:*** Geospatial technologies should use strong encryption techniques to protect location data during transmission and storage. Data confidentiality and protection from unauthorized access are guaranteed by encryption.
- ***User Data Control:*** Users should be able to manage their data using geospatial technologies. This covers choices about participation, access to concise and clear data usage policies, and the capacity to examine and remove their data.
- ***Data Access Controls:*** To guarantee that only individuals with permission can view and modify geospatial data, strict access controls must be in place. By doing this, the chance of illegal use and data breaches is reduced.
- ***Audits of data security:*** Perform routine security audits to identify weaknesses and guarantee data security in geospatial systems and applications. Proactively detecting and correcting security flaws is a hallmark of ethical practitioners.

- ***Transparency in Data Handling:*** Transparent Data Practices Continue to be open about how you collect and use data. Organizations and practitioners should be open and honest about how they handle data to promote accountability and trust.

In conclusion, protecting sensitive data, upholding people's rights, and adhering to ethical and legal norms all require a proactive and all-encompassing approach to privacy and security in geospatial technologies. These procedures are essential for preserving user confidence and safeguarding private location-based data.

Promoting Fair and Inclusive Geospatial Data Practices

In order to ensure that everyone has equal access to the advantages of location-based information and to reduce the possibility of bias and discrimination, fair and inclusive geospatial data practices are crucial. The ethical guidelines and practices that support equity and inclusivity in the use of geospatial data are described in this section.

- ***Bias Mitigation:*** To advance equity and equal representation, geospatial practitioners should proactively recognize and address biases in data, algorithms, and decision-making procedures.
- ***Fair Resource Distribution:*** Make sure that resources are distributed fairly and do not unfairly target specific communities when using geospatial data. Fairness is given priority in ethical behavior.
- ***Community Involvement:*** When it comes to geospatial projects that affect local communities' land, resources, and cultural heritage, particularly indigenous groups, they should be actively involved in the decision-making process.
- ***Ethical Data Collection:*** Respect cultural boundaries and customs by gathering geospatial data appropriately and conservatively. Steer clear of appropriation or deception.
- ***Environmental Equity:*** By guaranteeing fair access to green spaces and addressing differences in environmental quality, geospatial data can be used to address environmental justice issues.
- ***Fair Decision-Making:*** Ethical geospatial professionals should prioritize fair outcomes and avoid discrimination when making decisions, considering possible effects on various communities.
- ***Access to Vital Services:*** The availability of vital services, such as public transportation, healthcare, and education, is frequently assessed using

geospatial data. Ensuring access is fair and does not discriminate against underrepresented communities is a top priority for ethical practices.

- ***Environmental Stewardship***: Use geospatial data to encourage sustainable development that benefits all communities, ethical resource extraction, and environmental preservation.
- ***Legal and Ethical Compliance:*** Adherence to legal and ethical frameworks is essential to guarantee equitable and inclusive geospatial practices. International accords, regional legislation, and moral standards should all be followed by ethical practitioners.

To summarise, equitable access to location-based information benefits, mitigating biases, and addressing disparities depend on promoting fair and inclusive geospatial data practices. Ethical geospatial practitioners prioritize fairness and inclusivity when making decisions, actively engage with various communities, and respect cultural boundaries.

SHAPING THE FUTURE OF GEOSPATIAL DATA ETHICS AND FUTURE SCENARIOS

Emerging Technologies and Trends

The geographic data field is constantly evolving due to technological and fashion advancements. Understanding these developments is essential to influencing the future of geospatial data and ensuring its ethical use. This section examines recent advancements in geospatial data trends and technologies, highlighting their significance and implications.

- ***Learning Machines and Artificial Intelligence:*** Artificial intelligence (AI) and machine learning transform geospatial data analysis by enabling automated pattern recognition, predictive modeling, and real-time decision-making.
- ***Technological Advancements in Remote Sensing:*** High-resolution, real-time data for geospatial applications is made possible by developments in remote sensing technologies, such as satellite imagery and aerial photography.
- ***Blockchain Technology:*** Research is being done to see if blockchain technology can enhance data security, integrity, and decentralized ownership. It can provide a trustworthy, uncrackable ledger for location data. A few examples of ethical concerns are ensuring transparency and accessibility in blockchain-based geospatial data systems and addressing the environmental effects of energy-intensive blockchain networks.

- ***Real-time Data Streams:*** With the advent of real-time data streams from sensors, Internet of Things (IoT) devices, and mobile applications, the collection and use of geographic data is changing.
- ***Open Data Initiatives:*** These efforts encourage the public to benefit from sharing geospatial data. A diverse range of users can now access location-based information thanks to the efforts of governments, organizations, and private citizens. Respecting data privacy and ensuring that open data initiatives do not violate people's rights or reinforce prejudices are examples of ethical practices.
- ***Data from the Crowd and Citizen Science:*** Citizen science initiatives and crowdsourced data foster public participation in data collection, validation, and analysis. In addition to democratizing geospatial data, this can offer important insights for study and decision-making. Ensuring data privacy and quality in crowdsourced projects, recognizing citizen scientists' contributions, and resolving any biases in the data are all ethical considerations.
- ***Data Visualization and Storytelling:*** Geospatial data is becoming more readable and interesting through data visualization techniques. Using data to tell stories is becoming increasingly crucial in spreading knowledge to a large audience. Ensuring data visualizations are understandable, objective, and faithfully depict the underlying information is part of ethical best practices. When telling a story, the data's context and ethical issues should be respected.

In conclusion, there are many chances for creativity and solving difficult problems presented by new technologies and developments in geospatial data. Ethical considerations are necessary to navigate these trends responsibly, protect privacy, advance equity, and guarantee that the advantages of geospatial data are distributed fairly. The dynamic environment of geospatial data necessitates constant ethical examination and adjustment to fulfill societal demands while upholding individual liberties and cultural diversity.

Global Collaborations for Ethical Geospatial Data

In geospatial data, international collaboration is essential for establishing moral guidelines, exchanging knowledge, and tackling global issues. The importance of international cooperation in advancing moral geospatial data practices is discussed in this section, along with the field's ramifications.

- ***Harmonization and Compatibility:*** International cooperation makes creating uniform standards and geospatial data interoperability protocols easier. This guarantees the efficient integration and sharing of data from various sources and geographical areas, thereby augmenting the usefulness and significance of geospatial data. Promoting open standards and interoperability is a key component of ethical practices to avoid data fragmentation and silos. This facilitates efficient cross-border data sharing and fair access.

- ***Data Accessibility and Sharing:*** Through cooperative efforts, countries and organizations can share geospatial data, leading to a more thorough and knowledgeable grasp of global issues like public health, disaster relief, and climate change. Ensuring that data-sharing agreements uphold data security and privacy while encouraging accessibility and conscientious data management are examples of ethical considerations.
- ***Preservation of Indigenous and Cultural Knowledge:*** Respecting and preserving indigenous and cultural knowledge requires cooperative efforts. Traditional ecological knowledge, land rights, and cultural heritage can all be documented and protected using geospatial data. Respecting the rights and customs of indigenous communities and collaborating with them are key components of ethical practices. Cultural sensitivity and knowledge preservation should be given top priority in collaborative projects.
- ***Privacy and Data Protection:*** International agreements and norms about data privacy and protection must be defined through national collaboration. Since geospatial data frequently crosses national borders, international standards are required to safeguard people's rights. Engaging in global initiatives to create data protection and privacy accords is part of ethical behavior. To solve cross-border data issues and protect personal privacy, cooperation is crucial.
- ***Data Ownership and Rights:*** In cross-border and international contexts, collaborative efforts are especially helpful in addressing complex issues about data ownership. Ownership and rights over geospatial data must be established and upheld.

International cooperation is critical in developing transnational ethical geospatial data practices that support responsible data sharing, management, and protection. Ensuring equity, transparency, and respect for cultural diversity, privacy, and data ownership on a global scale are ethical considerations in such collaborations. In order to handle the many opportunities and challenges brought forth by the changing geospatial data landscape, international cooperation is essential.

The Role of Education and Awareness

Education and awareness greatly influence Ethical geospatial data practices, affecting practitioners and the general public. The importance of education and awareness in encouraging moral behavior, developing a culture of accountability, and guaranteeing the moral application of geospatial data is expounded upon in this section.

- ***Initiatives in Education:*** Geospatial ethics-focused educational programs, courses, and workshops equip practitioners with critical knowledge and abilities. These programs increase public awareness of best practices for managing and analyzing geospatial data, legal frameworks, and ethical issues.
- ***Moral Decision-Making Proficiency:*** Education provides individuals with the information and abilities to assess geospatial initiatives' ethical implications. Professionals who can make ethical decisions are better equipped to identify, assess, and resolve moral conundrums in their field of work.
- ***Diversity Awareness and Cultural Sensitivity:*** Education fosters diversity and cultural sensitivity awareness. When dealing with local communities and indigenous knowledge, geospatial professionals must learn to respect cultural boundaries and customs.
- ***Community Outreach and Involvement:*** Building relationships with local communities via outreach initiatives, workshops, and seminars fosters mutual respect and understanding. Participation from the community guarantees that geospatial projects respect cultural values, take ethical issues into account, and are in line with community needs.

To summarise, education and awareness campaigns are crucial in establishing moral values, encouraging conscientious conduct, and cultivating an ethical culture within the geospatial data industry. People who receive ethical education are better equipped to make moral decisions, interact with communities civilly and respectfully, and advance the moral development of geospatial technologies and practices. A sustainable and morally sound geospatial ecosystem requires constant awareness and learning.

Envisioning Ethical Geospatial Data in the Future

Predicting how ethical concerns will influence technological advancements, shape the field's growth, and affect societal outcomes is essential to the vision of geospatial data. This section examines the opportunities and challenges for ethical geospatial data practices in the coming years.

- ***Enhanced Privacy Technologies:*** Advancements in anonymization, differential privacy, and encryption for robust security of individuals' location data. Balancing privacy and essential uses like public safety requires ongoing efforts to respect individual rights.

- ***Bias Mitigation in Algorithms:*** Ethical emphasis on minimizing biases in geospatial algorithms and decision-making processes. Development of advanced AI models to identify and rectify biases, addressing challenges in achieving bias-free practices.
- ***Global Data Collaboration:*** Standardizing data sharing agreements for addressing global challenges like climate change and disaster response. Diplomatic efforts are needed to balance national interests and global needs amid geopolitical differences.
- ***Ethical AI Prioritization:*** Prioritization of ethical AI and transparent algorithms in geospatial technologies for public infrastructure, urban planning, and autonomous vehicles. Ongoing challenges in ensuring accountability for AI decisions and making them comprehensible to users.
- ***Cultural Sensitivity in Procedures:*** Ethical considerations becoming more attuned to cultural nuances, respecting indigenous and local communities. Continuous efforts are needed to balance technology demands with cultural preservation and foster cooperation and trust.
- ***Impact Assessment for Sustainability:*** Geospatial data's role in evaluating environmental and social impacts for responsible resource extraction and sustainable development, balancing development with social and environmental concerns, and acknowledging trade-offs in ethical evaluations.
- ***User Empowerment and Informed Consent:*** Increased user control with options for informed consent in geospatial data collection.
- ***Ethical Entrepreneurship for Innovation:*** Ethical entrepreneurship drives innovation in geospatial technologies, focusing on morality and responsible behavior.

In conclusion, the future of ethical geospatial data practices presents opportunities and challenges. Guided by ethical considerations, these practices will serve societal needs while upholding individual rights, cultural diversity, and environmental sustainability, shaping a responsible and thoughtful data ecosystem.

The Impact of Current Actions on Future Ethical Practices

Future ethical behavior is greatly influenced by the actions taken today. This section looks at how decisions, policies, and laws made in the geospatial data industry today will affect the ethical environment in the future.

- ***Regulations and Data Governance:*** The current initiatives to develop solid data governance and moral regulations will pave the way for future ethical geospatial data practices. Unambiguous rules and legislative frameworks will continue to impact data management and privacy protection. Finding the ideal balance between innovation and data regulation is crucial.

- ***Ethical Education and Training:*** Future practitioners' ethical competency will be determined by the caliber and scope of the moral education and training given to geospatial professionals today. Having a solid foundation in morality and judgment is essential. The ability of professionals to resolve challenging situations will be crucial to the future of geospatial data ethics. It will be essential to prioritize ethical awareness in training programs and to provide ongoing education.
- ***Public Awareness and Advocacy:*** The degree to which ethical geospatial data practices are currently understood and supported by the public will impact the demands made on businesses and governments to act responsibly.
- ***International Cooperation:*** By using geospatial data sharing to address global challenges, ongoing international collaborations will establish the standard for future cross-border cooperation.
- ***User Empowerment and Informed Consent:*** Present efforts to provide users authority over their data and informed consent will influence procedures in the future. The main focus of ethical data management will continue to be protecting user rights and privacy.
- ***Innovation and Ethical Entrepreneurship:*** Innovation in the future will be driven by ethical entrepreneurship, encouraging ethical and responsible behavior. Business decisions will continue to be influenced by the industry's ethical culture.

In conclusion, decisions and actions made today significantly impact future ethical behavior. The level of ethical integrity in the geospatial data field will depend on the current and future commitment to ethical governance, education, public advocacy, technological advancements, cultural respect, and responsible entrepreneurship. A sustainable and moral geospatial ecosystem requires balancing innovation and ethical responsibility.

Ethical Challenges and Opportunities in the Evolving Geospatial Landscape

The constantly changing geospatial environment presents various ethical opportunities and challenges that shape responsible data practices. These aspects are examined in more detail in this section.

- ***The Internet of Things and Data Privacy:*** There has never been more geospatial data thanks to the widespread use of Internet of Things (IoT) devices and sensors. During this data flood, maintaining people's privacy is a major challenge. Ensuring that private location information is not misused or disclosed without authorization is crucial.

- ***Data Protection in a Globalized Society:*** One major security challenge is the growing interconnectedness of geospatial data systems and increased cyber threats. Rebuilding the confidence of users and data providers requires constant efforts to protect data from breaches, cyberattacks, and unauthorized access.
- ***Cross-Border Data Governance:*** Addressing global issues requires international cooperation and data sharing. Harmonizing disparate legal and regulatory frameworks to create moral cross-border data governance is still difficult and delicate. In order to guarantee that data is used responsibly across borders, cooperation and diplomacy are needed.
- ***Rights of Indigenous and Local Communities:*** Respecting the rights of indigenous and local communities is morally challenging, particularly when it comes to their land, cultural heritage, and traditional wisdom. Collaborative projects must respect and engage these communities while navigating intricate cultural sensitivities, traditional boundaries, and knowledge preservation.
- ***User-Centric Data Empowerment:*** Encouraging users to take ownership of their geospatial data and obtaining informed consent opens doors to user involvement and responsible data management. This user-centric strategy respects both individual rights and ethical standards.
- ***Innovation in Geospatial Technologies:*** Ethical entrepreneurs have the potential to stimulate responsible practices within the industry and propel innovation in this field. These changes encourage an ethical business culture and moral decision-making.

A dynamic interaction between ethical opportunities and challenges defines the constantly changing geospatial environment. In order to shape a future where geospatial data practices remain responsible and ethically conscious, we must address these challenges while seizing the opportunities that present themselves. Geospatial technologies will continue to be developed and applied with ethical considerations in mind, guaranteeing that the rights of individuals, cultural diversity, and environmental sustainability are all respected while the technology advances society.

DATA ANALYSIS AND TECHNICAL INFORMATION IN GEOSPATIAL ETHICS AND DATA PRIVACY

The document provides an overview of geospatial data ethics, covering its historical evolution, current landscape, and future scenarios. It emphasizes the importance of ethical considerations in collecting, storing, and using geospatial data. The historical perspective section explores how geospatial data collection methods have evolved and discusses early geospatial data practices' positive and negative impacts. The current landscape section delves into modern data sources, technologies, and applications and has implications for geospatial data across various fields. It also highlights emerging ethical challenges arising from the rapid growth of geospatial data (Coetzee et al., 2020).

Ethics in geospatial data is examined through fundamental principles and legal frameworks, addressing privacy, security, data ownership, fairness, bias, and environmental impacts. It also discusses international agreements and standards governing geospatial data practices. The ethical dilemmas section explores privacy, bias, security, cultural heritage, and environmental responsibility challenges. It suggests governance best practices and ethical guidelines for promoting fair and inclusive geospatial data practices.

Looking toward the future, the document discusses emerging technologies, global collaborations, education, and awareness in shaping ethical geospatial data practices. It underscores the need for continued efforts to ensure responsible geospatial data handling. In conclusion, the document highlights the significance of moral considerations in the rapidly evolving field of geospatial data, advocating for ethical governance and responsible data practices to address present challenges and anticipate future scenarios.

The document encompasses a broad spectrum of technical information on geospatial data, spanning historical evolution, current applications, and future scenarios. It delineates the evolution of data collection methods, including advancements in satellite technology, remote sensing, and IoT sensors, which have revolutionized data acquisition. Moreover, it delves into the complexities of data processing, storage, and analysis, highlighting the integration of machine learning algorithms and AI in geospatial analysis. The document also underscores the significance of metadata standards, data interoperability, and quality assurance measures in ensuring the reliability and accuracy of geospatial datasets. Additionally, it discusses emerging technologies such as blockchain for enhancing data security and integrity. The document provides a comprehensive overview of the technical intricacies underpinning geospatial data management and analysis, emphasizing the need for continued innovation and ethical governance in this rapidly evolving domain.

CONCLUSION

The ethical dimensions of geospatial data form a dynamic tapestry woven through historical evolution, contemporary challenges, and prospects. From manual mapping in ancient civilizations to today's sophisticated technologies, ethical considerations have become inseparable from collecting, analyzing, and disseminating location-based information. Current ethical challenges encompass privacy, security, bias, environmental impact, and cultural sensitivity. Fundamental ethical principles guide responsible geospatial data practices, including privacy, security, fairness, and cultural and ecological impact considerations. Regulatory structures and international cooperation play pivotal roles in shaping the ethical landscape, acknowledging the importance of legal frameworks and global collaboration.

The ongoing ethical dialogue within the geospatial data community reflects its adaptive nature, responding to emerging technologies, evolving data governance, and cultural and environmental responsibilities. This continuous conversation raises awareness, promotes education, and fosters ethical business practices. Ethical considerations remain at the forefront as geospatial data drives innovation and societal progress. The dynamic discourse ensures that ethical frameworks evolve to align with changing norms and technological advancements. Through this commitment to moral principles and ongoing dialogue, the geospatial data community can navigate complexities, advance responsible practices, and contribute to a future where geospatial data is a force for societal benefit, respecting individual rights, cultural diversity, and environmental sustainability.

REFERENCES

Al Jawarneh, I. M., Bellavista, P., Corradi, A., Foschini, L., & Montanari, R. (2020). Big spatial data management for the Internet of Things: A survey. *Journal of Network and Systems Management*, *28*(4), 990–1035. doi:10.1007/s10922-020-09549-6

Bera, A., Taloor, A. K., Meraj, G., Kanga, S., Singh, S. K., Đurin, B., & Anand, S. (2021). Climate vulnerability and economic determinants: Linkages and risk reduction in Sagar Island, India; A geospatial approach. *Quaternary science advances, 4*, 100038.

Bertino, E., Thuraisingham, B., Gertz, M., & Damiani, M. L. (2008, November). Security and privacy for geospatial data: concepts and research directions. In *Proceedings of the SIGSPATIAL ACM GIS 2008 International Workshop on Security and Privacy in GIS and LBS* (pp. 6-19). ACM. 10.1145/1503402.1503406

Beuthner, C., Breuer, J., & Jünger, S. (2021). *Data linking—linking survey data with geospatial, social media, and sensor data*. GESIS Survey Guidelines.

Coetzee, S., Ivánová, I., Mitasova, H., & Brovelli, M. A. (2020). Open geospatial software and data: A review of the current state and a perspective into the future. *ISPRS International Journal of Geo-Information*, *9*(2), 90. doi:10.3390/ijgi9020090

Ekeu-Wei, I. T., & Blackburn, G. A. (2020). Catchment-scale flood modelling in data-sparse regions using open-access geospatial technology. *ISPRS International Journal of Geo-Information*, *9*(9), 512. doi:10.3390/ijgi9090512

Gómez, P. M., Santiago, A. R., Seco, G. G., Casanova, R., MacKenzie, D., & Tucker, C. (2022). Ethics in the use of geospatial information in the Americas. Technology in Society, 69, 101964.

Huang, H., Yao, X. A., Krisp, J. M., & Jiang, B. (2021). Analytics of location-based big data for smart cities: Opportunities, challenges, and future directions. Computers, Environment and Urban Systems, 90, 101712. Chang, V., Mou, Y., & Xu, Q. A. (2021). The ethical issues of location-based services on big data and IoT. In Modern Industrial IoT, Big Data and Supply Chain [Springer Singapore.]. *Proceedings of the IIoTBDSC, 2020*, 195–205.

Jafar, A., Dollah, R., Sakke, N., Mapa, M. T., Hua, A. K., Eboy, O. V., Joko, E. P., Hassan, D., & Hung, C. V. (2022). Assessing the challenges of e-learning in Malaysia during the pandemic of Covid-19 using the geo-spatial approach. *Scientific Reports*, *12*(1), 17316. doi:10.1038/s41598-022-22360-4 PMID:36243784

Kedron, P., & Trgovac, A. B. (2021). Assessing connections and tradeoffs between geospatial data ethics, privacy, and the effectiveness of digital contact tracing technologies. *Mapping COVID-19 in Space and Time: Understanding the Spatial and Temporal Dynamics of a Global Pandemic,* 115-136.

Kross, A., Kaur, G., & Jaeger, J. A. (2022). A geospatial framework for the assessment and monitoring of environmental impacts of agriculture. *Environmental Impact Assessment Review*, *97*, 106851. doi:10.1016/j.eiar.2022.106851

Ma, Q., Duan, Y., & Yao, Z. (2023). Meta-analysis of the impact of geospatial technologies on learning outcomes. *Education and Information Technologies*, *28*(12), 15739–15764. doi:10.1007/s10639-023-11712-w

Oluoch, I. (2024). Crossing Boundaries: The Ethics of AI and Geographic Information Technologies. *ISPRS International Journal of Geo-Information*, *13*(3), 87. doi:10.3390/ijgi13030087

Ricker, B., Schuurman, N., & Kessler, F. (2015). Implications of smartphone usage on privacy and spatial cognition: Academic literature and public perceptions. *GeoJournal*, *80*(5), 637–652. doi:10.1007/s10708-014-9568-4

Zhang, C., Wang, K., Yue, Y., Qi, X., & Zhang, M. (2023). Assessing Regional Ecosystem Conditions Using Geospatial Techniques—A Review. *Sensors (Basel)*, *23*(8), 4101. doi:10.3390/s23084101 PMID:37112442

Chapter 3
Navigating Ethical and Privacy Frontiers:
Geospatial Data in Collaborative Environments

R. Madhubala
University of Technology and Applied Science, Shinas, Oman

Saleem Raja Abdul Samad
https://orcid.org/0000-0002-7203-1426
University of Technology and Applied Science, Shinas, Oman

S. Thangam
Amrita Vishwa Vidyapeetham, India

ABSTRACT

Geospatial data maps people and objects on Earth, predominantly used for applications like emergency management, environmental monitoring, land planning, city planning, and military operations. These applications frequently require collaboration between multiple organizations, their respective data repositories, and users with varying responsibilities. These data can also be used to locate services such as schools and hospitals, develop local transportation and housing, monitor weather and traffic conditions, and design flood defenses. Addressing the ethical and privacy concerns of collecting and analyzing diverse data, particularly geospatial data, requires a delicate balance between individual rights and societal needs.

DOI: 10.4018/979-8-3693-6381-2.ch003

INTRODUCTION

Geospatial data, or geo-data, encompasses details about specific Earth locations. There are two types of Geospatial data: spatial data and attribute data. Spatial data consists of coordinates on the Earth's surface that indicate the location and shape of geographic features such as points, lines, and polygons. Lines depict linear components such as highways or rivers, whereas points represent countries or geographical units. However, attribute data includes these geographic features' population, temperature, land use, and other essential qualities. When combined, these two types of Geospatial data provide a full picture of both spatial relationships and the various attributes connected with the Earth's surface. Geographic data can be dynamic, such as the movement of a vehicle or the growth of an infectious disease, or static, such as the position of equipment, the occurrence of an earthquake, or the distribution of poor children. Geography is essential for urban planning, environmental monitoring, disaster management, agriculture, transportation, etc.

GIS manages, analyzes, and visualizes geographic data, enabling users to make informed decisions based on spatial relationships and patterns. Geospatial technologies include various tools and techniques to collect, process, store, and visualize geographical data. Geospatial technologies, like remote sensors, aid in visualizing inaccessible or hazardous areas, enabling safe assessment of damage from disasters or conflicts and monitoring recovery rates. They facilitate tracking the distribution of relief goods, analyzing temporary settlements, monitoring population displacement, and modeling environmental impacts for targeted disaster management planning (Manfre et al., 2012). Many Geographic Information Systems (GIS) software programs can analyze and visualize vast amounts of Geospatial data to find correlations, trends, and patterns. The most popular of these is ArcGIS, a popular Esri platform for mapping, spatial analysis, and reliable data management. QGIS is an open-source alternative with a simple interface and several spatial data analysis features. Google Earth Engine, a cloud-based environmental data analysis tool, excels at satellite image visualization and interpretation. Despite their benefits, Geospatial technologies bring ethical challenges like privacy and location-based stigma and discrimination.

Data ethics and privacy are of the utmost importance in geospatial data processing to protect individuals' location-related information and guarantee fair and transparent usage. Maintaining ethical procedures and privacy standards throughout the collection, analysis, and dissemination of Geospatial data is critical for building confidence, reducing the risk of misuse, and protecting individuals' rights. (Lee et al., 2022).

This chapter explores the ethical and privacy challenges inherent in Geospatial data, addressing key issues such as bias, discrimination, privacy concerns, and ownership dilemmas. A comprehensive mitigation plan is proposed to navigate these challenges, encompassing strategies for promoting fairness, protecting individual privacy, and clarifying data ownership. The chapter delves into legal frameworks governing privacy in Geospatial data, emphasizing the importance of compliance with regulations. Additionally, it discusses advanced techniques like anonymization and aggregation as essential tools for enhancing privacy while preserving the utility of Geospatial information.

ETHICS IN GEOSPATIAL DATA

Geospatial data ethics governs the ethical collection, analysis, dissemination, and use of spatial data. The primary objective of Geospatial data ethics is to tackle potential issues such as privacy concerns, discrimination, and biases that could emerge at any stage of the data lifecycle (Bertino et al., 2008). It is crucial to balance the benefits of utilizing location-based information and the rights of individuals (Shaham et al., 2019).

Bias and Discrimination in Geospatial Data

Bias and discrimination in Geospatial data refer to unfair or disproportionate effects on specific groups or individuals in the context of location-based information (Zhang et al., 2022). These biases might appear at different data collection, processing, and analysis phases, resulting in skewed or discriminating results.

- In Geospatial data gathering, sampling bias may disproportionately reflect certain demographics or places, resulting in an incomplete or biased dataset (Bystriakova et al., 2022).
- Geospatial datasets underrepresent particular populations, creating a visibility gap that may encourage resource allocation and decision-making biases (Freire et al., 2016).
- Biases in Geospatial algorithm training data can perpetuate or worsen inequities in predictions that affect urban planning, public services, and law enforcement (Manuel et al, 2023).
- Geospatial data bias can impair infrastructure development, education, and healthcare access. Biased Geospatial data may also cause environmental injustice in some communities (Cobb, 2020).
- Geospatial data may reveal sensitive information about gender identity or sexual orientation, risking privacy and even causing discrimination or harassment (Kress et al., 2021).

- Geospatial data gender and sexuality biases can disproportionately affect healthcare, social services, and public infrastructure allocation (Weber et al., 2021).

In order to promote fairness and equality in Geospatial data applications, it is important to deal with these biases.

Approaches to Mitigate Biases

- Use inclusive sampling methods to appropriately represent the population's variety and avoid underrepresentation or skewing. For example, in a municipality's urban planning program, inclusive sampling approaches were used to depict demographic diversity using Geospatial data. Rather than relying on traditional sampling that might inadvertently underrepresent certain communities, the project team actively sought input from residents across various neighborhoods, considering factors such as socioeconomic status, cultural backgrounds, and housing types (Cobb, 2003)). This inclusive approach provided a more accurate snapshot of the city's demographics and helped identify potential biases in the initial data collection.
- Use algorithms conscious of fairness to proactively address and correct biases and increase transparency by offering justifications for algorithmic outcomes. For instance, public transportation routes can be optimized using Geospatial data. The application can actively identify and mitigate biases that may disproportionately impact certain neighborhoods or demographic groups by integrating fairness-aware algorithms. These algorithms would rectify existing imbalances and explain the decision-making process, enabling city planners and stakeholders to understand how the algorithm considers different factors (Elwood & Cope, 2009). This approach enhances transparency and accountability and ensures that the Geospatial application contributes to equitable and unbiased urban planning decisions, promoting fairness across diverse communities.
- Foster cooperative alliances with community representatives, ensuring their input is duly considered throughout data acquisition, analysis, and application development. For instance, a city government is undertaking a Geospatial project to improve public transportation services. To foster cooperative alliances with community representatives, the government collaborates with local advocacy groups and residents (Dhand et al., 2021). Community representatives are invited to participate in workshops and meetings where they share insights about transportation needs, accessibility challenges, and preferred service locations.

- During the data acquisition phase, the government involves these representatives in defining key areas for data collection, ensuring that the Geospatial data is accurate and reflects the community's diverse needs. In the analysis stage, community representatives are consulted to interpret and validate findings, providing valuable context and nuances that might not be apparent solely from the data. When developing applications or implementing changes based on Geospatial insights, continuous feedback loops with community representatives are established to guarantee that the solutions align with the community's priorities and concerns. This collaborative approach ensures that the Geospatial project is technically sound and considers the real-world impact on the community it serves.
- Use anonymization and encryption to protect geographical data from inadvertent disclosure or discrimination. For instance, in a public health endeavor to map disease prevalence using Geospatial data, anonymization, and encryption protect sensitive geographical information (Jing et al., 2023). To protect participant identities, location data is anonymized by replacing IDs with randomized codes before analysis. To prevent accidental disclosure, the dataset is encrypted during transmission and storage.
- Establishing ethical frameworks for Geospatial data collection, analysis, and use that comply with industry standards and laws while educating data professionals on Geospatial data ethics, biases, and impacts promotes understanding and accountability in the responsible handling of location-based information (Cinnamon et al., 2020).

Privacy in Geospatial Data

Geospatial data privacy protects individuals against unauthorized tracking, disclosure, or exploitation of their location information. This includes protecting a person's geographic location, contextual data, and temporal features. To ensure competent and ethical handling of location-based data, it is critical to balance the benefits of Geospatial technology and the protection of individual privacy rights (. There are various geographical data privacy concerns:

- Geolocation privacy: protects people's precise coordinates or locations from unlawful tracking or disclosure (Colonna et al., 2020).
- Identity Protection: Protects individuals' identities during geographic data analysis, preventing personal details from being disclosed (Rubinstein, et al., 2015).

Figure 1. Geospatial data privacy concerns

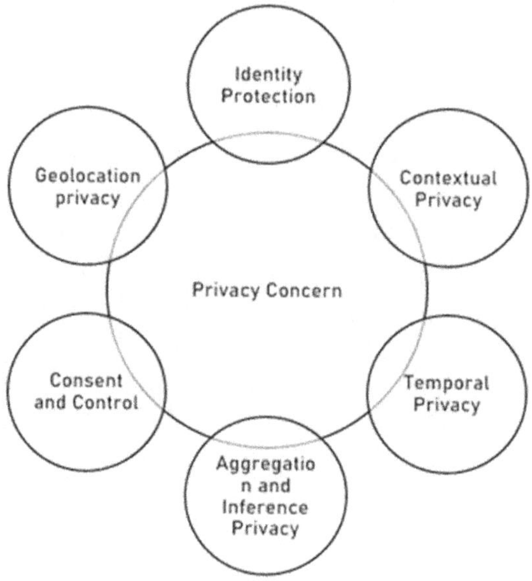

- Contextual Privacy: Protect location data's surrounding context, which includes people's actions and behaviors, from unauthorized access to their personal information (Garfinkel et al., 2015).
- Temporal Privacy: Protecting time-related geographical data, such as an individual's presence at a location, to prevent illegal tracking (Kerski, 2003)).
- Aggregation and Inference Privacy: Protecting people's sensitive information when data is aggregated, and conclusions are derived from it (Hoofnagle et al., 2019).
- Consent and Control: Emphasize the need for informed consent from persons before using geographical data. It also emphasizes controlling people's location data (Mei et al., 2022). The various Geospatial privacy concerns are given in Figure 1.

Approaches to Enhance Privacy in Geospatial Data

Implement methods such as anonymization and aggregation to eliminate or combine identifiers, thereby increasing the difficulty of attributing data to specific individuals. For example, a ride-sharing service that tracks user locations to improve efficiency. The service could replace user identifiers with generic or random identifiers to improve location privacy (Kuner, 2021). Conversely, rather than retaining individual users' precise GPS coordinates, the service could aggregate location data to generate density maps or heat maps that depict general patterns of user movement while safeguarding the anonymity of specific users. In this manner, the service ensures location privacy is prioritized while substantially reducing the risk of identifying and tracing individual users while retaining valuable insights into user behavior to enhance its operations.

Implement strong consent mechanisms to guarantee that individuals are aware and can grant or decline permission to collect and use their location data. Consider a smartphone application that makes personalized recommendations for nearby attractions or businesses based on their location. The app may employ a simple and user-friendly interface to tell users about location data collection and use it to build a strong consent mechanism. The first time users opened the app, a pop-up message explained location tracking and the benefits of personalized recommendations (Breunig et al., 2020). The notification would clearly state how and for how long location data will be used. Subsequently, users would be presented with the choice of granting or declining consent. The app might also have a settings option where users can change their location preferences and revoke or adjust their consent. It gives consumers informed options, respects their autonomy, and uses location data transparently and ethically.

Geospatial data should be encrypted during transmission and storage to prevent unauthorized access. Consider a Geospatial-based navigation app for real-time location. This app might use end-to-end encryption to improve security. It uses a strong cryptographic technique to encrypt geographical data connected with the route when a user enters their destination (Al-Hasani et al., 2023). The encrypted data would be transmitted via a secure connection to the app's servers for route calculation. Only authorized entities with decryption keys could access and understand geographical data. This protects users' geographical data from unauthorized access and tampering by making location information incomprehensible without cryptographic keys, even if intercepted during transport or storage.

Geo-fencing, a technique that entails the establishment of virtual boundaries, can be utilized to regulate and impede the gathering of location data within designated regions, provided that explicit consent is obtained (Schafer & Alexandro, 2023). Consider a mobile application that provides location-based promotions for a retail store. For geo-fencing, the app might set virtual boundaries around store locations. Users receive notifications on their devices regarding promotions and discounts when they enter geo-fenced regions. However, the app would not actively gather or store precise location data unless users consent to participate in the promotional program. Geo-fencing allows the app to provide a contextually relevant experience for users in specific places while protecting their privacy.

Adhere to privacy laws and regulations to guarantee that geographic data processing practices align with legal frameworks, giving a standardized approach to protecting location privacy (Kedron et al., 2020)[.

LEGAL FRAMEWORK FOR PRIVACY

Legal frameworks and industry standards shape geospatial data processing privacy practices (Cobb & Stephen, 2016). The legal frameworks governing privacy preservation in Geospatial data processing may differ depending on the jurisdiction. Some international guidelines and frameworks include the following:

- The US government regulates geographic data privacy through federal, state, and sector-specific laws. However, there is no all-encompassing federal legislation particularly addressing the privacy of Geospatial data. In some cases, the sectorial application of privacy regulations, such as the Health Insurance Portability and Accountability Act (HIPAA) and the Gramm-Leach-Bliley Act (GLBA), may include geographical data. State laws also matter, such as California's Consumer Privacy Act (CCPA), which gives residents rights over their data, including geographical data. The US Geospatial data privacy legal landscape evolves through existing regulations and new legislation as technology develops and location privacy concerns grow (Onsrud et al., 1994)).
- The General Data Protection Regulation (GDPR), which was established in 2016 and went into effect in 2018, is the fundamental legal framework for the protection of personal data in the European Union (EU) (Chris, 2019). Businesses handling personal data must follow rigorous GDPR data protection guidelines, such as:
    - Transparency: Businesses must explain their data processing purpose and legal basis.
    - Limited purpose: Personal data should only be processed for explicit objectives.
    - Data minimization: Companies should only collect and use the bare minimum of personal information essential to achieve their objectives.
    - Precision: Personal data should be complete and current.
    - Storage limit: Personal information should only be kept for as long as necessary.
    - Integrity and confidentiality: Unauthorized access, disclosure, or damage to personal data should be prevented.

- Accountability: Companies must appoint a Data Protection Officer (DPO) and educate customers about their rights and data protection.

The Canadian Personal Information Protection and Electronic Documents Act protects geographic data. Private organizations must comply with PIPEDA, a comprehensive federal law that regulates personal data collection, use, and disclosure. This includes Geospatial information that may disclose the location of an individual. PIPEDA requires organizations to seek informed consent before collecting such data and use reasonable security measures. Additionally, the law grants individuals the right to consult and contest the accuracy of their personal information held by organizations. PIPEDA balances technology improvements and privacy rights for Canadians in Geospatial data (Austin, 2006).

The Indian Penal Code of 1860 and the Information Technology Act of 2000 govern the privacy of geographical data in India. The Indian Penal Code criminalizes unauthorized sharing and exploitation of sensitive location data. The Information Technology Act regulates electronic communication, including geographical data. It protects personal data and offers legal remedies for unauthorized access or disclosure.

The United Arab Emirates (UAE) has enacted a comprehensive privacy law. The UAE Federal Decree-Law No. 1 of 2019 on the Privacy and the Use of Personal Information by Organizations, the E-Transaction Law, and the Decree Law No. 2 of 2020 on the General Provisions of the Cybercrimes Law govern personal data collection, use, and protection. Gulf countries, including Bahrain, Kuwait, Oman, and Qatar, have privacy laws that address personal data processing and privacy (Sami, 2009). These laws may have different criteria for data privacy and offer different degrees of protection.

Data Ownership in Geospatial Data

Geospatial data ownership determines who controls location-based data (Hu, 2017). Unlike real assets, Geospatial data ownership is complicated by legal, ethical, and practical factors.

- ***Producer vs. Subject:***

Geospatial data collectors may claim ownership. Government agencies, corporate companies, and people performing surveys or mapping exercises are examples (Breunig et al., 2020). Individuals whose personal data is subject to Geospatial data can claim their right to ownership or control over said information. So, it is imperative to institute explicit policies delineating data ownership from the beginning of data collection endeavors. Effectively convey ownership rights and obligations to data subjects and producers while addressing concerns regarding consent and control.

- ***Community or Public Ownership***

Geospatial data comes from government, business, and public sources. Coordinating data ownership across contributors might cause conflict (Ponnusamy, 2003). So, collaborative agreements should be created with data ownership and responsibility rights for each contributor. Prioritize consensus-building and transparent communication in order to manage potential conflicts effectively.

- ***Intellectual Property:***

Intellectual property rights and ownership may be disputed when Geospatial data incorporates unique map designs or proprietary datasets (Bédard et al., 2015). Utilize licensing agreements to delineate ownership and usage rights precisely. Preserve the rights of contributors by establishing the conditions under which intellectual property may be utilized.

- **Open Data Initiative:**

Ownership and control of Geospatial data disseminated via open data initiatives may present difficulties, particularly when contributors wish to maintain authority over their datasets (Bertino, 2008). So, it is important to establish all-encompassing data-sharing agreements that delineate ownership rights, authorized uses, and sharing conditions. Define contributor and data custodian roles in dataset maintenance and updates.

ANONYMIZATION AND AGGREGATION TECHNIQUES IN GEOSPATIAL DATA

In geographic data, anonymization and aggregation approaches have the combined benefit of protecting individual privacy while maintaining the data's analytical value.

Anonymization

Concealing or removing personally identifiable information from Geospatial data through anonymization makes it difficult to associate particular data points with specific individuals (Chrisman, 2002).

Anonymization Techniques

- Randomization:
  - Replace precise geographic coordinates with randomly generated values within a specified range (Araujo et al., 2013).
- Perturbation:
  - Introduce slight random perturbations to the original coordinates, adding noise to obscure specific locations (Rauch, 2022).
- Generalization:
  - Reduce the precision of coordinates by rounding or truncating values to a certain decimal place.
- Data Swapping:
  - Swap the geographic coordinates of different individuals, maintaining overall spatial patterns (Holt, 2010).

Aggregation

Aggregation involves combining individual data points into groups or summaries, preserving patterns without disclosing precise details. Aggregation can derive important insights from geographical data while resolving privacy concerns and following ethical data practices (Zhao, et al., 2015).

Aggregation Techniques:

- Grid-based Aggregation:
  - Divide the geographic area into a grid and aggregate data within each grid cell (Sarang, 2023).
- Spatial Clustering:
  - Group nearby data points into clusters, replacing individual coordinates with the centroid of each cluster (Iftikhar & Pedersen, 2010).
- Temporal Aggregation:
  - Aggregate data over time intervals, such as hours or days, to reduce temporal resolution (Bendechache, 2016).
- Hierarchy-based Aggregation:
  - Aggregate data at different hierarchical levels, such as city, district, or country (Eckman & Himelein, 2019).
- Statistical Aggregation:
  - Calculate statistical measures (mean, median) for geographic coordinates within specified regions (Yoba et al., 2024).

CASE STUDIES

Geospatial Data can be used to uncover important location-based insights through a variety of industry-specific applications.

Healthcare

The goal of a medical organization is to monitor diseases and assess public health measures using geographical data. The company uses cutting-edge technologies to gather and examine data that includes several facets of people's geographic information. This covers their travel patterns, health issues, and other potentially private information.

Healthcare organizations can use geospatial data to monitor the geographic spread of infectious diseases, locate hotspots, and comprehend transmission patterns. This data and demographic and health-related information allow the organization to evaluate the success of public health campaigns, outreach programs, and quarantine regulations.

Healthcare providers may find it useful to use this geographical data to pinpoint high-infection areas so they may better allocate resources and carry out focused therapies. Nonetheless, significant ethical questions are raised by gathering such data. The company must protect the privacy and confidentiality of people's data, particularly when handling delicate information about their medical issues.

A well-known exercise and health app inadvertently revealed private data, emulating the Strava example from 2018 (Starva, 2018). The program, extensively used to monitor users' workout schedules and locations, unintentionally revealed military personnel's whereabouts and habits. This incident attracted attention to the security flaws in managing Geospatial data, especially when people in sensitive professions are involved.

The unintentional release of military personnel's whereabouts and daily activities via the fitness app highlighted the continued requirement to examine privacy considerations in geospatial data operations closely. The event highlighted the possible dangers of gathering, keeping, and disclosing location-based data, particularly for those working in fields where secrecy is crucial.

This event should remind application developers and users about the importance of implementing robust security mechanisms and privacy safeguards when working with geographical data.

Education

A school district decided to optimize school bus routes using a Geospatial data system to increase productivity and lower transportation costs. In addition to real-time traffic patterns and bus stop locations, the system gathers data from various sources, including the home addresses of the pupils. The goal is to develop more efficient and economical bus routes so that the district can distribute resources more effectively and children will spend less time traveling.

There are serious privacy concerns when bus routes are optimized using geographical data. Sensitive data, such as the exact addresses of kids' houses and the routes they travel to school, might be included in the data collection. Parents and students who may not have been properly informed about the extent to which their location data would be employed may find this breach of privacy to be especially concerning.

The school district unintentionally gives third-party advertising looking to target the student population access to geolocation data, including the travel habits of the pupils. Since students' information is disclosed to outside parties without their knowledge or approval, this unlawful data sharing raises major ethical concerns. There is a significant possibility of commercial exploitation or misuse of this information, which might betray trust between the school system, kids, and their families.

The risk of profiling and targeting vulnerable groups, such as students from lower-income households, increases when geolocation data is shared with third-party advertising (Source: Medium.com). Marketers might utilize the information to customize their ads according to perceived socioeconomic characteristics, which could exacerbate already-existing inequalities. This raises concerns about the possibility of inadvertent discrimination and the development of unfair preconceptions based on geographical data, in addition to undermining the ethical norms of data use. School districts need to prioritize informed consent, create clear policies for data sharing, and put strong security measures in place to safeguard sensitive student information to solve these ethical issues.

Environment

Using geographic data to monitor migratory patterns, environmental experts begin a study to comprehend and safeguard species. They equip endangered species with satellite tags and GPS collars to gather vital data about their movements and habitats. Nevertheless, the researchers unintentionally reveal the exact locations of these endangered species while disseminating their findings to the larger scientific community.

Various ethical questions are raised by the inadvertent disclosure of the whereabouts of endangered animals using Geospatial data (Source:medium.com). First, there are the unintended repercussions of data sharing, which occur when the intentions behind scientific study unintentionally cause harm. In this instance, individuals looking to exploit and jeopardize endangered species use the very data gathered to conserve wildlife as a tool.

It is imperative for researchers to exercise caution when sharing Geospatial data and to have strong security measures in place to protect their study subjects. This entails finding a careful balance between the ethical duty to ensure that sharing scientific knowledge does not unintentionally lead to the injury or exploitation of the same people being studied and the necessity to share knowledge for the benefit of society.

Smart Cities

The ascent of Smart Cities has witnessed a growing prevalence of Geospatial data integration into urban infrastructure, aiming to augment efficiency and elevate the standard of living for inhabitants. One common scenario is implementing a smart traffic management system that uses Geospatial data to optimize traffic flow and lessen congestion. However, ethical questions are raised when this data is used for widespread surveillance without getting permission from the public.

Significant ethical concerns are raised by the opaqueness of repurposing geographical data for surveillance purposes (Fabrègue & Bogoni, 2023). There is a breach of confidence between the people and their government since citizens are frequently unaware of how much their movements and activities are being watched. Furthermore, when people face unjustified intrusions into their private lives, the fundamental concepts of privacy and personal freedom are jeopardized, making violating civil liberties an urgent problem. These worries are heightened by the possibility that government surveillance programs may be overreached, as citizens' rights and liberties may be violated by using the geolocation data for purposes other than traffic control.

Transparent communication using Geospatial data between local officials and residents is necessary to solve these ethical challenges. Achieving a balance between protecting individual privacy and advancing technology can be facilitated by establishing clear standards, receiving informed consent, and putting strong data protection mechanisms in place. To guarantee that Smart Cities emphasize ethical considerations and protect the rights and privacy of their residents, thorough regulatory frameworks must be developed.

SUMMARY

In the rapidly evolving landscape of Geospatial technology, the intersection of privacy and ethics has emerged as a critical and complex terrain. The dynamic nature of location-based data collection, analysis, and utilization necessitates a thoughtful and ethical approach to ensure responsible practices and protect individual privacy. Ethical considerations demand responsible data stewardship, equitable access, and transparent decision-making, emphasizing the need for accountability in Geospatial data practices. Privacy concerns, particularly in individual privacy and consent, highlight the imperative to implement robust measures such as anonymization and encryption. The legal landscape and community perspectives further contribute to the nuanced nature of Geospatial data ownership and usage. Achieving an equilibrium between the utility of Geospatial data for valuable insights and the imperative to safeguard individual privacy is at the heart of ethical considerations. Striking this balance requires nuanced strategies such as anonymization, aggregation, and robust consent mechanisms.

REFERENCES

Joseph, S. (2023). AI in Wildlife Conservation: Poaching Prevention and Ethics. *Medium.* https://medium.com/@staneyjoseph.in/ai-in-wildlife-conservation-poaching-prevention-and-ethics-03e2076f1a3a

Al-Hasani, B. (2023), *Integrated Geospatial approach for adaptive rainwater harvesting site selection under the impact of climate change.*

Araujo, V., Pacifico, M. J., & Pinheiro, M. (2013). Adapted random perturbations for non-uniformly expanding maps. *Stochastics and Dynamics, 14*(4), 1450007. doi:10.1142/S0219493714500075

Austin, L. M. (2006). *Reviewing PIPEDA: Control.* Privacy and the Limits of Fair Information Practices.

Bédard, Y., Wright, E., & Rivest, S. (2015), *A Guide to Geospatial Data Quality.*

Bendechache, M., Le-Khac, N., & Kechadi, M. T. (2016). Hierarchical Aggregation Approach for Distributed Clustering of Spatial Datasets. *2016 IEEE 16th International Conference on Data Mining Workshops (ICDMW),* (pp. 1098-1103). IEEE.

Bertino, E., Thuraisingham, B., & Gertz, M. (2008). *Security and privacy for Geospatial data: Concepts and research directions.* Cyber Center Publications. doi:10.1145/1503402.1503406

Bertino, E., Thuraisingham, B. M., Gertz, M., & Damiani, M. L. (2008). Security and privacy for Geospatial data: concepts and research directions. *International Workshop on Security and Privacy in GIS and LBS*. ACM. 10.1145/1503402.1503406

Breunig, M., Bradley, P., Jahn, M., & Kuper, P. (2020). Geospatial Data Management Research: Progress and Future Directions. *ISPRS International Journal of Geo-Information*.

Breunig, M., Bradley, P. E., Jahn, M. W., Kuper, P. V., Mazroob, N., Rösch, N., Al-Doori, M., Stefanakis, E., & Jadidi, M. (2020). Geospatial Data Management Research: Progress and Future Directions. *ISPRS International Journal of Geo-Information*, *9*(2), 95. doi:10.3390/ijgi9020095

Bystriakova, N., Peregrym, M., Erkens, R., Bezsmertna, O., & Schneider, H. (2012). Sampling bias in geographic and environmental space and its effect on the predictive power of species distribution models. *Systematics and Biodiversity*, *10*(3), 305–315. doi:10.1080/14772000.2012.705357

Chrisman, N. (2002). *Exploring geographic information systems* (2nd ed.). John Wiley & Sons.

Cinnamon, J. (2020). *Geographic Information Systems; Ethics*. International Encyclopedia of Human Geography. doi:10.1016/B978-0-08-102295-5.10554-2

Cobb, C. D. (2003). GEOGRAPHIC METHODS & POLICY: Using Geographic Information Systems to Inform Education Policy. *Educational Research Quarterly*, *27*, 28.

Cobb, C. D. (2020). Geospatial Analysis: A New Window Into Educational Equity, Access, and Opportunity. *Review of Research in Education*, *44*(1), 129–197. doi:10.3102/0091732X20907362

Cobb, S. (2016). *Data privacy and data protection: US law and legislation.*

Colonna, L. (2020). Privacy, Risk, Anonymization and Data Sharing in the Internet of Health Things. *Pittsburgh Journal of Technology Law & Policy*, *20*(1), 148–175. doi:10.5195/tlp.2020.235

Dhand, S., Singh, A., Litoria, P., & Pateriya, B. (2021). *Geo-smart City Planning Using Geospatial Techniques.*

Eckman, S., & Himelein, K. (2019). *Methods of Geospatial Sampling*. Data Collection in Fragile States.

Elwood, S., & Cope, M. (2009). *Introduction: Qualitative Gis: Forging Mixed Methods Through Representations*. Analytical Innovations, and Conceptual Engagements.

Fabrègue, B. F., & Bogoni, A. (2023). *Privacy and Security Concerns in the Smart City*. Smart Cities. doi:10.3390/smartcities6010027

Freire, S., Macmanus, K., Pesaresi, M., Doxsey-Whitfield, E., & Mills, B. (2016). *Development of new open and free multi-temporal global population grids at 250 m resolution*.

Garfinkel, S.L. (2015). De-Identification of Personal Information.

Holt, D., Steel, D. G., Tranmer, M., & Wrigley, N. (2010). Aggregation and Ecological Effects in Geographically Based Data. *Geographical Analysis*, *28*(3), 244–261. doi:10.1111/j.1538-4632.1996.tb00933.x

Hoofnagle, C. J., van der Sloot, B., & Borgesius, F. J. (2019). The European Union general data protection regulation: What it is and what it means*. *Information & Communications Technology Law*, *28*(1), 65–98. doi:10.1080/13600834.2019.1573501

Hoofnagle, C. J., van der Sloot, B., & Borgesius, F. Z. (2019). The European Union general data protection regulation: What it is and what it means*. *Information & Communications Technology Law*, *28*(1), 65–98. doi:10.1080/13600834.2019.1573501

Hu, Y. (2017). *Spatial Data Infrastructures*. Geographic Information Science & Technology Body of Knowledge. doi:10.22224/gistbok/2017.2.1

Iftikhar, N., & Pedersen, T. (2010). Using a Time Granularity Table for Gradual Granular Data Aggregation. *Fundamenta Informaticae*.

Jing, F., Li, Z., Qiao, S., Zhang, J., Olatosi, B., & Li, X. (2023). Using Geospatial social media data for infectious disease studies: A systematic review. *International Journal of Digital Earth*, *16*(1), 130–157. doi:10.1080/17538947.2022.2161652 PMID:37997607

Kande, Y., Diogoul, N., Brehmer, P., Dabo-Niang, S., Ngom, P., & Perrot, Y. (2024). Demonstrating the relevance of spatial-functional statistical analysis in marine ecological studies: The case of environmental variations in micronektonic layers. *Ecological Informatics*, *81*, 102547. doi:10.1016/j.ecoinf.2024.102547

Kedron, P. J., Li, W., Fotheringham, S. A., & Goodchild, M. F. (2020). Reproducibility and replicability: Opportunities and challenges for Geospatial research. *International Journal of Geographical Information Science*, *35*(3), 427–445. doi:10.1080/13658816.2020.1802032

Kerski, J. J. (2003). The Implementation and Effectiveness of Geographic Information Systems Technology and Methods in Secondary Education. *The Journal of Geography*, *102*(3), 128–137. doi:10.1080/00221340308978534

Kress, A. C., Asberry, A., Taillepierre, J. D., Johns, M. M., Tucker, P., & Penman-Aguilar, A. (2021). Collection of Data on Sex, Sexual Orientation, and Gender Identity by US Public Health Data and Monitoring Systems, 2015–2018. *International Journal of Environmental Research and Public Health*, 18. PMID:34831945

Kuner, C., Bygrave, L., Docksey, C., Tosoni, L., & Drechsler, L. (2021, May 10). *The EU General Data Protection Regulation: A Commentary/Update of Selected Articles*. Oxford University Press.

Lee, W.W., Zankl, W.D., & Chang, H.K. (2016). *An Ethical Approach to Data Privacy Protection*.

Majeed, A., & Lee, S. (2021). Anonymization Techniques for Privacy Preserving Data Publishing: A Comprehensive Survey. *IEEE Access : Practical Innovations, Open Solutions*, *9*, 8512–8545. doi:10.1109/ACCESS.2020.3045700

Manfre, L. A., Hirata, E., Silva, J. B., Shinohara, E. J., Giannotti, M. A., Larocca, A. P. C., & Quintanilha, J. A. (2012). An Analysis of Geospatial Technologies for Risk and Natural Disaster Management. *ISPRS International Journal of Geo-Information*, *1*(2), 166–185. doi:10.3390/ijgi1020166

Manuel, A., Delgado, J., Parra Jounou, I., Ausín, T., Casacuberta, D., Cruz, M., Guersenzvaig, A., Moyano, C., Rodríguez-Arias, D., Rueda, J., & Puyol, Á. (2023). Ethical assessments and mitigation strategies for biases in AI-systems used during the COVID-19 pandemic. *Big Data & Society*, *10*(1), 10. doi:10.1177/20539517231179199

Mei, R., Yan, H., He, Y., Wang, Q., Zhu, S., & Wen, W. (2022). *Considerations on Evaluation of Practical Cloud Data Protection*. CNCERT. doi:10.1007/978-981-19-8285-9_4

Onsrud, H., Johnson, J., & Lopez, X. (1994). Protecting personal privacy in using geographic information systems. *Photogrammetric Engineering and Remote Sensing*, *60*, 1083–1096.

Ponnusamy, J. & Venugopal, K. (2003). GIS Data Dissemination and Intellectual Property Rights. *VII, 336.*

Rauch, S. (2022). *Analysing Long Term Spatial Mobility Patterns of Individuals and Large Groups Using 3D-GIS: A Sport Geographic Approach.* Tijdschrift voor Economische en Sociale Geografie.

Rubinstein, I., & Hartzog, W. (2016). Anonymization and Risk (August 17, 2015). *91 Washington Law Review.*

Sami, K. (2009). *Georeferencing System for Abu Dhabi Spatial Data.*

Sarang, P. (2023). Centroid-Based Clustering. In *Thinking Data Science. The Springer Series in Applied Machine Learning*. Springer.

Schafer, A. (2023). *Geospatial Technology, GeoAI and Spatial Data Science applied to Engineering.*

Shaham, S., Rafieian, S., Ding, M., Shirvanimoghaddam, M., & Lin, Z. (2019). *On the Importance of Location Privacy for Users of Location Based Applications.* ArXiv, abs/1911.01633.

Weber, A. M., Gupta, R., Abdalla, S., Cislaghi, B., Meausoone, V., & Darmstadt, G. L. (2021). Gender-related data missingness, imbalance and bias in global health surveys. *BMJ Global Health*, *6*(11), 6. doi:10.1136/bmjgh-2021-007405 PMID:34836912

Zhang, G. (2022). *Mitigating Spatial Bias in Volunteered Geographic Information for Spatial Modeling and Prediction.*

Zhao, Q., Shi, Y., Liu, Q., & Fränti, P. (2015). A grid-growing clustering algorithm for Geospatial data. *Pattern Recognition Letters*, *53*, 77–84. doi:10.1016/j.patrec.2014.09.017

Chapter 4
Geospatial Data Analysis: A Comprehensive Overview of Python Libraries and Implications

Tewodros Demise Bezabih
https://orcid.org/0009-0005-4599-316X
Bule Hora University, Ethiopia

Mohammad Gouse Glaety
https://orcid.org/0000-0003-1666-2001
Samarkand International University of Technology, Uzbekistan

Desalegn Aweke Wako
https://orcid.org/0000-0003-2803-7959
Bule Hora University, Ethiopia

Senait Girma Worku
Bule Hora University, Ethiopia

ABSTRACT

This chapter examines the introduction to geospatial data, Python programming, and the implications of Python in geospatial data analysis. There are different forms of geospatial data: tabular data, raster data, and vector data. The package or library is essential when using geospatial data in Python programming. Geospatial data is huge and complex because of the nature of geospatial data computing Python is used for. This chapter briefly introduced the Role of Python libraries in Geospatial Analysis such as Arcpy, Basemap, Cartopy, EarthPy, Fiona-GO, Folium, GDAL and OGR, GEE-Py, GeoAlchemy, Geocoder, Geodaisy, Geopandas, Geoplot, Geopy, Geopyspark, GeospatialPDF, GeostatsPy, GPSBabel, 3-Py,ipyleaflet, KeplerglPandas, Plotl, Plotly Express, Plotnine, PyGeos, SentinelHub-Py, Shapely, SpatialPandas, Turfpy. Python has emerged as an indispensable tool in geospatial data analysis.

DOI: 10.4018/979-8-3693-6381-2.ch004

INTRODUCTION TO GEOSPATIAL

Geospatial refers to information that identifies the location of specific features on the Earth's surface, such as oceans and mountains. It involves data that describes objects, events, or features with a precise geographic location. Geospatial data can be static or dynamic, providing insights into relationships between variables and revealing patterns and trends (Pebesma, 2024) Geospatial refers to information located on the Earth's surface using coordinates. Geospatial data often associates some piece of information with a particular location. Geospatial data are facts, figures, files, pixels, and other resources that contain locational data referenced to the Earth via geographic coordinates. They are captured and collected by sensors and people, stored in systems, and exposed to hardware and software where practitioners can conduct geospatial analysis (Martin Werner, 2023). This geospatial data becomes part of the research inquiry and is eventually transformed into useful information for an audience through the data-to-information refinement process. The term spatial data is commonly referred to as data that can be mapped and is often used interchangeably with geospatial data when referring to the portion of geospatial data that can be mapped.

It's important to note that the overall category of geospatial data includes both spatial data, which records the relationship among and about geographically distinguishable features, and imagery data, which is not generally referred to as spatial data (Cartography: Visualization of Geospatial Data, 2013). Spatial data, for instance, can record the location of a rain gauge, the area submerged under flood, the route a delivery truck takes, and the extent of damage from a forest fire or a tourist place. Like other kinds of data, spatial data can be categorized into primary and secondary spatial data.

Representation of real-world features as discrete objects is done through two modes of data representation, which embody the linkage between the real-world domain of geographic data and the computer representation of these features (Martin Werner, 2023). Spatial data, Geospatial data, GIS data, or geodata are names for numeric data that identify the geographical location of a physical object, such as a building, a street, a town, a city, a country, etc., according to a geographic coordinate system. The spatial data lets you find any object's location, length, size, area, or shape. An example of spatial data you can get is the coordinates of an object, such as latitude, longitude, and elevation (2023). Python is used in various domains to predict accurate results. According to Gupta and Arulkumar (2023), Python has played a significant role in advancing cryptocurrency and blockchain ecosystems, as highlighted in their exploratory study within the book "Advanced Applications of Python Data Structures and Algorithms"

GEOSPATIAL DATA

Geospatial data refers to information that identifies the location of objects, events, or phenomena on the Earth's surface. This data includes coordinates like latitude and longitude, enabling the visualization of features on maps and facilitating spatial analysis in Geographic Information Systems (GIS) (Meenakshi Kandpal, 2024). Geospatial data can be categorized into two primary models: vector and raster. Vector data represents discrete features using points, lines, and polygons, while raster data consists of a grid of cells where each cell holds a value. Geospatial datasets often include attribute tables that provide additional information about the data, allowing for queries, analyses, and visualization of specific variables. Common file formats in GIS applications include .shp, .dbf, .prj, and .shx. Geospatial data is crucial in various fields, offering insights into patterns, relationships, and trends that help in decision-making processes and spatial analysis (A. Voisard, 2024).

Tabular Data

Tabular data is data in a table, usually a system or a spreadsheet made up of columns and rows, similar to the parallels and meridians of the geographic grid. Some tabular datasets contain relative locations such as street addresses presented along with additional attribute data. In a tabular dataset, the horizontal data are referred to as records or rows, and vertical data are referred to as fields or columns. To become a geospatial dataset, all relative locations must be linked to geo coordinates (Aaron Jabbour, 2023), Geospatial datasets may also contain additional attribute data that provides context. The more attribute data the practitioner can connect to specific locations, the more relations and context will be revealed during geospatial analysis.

Locational data are the key fields that allow the practitioner to solve for where, while additional attribute fields provide context that may help solve for who, what, why, when, and how. Tabular data can also be extracted from a larger source system that houses many tables, and then separately joined to create a customized, coherent dataset (Aaron Jabbour, 2023). These datasets take a larger level of effort to assemble and then either connect to or extract. However, this tabular data, especially when created into an automated map service, can be especially valuable for organizations that need to visualize and analyze large volumes of their own data, partner data, and other open data on a recurring basis (Aaron Jabbour, 2023)

Raster Data

Raster data consists of a matrix of cells (or pixels) organized into rows and columns (or a grid), where each cell contains a value representing data such as temperature or tone For example, raster data is represented by a grid of pixels of varying tone geospatial raster data origins include satellites, airplanes, drones, sensors, and people with any form of digital cameras. Raster data include digital photographs and images, videos, and heat maps. Raster data can be exploited on a GIS or ELT (Silma, 2024). For example, a sensor can collect raster data and send it for processing to a computer that will reorganize it into a matrix of cells that contain various tonal differences, elevations, and locations. The resulting digital image can be analyzed by a practitioner on an ELT (ESRI, n.d.). In the raster data type, real-world features are represented as grids. Raster data uses a fixed grid dimension and records information about each grid. One or more features are associated with each grid cell. One set of cells and associated values is known as a layer. Digital satellite images are one of the commonly used forms of raster data. If you are familiar with digital photography, you can recognize the raster graphics pixel as the smallest individual grid unit, a building block of an image (Silma, 2024).

Each cell (also known as a pixel) has a value in raster datasets. The cell values represent the phenomenon the raster dataset portrays, such as a category, magnitude, height, or spectral value. The category could be a land-use class such as grassland, forest, or road. A magnitude might represent gravity, noise pollution, or percent rainfall. Height (distance) could represent surface elevation above mean sea level, which can be used to derive slope, aspect, and watershed properties. Spectral values are used in satellite imagery and aerial photography to represent light reflectance and color. Cell values can be either positive or negative, integer, or floating point. Integer values are best used to represent categorical (discrete) data, and floating-point values to represent continuous surfaces. Additional information on discrete and continuous data is given in work done by Laube (2023).

Vector Data

Vector data is a coordinate-based data model representing geographic features as points, lines, and polygons (also called areas). Points are locations on Earth represented by a single geographic coordinate, such as a latitude/longitude. Lines are representations of the distance between points that connect them (Karamelaki, 2023). Polygons are enclosed lines that form a contiguous boundary or the outline of an area around points. Vector data are most commonly seen on maps, although vector files can be created on imagery and shared between imagery and GIS environments. A common example of a vector dataset, often called spatial data, is one of the features, single entities on a GIS that contain geometry and attributes. Geometry refers to only the spatial aspects of the feature, such as the measurement of the points, lines, and polygons in spatial data. Attributes are the descriptive data and information linked to the spatial data, such as count of events, identification of an entity, data source, or date and time (Karamelaki, 2023). In the vector data, spatial locations of features are defined based on coordinate pairs. In this data format, real-world features are represented in the form of:

- Lines (arc or line data)
- Points (point or node data),
- Polygons (closed boundary encompassing area).

Points (or nodes) are the fundamental building blocks of spatial data. They refer to a specific place, generally in a two-dimensional space. Points are used to record the locations of objects, such as tourist places or overhead bridges. Vectors are lines between the points and are generally stored as an ordered series of two or more nodes.

Linear features like pipelines, roads, and rivers are usually represented as vector data. A polygon is a closed figure made from several vector lines where the first and last node in the series is the same point. Polygons represent features of an areal extent, such as lakes, political boundaries, and soil type (Karamelaki, 2023). A common example of a vector dataset, often called spatial data, is one of the features, single entities on a GIS that contain geometry and attributes. Geometry refers to only the spatial aspects of the feature, such as the measurement of the points, lines, and polygons in spatial data. Attributes are the descriptive data and information linked to the spatial data, such as count of events, identification of an entity, data source, or date and time (Karamelaki, 2023).

The Case for Geospatial Data Analysis

A compelling case study demonstrating the application of geospatial data analysis in ensuring public safety is the integration of Geographic Information Systems (GIS) in law enforcement for crime mapping and predictive policing. By harnessing geospatial technologies, law enforcement agencies can analyze crime patterns, identify high-risk areas, and optimize resource allocation to prevent criminal activities effectively (David, 2023).

CASE STUDY

Crime Mapping and Predictive Policing

Overview: Law enforcement agencies leverage geospatial data analysis to map crime incidents, identify crime hotspots, and predict potential criminal activities through advanced analytical techniques. Implementation: By analyzing historical crime data and combining it with geospatial information, law enforcement agencies can create interactive maps that provide actionable insights for crime prevention and response. Predictive policing techniques enable informed decisions on resource deployment based on spatial patterns and trends. Benefits: Crime Prevention: Geospatial analysis helps predict crime hotspots, enabling law enforcement to increase patrols in high-risk areas and respond quickly to incidents. Resource Optimization: Using GIS tools, law enforcement agencies can efficiently manage workforce deployment, plan field visits, optimize routes, and track police vehicles in real time. Prison and Parole Management: GIS aids in identifying areas prone to inmate violence, assigning probation officers geographically, and making informed decisions on facility placements within communities (David, 2023).

Outcome: Integrating geospatial data analysis in law enforcement enhances public safety by enabling smarter policing strategies, proactive crime prevention measures, and efficient resource management. By leveraging GIS technologies, law enforcement agencies can make data-driven decisions, improve response times, and enhance community safety. This case study exemplifies how geospatial data analysis plays a pivotal role in ensuring public safety through smart policing practices, demonstrating the tangible benefits of integrating geospatial technologies in law enforcement for crime prevention and effective resource allocation (David, 2023).

Geospatial Data: Embedded in Our Everyday

The exposure and accessibility of geospatial data from individual users are some of the greatest transformations of the human experience in the twenty-first century. For example, smartphones and some vehicles automatically record the geospatial data tracking the device's location, which often reveals the device user's location as well. This data has emerged as a primary source for tracking users on the electronic grid, as evidenced in the Danville murder case. In these cases, geospatial data is the input that allows practitioners to conduct geospatial analysis that reveals the patterns in which humans carry out their lives, known in the field as pattern of life. Such mobile data is available to private companies and government entities to collect and process (Haining, 2022).

PYTHON'S ROLE IN GEOSPATIAL DEVELOPMENT

Python programming plays a significant rolc in Gcospatial dcvclopment, bridging the gap between geospatial technology and software development. Python is a commonly used programming language in GIS due to its versatility and extensive libraries for geospatial data processing. Python developers in Geospatial roles are responsible for tasks such as developing custom web-based GIS applications, maintaining and enhancing existing GIS applications, creating system documentation, and providing support for GIS tools and services. Proficiency in Python and other languages like JavaScript, Java, or C# is essential for GIS developers to create custom applications, tools, and services that analyze and visualize geospatial data effectively.

Python developers in Geospatial roles are expected to have strong software development skills, adaptability to learn new technologies quickly, and the ability to write, test, and maintain code efficiently to meet the demands of the geospatial industry (Westra, 20123). Python plays a significant role in geospatial analysis due to its versatility, rich ecosystem of libraries, and ease of use. Here are some critical aspects of Python's role in geospatial analysis (Westra, 20123). In the realm of geospatial analysis, the integration of Python programming presents a powerful toolset for addressing complex spatial problems. However, for many individuals new to this field, bridging the gap between geospatial concepts and Python coding proficiency is the challenge. This gap hinders the effective utilization of geospatial data and tools, limiting the potential for insightful analysis and decision-making.

Therefore, the need arises for a comprehensive introduction that not only elucidates the fundamental principles of geospatial analysis but also equips learners with the necessary Python programming skills to manipulate, analyze, and visualize spatial data effectively. This statement of the problem underscores the critical importance of developing a cohesive educational framework that empowers individuals to harness the synergy between geospatial analysis and Python programming for enhanced spatial understanding and problem-solving capabilities (Westra, 20123).

Data Access and Manipulation

Python provides libraries like GDAL, Fiona, and Rasterio for reading, writing, and manipulating geospatial data in different formats, including shapefiles, GeoTIFFs, and more. These libraries enable users to access and work with geospatial datasets seamlessly (M. Kanevski, 2024).

Data Visualization

Python libraries such as Matplotlib, Seaborn, and Plotly are widely used to create interactive and informative geospatial visualizations. These tools allow for the creation of maps, charts, and graphs to represent geographic data effectively (M. Kanevski, 2024).

Geospatial Analysis Libraries

Python offers specialized geospatial analysis libraries like GeoPandas, Shapely, and Pyproj that facilitate operations on geometric objects, spatial relationships and coordinate transformations. These libraries simplify conducting complex spatial analyses (M. Kanevski, 2024).

Web Mapping

Python libraries like Folium and Bokeh allow developers to create interactive web maps and applications. These tools can integrate with web mapping services like Leaflet and OpenLayers, making it easier to visualize and share geospatial data online (Lee, 2023).

Machine Learning and AI

Python's extensive machine learning libraries, such as scikit-learn and TensorFlow, enable geospatial analysts to apply machine learning techniques to remote sensing data, land use classification, and other geospatial tasks. This is valuable for predictive modeling and pattern recognition (M. Kanevski, 2024).

Geospatial Data Science

Python is the preferred language for data scientists working with geospatial data. It supports data preprocessing, feature engineering, and model building, making it an ideal choice for solving real-world geospatial problems (Lee, 2023).

Integration with GIS Software

Python can seamlessly integrate with popular GIS software like ArcGIS, QGIS, and GRASS GIS. This enables users to extend the functionality of these tools, automate repetitive tasks, and customize workflows (Lee, 2023).

GEOSPATIAL PYTHON LIBRARIES

Arcpy

Arcpy is a Python library developed by Esri for automating and customizing tasks within ArcGIS, a popular geospatial software. It provides access to ArcGIS functionality, allowing users to script and extend its capabilities. Arcpy offers tools for geoprocessing, map automation, and spatial analysis. Users can create and manage geospatial data, perform spatial queries, and automate complex GIS workflows. It's a valuable resource for ArcGIS users and GIS professionals (Kempeneers, 2023).

Basemap

Basemap, though deprecated in favor of Cartopy, was a Python library for creating static, interactive, and animated maps. It enabled the visualization of geospatial data on various map projections. Basemap allowed users to plot data on different map projections, add geographic features, and customize map layouts. While it's no longer actively maintained, it was once a widely used tool for geospatial visualization (Kempeneers, 2023).

Cartopy

Cartopy is a Python library for geospatial data visualization. It's a more modern and actively maintained alternative to Basemap, offering various map projections and customization options. Cartopy supports the creation of maps, data visualization, and integration with multiple map data sources. It's used for scientific and environmental data visualization, making it suitable for various applications (Kempeneers, 2023).

EarthPy

EarthPy is a Python package designed for geospatial data analysis in the context of environmental science. It focuses on working with satellite and aerial imagery. EarthPy provides tools for processing, analyzing, and visualizing geospatial data. It's beneficial for land cover analysis, time series data, and the manipulation of raster data (Kempeneers, 2023).

Fiona-GO

Fiona-GO is a lightweight wrapper around the Fiona library, simplifying access to geospatial data. It enhances the convenience of working with vector data formats, such as Shapefiles, in Python. Fiona-GO simplifies tasks like reading, writing, and manipulating vector geospatial data. It streamlines working with formats like Shapefile, making it easier for Python developers (Kempeneers, 2023).

Folium

Folium is a Python library for creating interactive maps. It allows users to embed Leaflet maps into web applications and customize them with various data overlays. Folium is user-friendly and suitable for web developers. It simplifies map creation, adding markers, popups, and other interactive features. It's a versatile tool for data visualization and location-based applications (Kempeneers, 2023).

GDAL and OGR

GDAL (Geospatial Data Abstraction Library) and OGR (Simple Feature Library) are powerful tools for geospatial data processing. Geospatial Data Abstraction Library or GDAR handles raster data, while OGR is responsible for vector data. GDAL/OGR provides extensive capabilities for data conversion, analysis, and manipulation. Users can read and write various geospatial data formats, perform geoprocessing tasks, and manage data efficiently (Madhugiri, 2023).

GEE-Py

GEE-Py is a Python package for interacting with Google Earth Engine (GEE). GEE is a platform for analyzing and visualizing geospatial data on a global scale. GEE-Py allows users to access and analyze Earth Engine data using Python. It simplifies tasks like data retrieval, processing, and visualization. It's an essential tool for leveraging GEE's capabilities (Madhugiri, 2023).

GeoAlchemy

GeoAlchemy is a library that integrates geospatial functionality into SQLAlchemy, a popular Python library for database interaction. It enables the storage and querying of geospatial data within relational databases. It supports spatial data types and provides a seamless way to work with geospatial data in a database context (Madhugiri, 2023).

Geocoder

Geocoder is a Python library for geocoding, converting addresses or place names into geographic coordinates and vice versa. It offers a straightforward and consistent interface for geocoding tasks. It supports various geocoding services, making it easy to work with location-based data and applications (Madhugiri, 2023).

Geodaisy

Geodaisy is a toolset that provides functionalities for geospatial data analysis and visualization. It simplifies working with spatial data, making it accessible to a broader audience. Geodaisy offers tools for data processing, mapping, and geospatial analytics. It supports various data formats and enables users to create custom geospatial applications and visualizations (Madhugiri, 2023).

GeoDjango

GeoDjango is an extension of Django, a popular web framework for Python, designed to handle geospatial data. It empowers developers to build web applications with geospatial features. GeoDjango integrates geospatial data types, spatial queries, and mapping capabilities into web applications. It simplifies the development of location-based services and geospatial web applications (Madhugiri, 2023).

Geopandas

Geopandas-Tools likely refers to additional tools or extensions for the Geopandas library. In Python, Geopandas is itself used for geospatial data manipulation. Extensions for Geopandas could enhance its functionality for data processing, analysis, and visualization in geospatial applications (Madhugiri, 2023).

Geoplot

Geoplot is a Python library that provides a high-level interface for creating various map types. It simplifies the process of visualizing geospatial data. Geoplot offers an easy way to create choropleth maps, scatter plots on maps, and other geospatial visualizations. It is suitable for data exploration and presentation in geospatial analysis (Madhugiri, 2023).

Geopy

Geopy is a Python library for geocoding, converting addresses or place names into geographic coordinates and vice versa. It supports various geocoding services, making it a versatile tool for location-based data applications. It simplifies working with geospatial coordinates and addresses (Madhugiri, 2023).

Geopyspark

Geopyspark is a Python library designed for distributed geospatial analytics. It leverages PySpark, a powerful tool for large-scale data processing. Geopyspark enables geospatial data analysis on distributed systems, making it suitable for handling big geospatial datasets. It supports operations like raster data processing and spatial analytics at scale (Madhugiri, 2023).

GeospatialPDF

GeospatialPDF is a tool that empowers users to embed geospatial data within PDF documents. It is a valuable solution for integrating spatial information into reports, maps, and presentations. GeospatialPDF simplifies the process of adding spatial context to PDF files. It allows users to include maps, geographic coordinates, and other location-based data within PDFs, enhancing the visual representation of information (Madhugiri, 2023).

GeostatsPy

GeostatsPy is a Python library that specializes in geostatistical analysis for spatial data. It is designed to handle the statistical aspects of geospatial datasets. GeostatsPy offers a range of geostatistical tools, including variogram modeling, kriging, and spatial interpolation. It is a valuable resource for geospatial analysts looking to perform advanced statistical analysis on their spatial data (Madhugiri, 2023).

GPSBabel

GPSBabel is a versatile program for converting and transferring GPS data. It facilitates the interoperability of various GPS file formats and simplifies data exchange. GPSBabel supports a wide range of GPS data formats and allows users to convert data between formats, making working with GPS data from different sources easier. It is a helpful tool for GPS enthusiasts and professionals (Madhugiri, 2023).

H3-Py

H3-Py is a Python binding for the H3 geospatial indexing system. H3 is a popular spatial indexing system developed by Uber, and H3-Py provides Python access to its functionality. H3-Py enables users to perform geospatial indexing, hexagonal binning, and spatial analysis using the H3 system. It is handy for applications involving location-based data and spatial aggregation (Madhugiri, 2023).

ipyleaflet

ipyleaflet is a Python library for interactive, browser-based mapping. It is designed to create interactive and visually appealing maps in Jupyter notebooks. It offers a range of mapping tools and widgets for Jupyter environments. Users can create interactive maps, add markers, and visualize geospatial data, making it an excellent choice for data exploration and presentation (Madhugiri, 2023).

Kepler.gl

Kepler.gl is an open-source geospatial analysis tool tailored for large-scale datasets. It is designed to simplify visualizing and analyzing complex geospatial information. Kepler.gl provides a user-friendly interface for building customizable maps and analyzing geospatial data. It can handle large datasets and offers data filtering, styling, and sharing features it a valuable resource for geospatial professionals (Madhugiri, 2023).

Pandas

Pandas is a widespread data manipulation and analysis library in Python. While not exclusively a geospatial tool, it is widely used for processing and analyzing tabular and structured data, including geospatial data. Pandas offers data structures and functions for data cleaning, transformation, and analysis. It is a versatile library for handling and preparing geospatial datasets for analysis (Madhugiri, 2023).

Plotly and Plotly Express

Plotly and Plotly Express are Python libraries for interactive data visualization. They can create various charts and graphs, including geospatial visualizations. Plotly and Plotly Express provide high-quality, interactive plotting capabilities. They allow users to develop geospatial visualizations, such as maps, scatter plots, and heat maps, with ease (Madhugiri, 2023).

Plotnine

Plotnine is a Python library that brings the concept of a grammar of graphics to geospatial data visualization. It allows users to create custom and complex geospatial visualizations with a structured and consistent approach. Plotnine offers a powerful and flexible framework for creating geospatial visualizations. It enables users to define the aesthetics and components of their visualizations, making it a valuable resource for advanced geospatial data visualization (Madhugiri, 2023).

PyGeos

PyGeos is a Python library designed to perform efficient geometric operations using the GEOS library (Geometry Engine – Open Source). It finds application in advanced geospatial calculations. PyGeos offers high-performance geometric operations, such as buffering, intersections, and overlays. It is optimized for speed and memory efficiency, making it a valuable tool for geospatial analysis (Madhugiri, 2023).

SentinelHub-Py

SentinelHub-Py is a Python library designed to work with satellite imagery from the Sentinel series of Earth observation satellites. It offers powerful tools for accessing, processing, and analyzing satellite data, making it a valuable resource for remote sensing applications. Key features include access to Sentinel Hub services, custom band combinations, and creating time series analyses for environmental monitoring (Madhugiri, 2023).

Shapely

Shapely is a Python library for geometric operations and manipulations. It facilitates creating and analyzing geometric shapes, such as points, lines, and polygons. Many GIS (Geographic Information Systems) applications widely use Shapely for spatial data processing and integration. Key features include spatial predicates, geometric operations, and the ability to check for geometric relationships (Madhugiri, 2023).

SpatialPandas

SpatialPandas extends the functionality of the Pandas library to handle geospatial data efficiently. It provides data structures and operations for working with geospatial data like points, lines, and polygons. Key features include spatial indexing, geographic transformations, and seamless integration with existing Pandas workflows, making it easier to manage and analyze large geospatial datasets (Madhugiri, 2023).

Turfpy

Turfpy is a Python port of Turf.js, a geospatial engine that offers a wide range of geospatial analysis functions. In Python, it enables users to perform geospatial calculations, such as distance measurement, intersection detection, and buffer operations. Turfpy is a valuable resource for geospatial professionals and developers who require powerful geospatial processing capabilities in their applications (Madhugiri, 2023).

WORKING WITH GEOSPATIAL DATA IN PYTHON

Spatial data, also known as geospatial data, GIS data, or geodata, is a type of numeric data that defines the geographic location of a physical object, such as a building, a street, a town, a city, a country, or other physical objects, using a geographic coordinate system. You may determine not just the position of an object but also its length, size, area, and shape using spatial data (Lawhead, 2023). To work with geospatial data in Python, use the GeoPandas and GeoPlot library. GeoPandas is an open-source project to make working with geospatial data in Python easier. GeoPandas extends the data types used by pandas to allow spatial operations on geometric types. Geometric operations are performed shapely. Geopandas further depends on Fiona for file access and matplotlib for plotting. GeoPandas depends on its spatial functionality on a large geospatial, open-source stack of libraries (GEOS, GDAL, and PROJ). See the Dependencies section below for more details. Geoplot is a geospatial data visualization library for data scientists and analysts wanting to get things done quickly (Lawhead, 2023).

Installation

Please install all the dependencies and modules to function the given codes properly. Installing can be done through Anaconda:

Syntax:

```
conda install geopandas
conda install geoplot
```

conda-forge is a community effort that provides conda packages for a wide range of software. It provides the conda-forge package channel for conda from which packages can be installed, in addition to Anaconda's "defaults" channel. GeoPandas and all its dependencies are available on the conda-forge channel and can be installed as:

Syntax:

```
conda install --channel conda-forge geopandas
conda install geoplot -c conda-forge
```

Figure 1. Output of first few rows of the country

	GEO_ID	STATE	NAME	LSAD	CENSUSAREA	geometry
0	0400000US01	01	Alabama		50645.326	(POLYGON ((-88.124658 30.28364, -88.0868119999...
1	0400000US02	02	Alaska		570640.950	(POLYGON ((-166.10574 53.988606, -166.075283 5...
2	0400000US04	04	Arizona		113594.084	POLYGON ((-112.538593 37.000674, -112.534545 3...
3	0400000US05	05	Arkansas		52035.477	POLYGON ((-94.042964 33.019219, -94.043036 33....
4	0400000US06	06	California		155779.220	(POLYGON ((-122.421439 37.869969, -122.421341 ...

```
pip install geopandas
pip install geoplot
```

GeoPandas can also be installed with pip if all dependencies can be installed as well:

```
# Load all importance packages
import geopandas
import numpy as np
import pandas as pd
from shapely.geometry import Point
import missingno as msn
import seaborn as sns
import matplotlib.pyplot as plt
% matplotlib inline
```

First, let's look at the first geospatial dataframe: US States Geodata

```
# Getting to know GEOJSON file:
country = geopandas.read_file("data/gz_2010_us_040_00_5m.json")
country.head()
```

This code uses the geopandas library to read a GEOJSON file named "gz_2010_us_040_00_5m.json" in the "data" folder. The read_file() function from geopandas is used to read the file and store it in a variable named country. The head() function is then used to display the first few rows of the country data frame This allows the user to get a quick overview of the data and its structure (Figure 1).

Checking the type of the dataframe that you just load in, you can see that it's Geo Data Frame, which has all the regular characteristics of a Pandas DataFrame.

```
type(country)
```

This code is written in Python. The type() function is used to determine the data type of a variable or object in Python. In this code snippet, the type() function is being used to determine the data type of the variable country. When this code is executed, it will return the data type of the country variable. For example, if country is a string, the output will be. If country is an integer, the output will be.

```
geopandas.geodataframe.GeoDataFrame
```

This code creates an instance of the GeoDataFrame class from the geopandas module. The GeoDataFrame class is a specialized version of the pandas DataFrame that is designed to handle spatial data. It allows for the storage and manipulation of geospatial data, such as points, lines, and polygons, and provides methods for performing spatial operations and analysis. The geopandas.geodataframe part of the code specifies the module and sub-module where the GeoDataFrame class is located, while GeoDataFrame is the actual class being instantiated.

```
type(country.geometry)
```

This code uses the type() function in Python to determine the data type of the geometry attribute of the country object. The geometry attribute is likely a spatial data type, such as a point, line, or polygon, which is commonly used in geographic information systems (GIS) and mapping applications. The output of this code will be the data type of the geometry attribute, which could be a class or a string representation of the data type.

```
geopandas.geoseries.GeoSeries
```

This code snippet is importing the GeoSeries class from the geoseries module of the geopandas library. GeoSeries is a class in geopandas that represents a series of geometric objects, such as points, lines, or polygons, with associated coordinate reference system (CRS) information. It is similar to a pandas Series, but with additional spatial functionality. By importing this class, the user can create and manipulate GeoSeries objects in their code.

Each value in the GeoSeries is a Shapely Object. It can be: Point Line, Polygon, MultiPolygon

Each object can be used for a different type of physical object such as: Point for building, Line for Street, Polygon for city, and MultiPolygon for country with multiple cities inside.

Geospatial Data Model

The fundamental types of geometric objects implemented by Shapely are points, curves, and surfaces. Each is associated with three sets of (possibly infinite) points in the plane. The *interior*, *boundary*, and *exterior* sets of a feature are mutually exclusive and their union coincides with the entire plane. A Point has an interior set of exactly one point, a boundary set of exactly no points, and an exterior set of all other points. A Point has a topological dimension of 0. A Curve has an interior set consisting of the infinitely many points along its length (imagine a Point dragged in space), a boundary set consisting of its two end points, and an exterior set of all other points.

A Curve has a topological dimension. A Surface has an interior set consisting of the infinitely many points within (imagine a Curve dragged in space to cover an area), a boundary set consisting of one or more Curves, and an exterior set of all other points including those within holes that might exist in the surface. A Surface has a topological dimension (Toms, 2022)

Geometric Objects

Geometric objects are created in the typical Python fashion, using the classes themselves as instance factories. A few of their intrinsic properties will be discussed in this sections, others in the following sections on operations and serializations.

General Attributes and Methods

```
object.area
Returns the area (float) of the object.
object.bounds
Returns a (minx, miny, maxx, maxy) tuple (float values) that
bounds the object.
object.length
Returns the length (float) of the object.
object.minimum_clearance
```

Returns the smallest distance by which a node could be moved to produce an invalid geometry. This can be thought of as a measure of the robustness of a geometry, where larger values of minimum clearance indicate a more robust geometry. If no minimum clearance exists for a geometry, such as a point, this will return *math.infinity*.

```
type(country.geometry[0])
```

This code uses Python to determine the data type of the first element in the "geometry" attribute of the "country" object. The "type()" function is used to return the data type of the specified object. In this case, the object is "country.geometry[0]", which refers to the first element in the "geometry" attribute of the "country" object. The output of this code will be the data type of the first element in the "geometry" attribute, which could be a point, line, polygon, or other geometric shape depending on the data being analyzed.

```
hapely.geometry.multipolygon.MultiPolygon
```

This code imports the MultiPolygon class from the multipolygon module of the shapely.geometry package. The MultiPolygon class is used to represent a collection of polygons in a single object. This can be useful for working with complex geometries that consist of multiple polygons, such as a country with multiple islands or a city with multiple neighborhoods. The shapely library is a popular Python package for working with geometric objects and performing geometric operations. Similar to a Pandas DataFrame, a GeoDataFrame also has attribute plot, which makes use of the geometry character within the dataframe to plot a map:

```
country.plot()
```

This code snippet is written in Python and it uses the plot() function to create a plot of a variable. named country. The plot() function is a part of the matplotlib library, which is a popular data visualization library in Python. • The plot() function is used to create line plots, scatter plots, bar plots, and many other types of plots. • In this case, it is used to create a plot of the country variable.. The plot will be displayed in the output window or saved to a file, depending on the configuration of the environment. Without more context, it is difficult to say what the plot will look like or what information it will convey. However, the plot() function is a powerful tool for visualizing data and can be customized in many ways to create informative and visually appealing plots.

```
# Exclude Alaska and Hawaii for now
country[country['NAME'].isin(['Alaska','Hawaii'])==False].
plot(figsize=(30,20), color='#3B3C6E');
```

This code is written in Python. • The code is plotting a map of the United States with the states colored in. However, it is excluding Alaska and Hawaii from the plot. The first line of code is a comment, which is ignored by the interpreter. The second line of code is using the isin() method to check if the NAME column of the country dataframe contains the values 'Alaska' or 'Hawaii'. The == False at the end is used to invert the boolean values returned by isin(), so that it returns True for all states that are not Alaska or Hawaii. The resulting boolean series is then used to subset the country dataframe using square brackets. This creates a new dataframe that only contains the rows where the NAME column is not 'Alaska' or 'Hawaii'. Finally, the plot() method is called on this new dataframe to create the map. The figsize parameter is used to set the size of the plot, and the color parameter is used to set the color of the states.

CONCLUSION

Most of this chapter devoted to explorer and examining the introduction to geospatial data, python programming and implication of Python in geospatial data analysis. There are different form of geospatial data, Tabular Data Raster Data Vector Data. When using geospatial data using Python programming the package or library are very essential. Geospatial data is very huge and complex because of this the nature of geospatial data computing python is used for. This chapter briefly introduced. Role of Python in Geospatial Analysis and Geospatial Python Libraries such as Arcpy,Basemap, Cartopy, EarthPy, Fiona-GO, Folium, GDAL and OGR, GEE-Py, GeoAlchemy, Geocoder, Geodaisy, Geopandas, Geoplot, Geopy, Geopyspark, GeospatialPDF, GeostatsPy, GPSBabel,H3-Py,ipyleaflet,Kepler.glPandas,Plotl, Plotly Express, Plotnine,PyGeos,SentinelHub-Py,Shapely, SpatialPandas, Turfpy.

In addition Python has emerged as an indispensable tool in geospatial analysis. The versatility, extensive library ecosystem, and user-friendly nature of this technology have revolutionized the way people access, process, and visualize geospatial data. Python facilitates seamless data manipulation with libraries like GDAL, Fiona, and Rasterio, allowing users to work with various geospatial formats effortlessly. It empowers geospatial analysts to create interactive and informative visualizations using libraries such as Matplotlib, Seaborn, and Folium, while specialized tools like GeoPandas and Shapely simplify complex spatial operations. In essence, Python has transformed geospatial analysis by providing a comprehensive, user-friendly, and powerful platform that empowers analysts and data scientists to harness the full potential of geographic data, ultimately contributing to better decision-making in various fields, from urban planning to environmental science and disaster management.

REFERENCES

Aaron Jabbour, R. B. (2023). Geospatial Data, Information and Intelligence. In R. B. Aaron Jabbour, Geospatial Data, Information and Intelligence (pp. 20-50). USA.

Cartography: Visualization of Geospatial Data. In (2013). *M.-J. K. ORMELING, Cartography: Visualization of Geospatial Data* (pp. 34–70). Pearson Education Limited.

David. In (2023). *Designing Geodatabases: Case Studies in GIS Data Modeling* (pp. 300–350). ESRI.

Gupta, A., & Arulkumar, N. (2023). An Exploratory Study of Python's Role in the Advancement of Cryptocurrency and Blockchain Ecosystems. In *Advanced Applications of Python Data Structures and Algorithms* (pp. 236–252). IGI Global. doi:10.4018/978-1-6684-7100-5.ch012

Haining, R. (2022). *Spatial Data Analysis Theory and Practice*. University of Cambridge.

Heinold, B. (2018). *A Practical Introduction to Python Programming*.

Kanevski, A. P. (2024). *Machine learning algorithms for geo spatial data. Applications and software*. Research Gate.

Karamelaki. (2023). *Guide on Geospatial Data Integration*. ESRI. https://support.esri.com/en/otherresources/gis-dictionary/term/7cbd3f7c-e17f-4bb0-a51a-318ccf5b68f1#:~:text=and%20

Kempeneers, P. (2023). A Python Package for the Analysis of Geospatial Data. *International Journal of Geo-Information*.

Laube, S. T. (2023). Advances in Geographic Information Science. In *Advances in Spatial Data Handling* (pp. 120–160). Springer.

Lawhead, J. (2023). *Learning Geospatial Analysis with Python*.

Lee, P. D. (2023). *Machine Learning Techniques Applied to Geospatial Big Data*. MDPI.

Madhugiri, D. (2023). *Python Packages For Geospatial Data Analysis*. ResearchGate.

Martin Werner, Y.-Y. C. (2023). Handbook of Big Geospatial Data. In *Y.-Y. C. Martin Werner, Handbook of Big Geospatial Data* (pp. 60–90). Springer.

McClain, B. P. (2022). *Python for Geospatial Data Analysis*. Theory, Tools, and Practice.

McClain, B. P. (2023). *Advanced Python for Geospatial Data Analysis*. O'Reilly Media.

Meenakshi Kandpal, S. B. (2024). Geospatial Data Analytics using Machine Learning. *International Journal of Geo-information* .

Pebesma, E. (2024). *Applied Geospatial data Science with python.*

Ramalho, L. (2019). *Fluent Phyton, Advance Praise for Head First Python*. Orielly.

Silma, D. (2024). Applied Geospatial Data Science with Deep learning.

Toms, S. (2022). *ArcPy and ArcGIS – Geospatial Analysis with Python*. Packt Publishing Ltd.

Voisard, A. B. D. (2024). geospatial data modeling using Deep learning approach. IEEE.

Westra, E. (20123). Python Geospatial Develoment. In Python Geospatial Develoment (pp. 160-180).

Chapter 5
Python for Geospatial Data Analysis

Gurram Sunitha
https://orcid.org/0000-0002-3305-8167
Mohan Babu University, India

K. G. Suma
VIT-AP University, India

Mohammad Gouse Galety
https://orcid.org/0000-0003-1666-2001
Samarkand International University of Technology, Uzbekistan

Ganesh Davanam
Sree Vidyanikethan Engineering College, India

Chinthapatla Pranay Varna
Sree Vidyanikethan Engineering College, India

ABSTRACT

This chapter serves as a valuable resource for computer science enthusiasts, researchers, and practitioners seeking an understanding of geospatial analysis with Python. The chapter begins with an introduction to geospatial analysis, highlighting the significance of geospatial data across various domains. It establishes Python's stand in this field, positioning it as a powerful tool for geospatial analyses. The subsequent sections explore fundamental concepts, such as vector versus raster data, coordinate reference systems and projections, geometric objects, topological relationships, and spatial operations. Further, prominent Python libraries for geospatial analysis are explored. GeoPandas is introduced, detailing its capabilities in working with geospatial data, handling geometric data structures, and leveraging spatial operations. Shapely is examined for its role in geometric manipulations. Fiona is explored as a library for handling geospatial data. Discussion on Folium showcases its utility in creating interactive and customized maps.

DOI: 10.4018/979-8-3693-6381-2.ch005

INTRODUCTION

Geospatial analysis involves studying data associated with specific geographical locations on Earth's surface (Lawhead, 2015). It includes a range of techniques to interpret, analyze and visualize geographical information (Andrienko et al., 2011). This data could be anything from the location of trees in a forest to the spread of diseases across continents. The primary aim of geospatial analysis is to derive meaningful insights into spatial patterns and relationships in order to make informed decisions (Lee and Kang, 2015). The importance of geospatial data spans across various domains (Cobb, 2020). It impacts application areas like urban planning, environmental science, transportation, agriculture, disaster management etc (Roy et al., 2021; Reece and Hulse, 2020;). In urban planning domain, geospatial analysis assists in designing cities by analyzing population density, traffic patterns, and infrastructure development (Dano et al., 2020). Environmental scientists rely on geospatial data to monitor deforestation, assess biodiversity, and study climate change effects (Aksha et al., 2020). Transportation industries optimize routes and logistics through geospatial analysis, reducing costs and enhancing efficiency (Zhang et al., 2020). Agriculture benefits from precision farming techniques derived from analyzing soil types, weather patterns, and crop yield data (Emam and Soliman, 2022).

Python emerged as a formidable tool for geospatial analysis due to its versatile and extensive libraries (Sunitha et al., 2023). These libraries empower users to handle, process, and visualize geospatial data efficiently. Python's derives its popularity due to its user-friendly syntax and the availability of powerful and vast open-source geospatial libraries. The Python ecosystem offers specialized libraries tailored for geospatial tasks. "GeoPandas" Python library streamlines the handling of geospatial datasets by introducing data structures like GeoDataFrames. "Shapely" Python library provides powerful geometric operations, thereby allowing users to manipulate spatial objects effortlessly. "Fiona" Python library facilitates reading and writing of various geospatial file formats. It eases data interchange. "Pyproj" Python library provides efficient coordinate system transformations. It ensures accurate spatial analysis. "Folium" enables creation of interactive maps, thus enhancing visualization capabilities. "Rasterio" Python library specializes in managing raster datasets. "Cartopy" Python library aids in geospatial data visualization. These libraries empower analysts, researchers, and developers to conduct intricate geospatial analysis tasks within Python's programming environment. Their ease of use and flexibility have made Python the language of choice for professionals performing geospatial analysis.

Geospatial analysis plays a pivotal role across multiple domains. Python as a programming language, coupled with its robust geospatial libraries, offers a potent combination for conducting insightful and impactful geospatial analysis (Coetzee et al., 2020). Python's adaptability in handling geospatial data is evident in its ability to integrate seamlessly with other scientific libraries and tools. This interoperability allows to combine geospatial analysis with statistical analysis, machine learning, and data visualization (Jiang, 2015). Thereby, unlocking new possibilities for deriving insights from geospatial datasets. The simplicity and accessibility of Python have democratized geospatial analysis. Thereby, enabling a broader spectrum of users, including researchers, data scientists, geographers etc to leverage geospatial data. The language's robustness allows to dive into complex geospatial analysis tasks with relative ease. This facilitates innovation and problem-solving across various disciplines.

Python's open-source nature fosters a collaborative environment. Developers continuously contribute to enhancing existing libraries. New tools tailored to evolving geospatial challenges are being created. This collaborative ecosystem ensures that Python geospatial landscape remains dynamic and adaptive. This allows to tackle emerging spatial analysis needs timely (Patroumpas et al., 2014). As technology advances, the volume, variety, and velocity of geospatial data continue to grow exponentially (Breunig et al., 2020; Liu et al., 2022). Python's role in handling and analyzing this data becomes increasingly pivotal. The language demonstrates scalability and efficiency in processing large datasets. Its extensive library support, positions it as a go-to choice for addressing geospatial challenges. Its technical capabilities, community-driven development, adaptability, and versatility makes it dominant in geospatial analysis.

A JOURNEY THROUGH SPACE AND TIME: THE EVOLUTION OF GEOSPATIAL ANALYSIS AND EMERGING TRENDS

The analysis of geographic information has a rich history, constantly evolving alongside technological advancements.

Early Beginnings (Pre-1960s)

- **Paper Maps and Overlays:** The foundation of geospatial analysis lies in traditional paper maps. People used overlays to visually compare different geographic features. For example, Dr. John Snow's famous cholera map in 1854, which linked cholera outbreaks to contaminated water sources, is a prime example of this technique.
- **Limited Tools and Techniques:** Early analysis relied on manual calculations and rudimentary statistical methods. Distance and area measurements were laborious tasks.

The Rise of Geographic Information Systems (GIS) (1960s-1990s)

- **The Dawn of Automation:** The 1960s saw the birth of GIS, a revolution in geospatial analysis. Advances in computer technology allowed for digital storage, manipulation, and analysis of spatial data.
- **Software Development and Standardization:** Early GIS systems were complex and expensive. Over time, software became more user-friendly and accessible, leading to standardization and wider adoption.
- **Focus on Data Management and Visualization:** GIS revolutionized how we store, manage, and visualize spatial data. Thematic maps, overlays, and spatial queries became central tools for uncovering patterns and relationships.

The Modern Era (2000s-Present)

- **The Age of Big Geospatial Data:** The explosion of data from satellites, sensors, GPS devices, and social media has created a new era of "Big Geospatial Data."
- **Advanced Analytics and Artificial Intelligence (AI):** With massive datasets, advanced statistical methods, machine learning, and AI are being employed for spatial modeling, predictive analysis, and automated feature extraction.
- **Cloud Computing and Web-based GIS:** The rise of cloud computing has enabled access to powerful GIS capabilities without expensive hardware investments. Web-based GIS platforms allow for collaborative and real-time geospatial analysis.

Emerging Trends in Geospatial Analysis

The future of geospatial analysis is brimming with exciting possibilities:

- **Spatial Data Integration:** Integrating geospatial data with other forms of data like weather, social media, and economic indicators will lead to deeper insights and holistic understanding.
- **Real-time Geospatial Analysis:** The growing availability of real-time data from sensors and citizen science initiatives will enable real-time decision making in areas like traffic management, disaster response, and environmental monitoring.
- **3D Visualization and Immersive Technologies:** Integrating 3D geospatial data with virtual reality (VR) and augmented reality (AR) holds immense potential for creating immersive and interactive visualizations for planning, education, and public engagement.

- **Location Intelligence and Citizen Science:** Citizen science initiatives and crowdsourced data collection platforms will play a bigger role in enriching geospatial datasets, leading to more comprehensive analysis.
- **Ethical Considerations and Privacy Concerns:** As geospatial data becomes more personal and detailed, ethical considerations regarding data privacy and security will need to be carefully addressed.

The field of geospatial analysis continues to evolve at a rapid pace. By harnessing the power of new technologies and embracing emerging trends, we can unlock the true potential of location-based data to solve complex challenges and build a more informed and sustainable future.

GEOSPATIAL POWERHOUSE: PYTHON LIBRARIES LEADING THE CHARGE IN TODAY'S DATA-DRIVEN WORLD

The world is awash with data, and a significant portion of it has a spatial dimension. From tracking environmental changes to analyzing disease outbreaks, geospatial analysis has become crucial across various fields. This is where Python's robust geospatial ecosystem shines, with libraries like GeoPandas, Shapely, Fiona, and Folium playing a vital role in unlocking the power of location-based information.

Significance of Geospatial Libraries in the Current Context

- **Democratization of Geospatial Data**: Open-source data platforms and citizen science initiatives are generating vast amounts of geospatial data. These libraries, with their user-friendly interfaces and powerful functionalities, empower researchers, analysts, and even students to analyze and visualize this data more efficiently.
- **Real-time Decision Making**: The ability to work with real-time geospatial data streams (e.g., sensor data, traffic information) is crucial in areas like disaster response, environmental monitoring, and smart cities. Libraries like GeoPandas, with their efficient data manipulation capabilities, facilitate this by enabling rapid analysis and visualization of constantly updating information.
- **Communication and Public Engagement**: Interactive web maps created with Folium can effectively communicate complex geographical patterns and relationships to a broad audience. This is invaluable for raising awareness about environmental issues, public health concerns, and socio-economic trends.

Synergy in Action: Addressing Contemporary Challenges

The following are the real-time applications in today's world that can be solved using geospatial applications.

- Climate Change Analysis: GeoPandas excels at analyzing geospatial data related to climate change, such as temperature variations, sea level rise, and deforestation patterns. It enables researchers to identify trends, model future scenarios, and inform mitigation strategies.
- Precision Agriculture: Shapely's geometric calculations and spatial analysis capabilities are valuable for optimizing farm layouts, analyzing soil health data, and creating efficient irrigation plans, contributing to sustainable agricultural practices.
- Urban Planning and Management: Fiona plays a role in integrating various geospatial data sources (e.g., building footprints, traffic data, demographic information) for informed urban planning decisions. It facilitates analyzing population density, identifying potential transportation bottlenecks, and optimizing resource allocation.

CORE CONCEPTS IN GEOSPATIAL ANALYSIS

This section introduces core concepts of geospatial analysis - vector vs. raster data, Coordinate Reference Systems (CRS) and projections, geometric objects, topological relationships, and spatial operations.

Vector vs. Raster Data

Two primary types of formats exist in geospatial data: vector and raster data. These formats differ in how they represent and store spatial information (Grekousis, 2020). Each of the formats have its unique characteristics and applications (Figure 1).

Vector data represents geographic features using points, lines, and polygons. These elements store location information explicitly as coordinates, along with attributes describing the features they represent. Points denote individual locations, lines represent linear features like roads or rivers, and polygons delineate areas such as forests or administrative boundaries. Vector data is highly precise and suitable for representing discrete objects with well-defined boundaries. It allows for storing attribute information linked directly to specific spatial elements, enabling detailed analysis based on these attributes. Examples of vector data include street networks, parcel boundaries, administrative boundaries etc.

Figure 1. Vector vs. raster data comparison

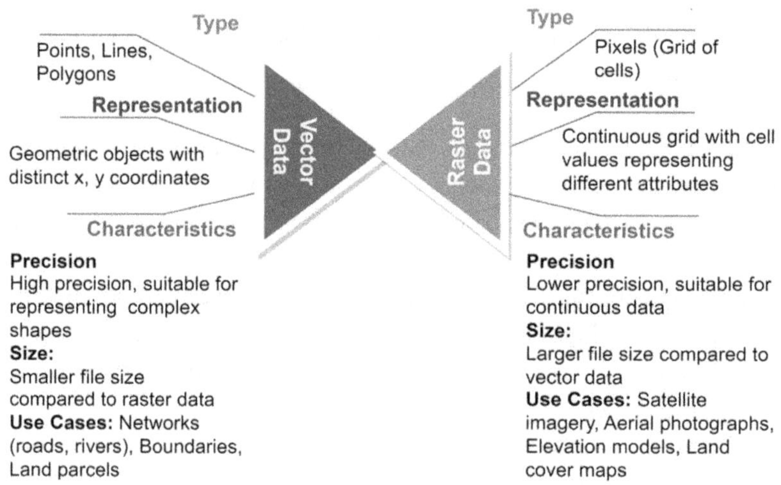

Raster data represents spatial information as a grid of cells. Each cell contains a value representing a specific attribute. In this format, each pixel holds information about a particular location on the Earth's surface. Raster data is suitable for continuous phenomena, such as elevation models, satellite imagery, land cover classification etc. It allows to represent phenomena that varies continuously across the geographical space. Thereby, providing a comprehensive view of the geographical landscape. However, raster data might lack the precision in representing detailed boundaries that vector data offers.

The fundamental difference between vector and raster data lies in their data structures. Vector data relies on discrete geometric objects, making it ideal for representing distinct features and precise boundaries. Raster data, uses a grid-based structure, suitable for representing continuous surfaces and phenomena. Vector data excels in applications requiring precise boundaries and detailed attribute information. Raster data finds its strength in applications involving continuous spatial phenomena.

In practice, choosing between vector and raster data often depends on the specific analysis requirements, the nature of the data, and the desired output. Both formats have their advantages and are complementary in many applications.

Coordinate Reference Systems (CRS) and Projections

CRS serves as frameworks to define locations on the Earth's surface. They use coordinates to pinpoint positions accurately. A CRS includes two primary components: the datum and the projection. The datum defines the reference point, shape, and

Figure 2. Examples of different EPSG codes

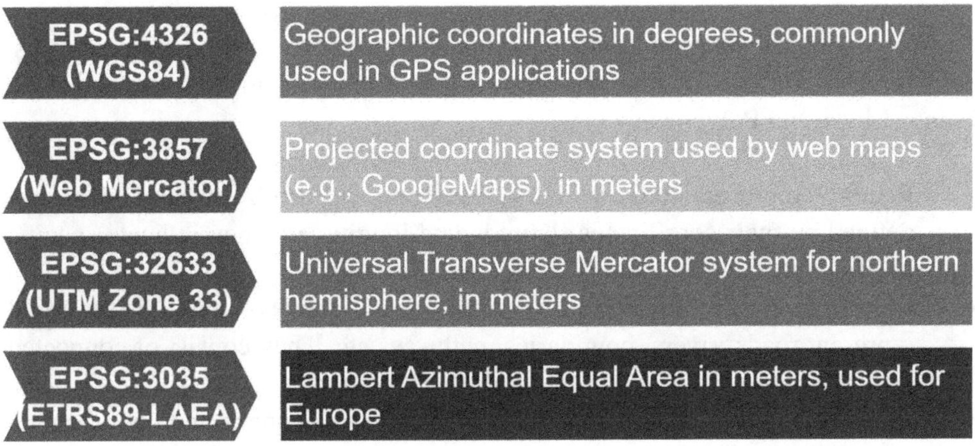

size of the Earth. It acts as a starting point to measure locations. There are various datums, like WGS84, NAD83 etc. Each is tailored for specific regions or purposes. A projection method transforms the Earth's spherical or ellipsoidal surface onto a plane. It is needed because representing a round Earth on a flat surface lead to distortions. Different projections have unique properties suited for various uses, like preserving shapes, areas, distances etc (Hu et al., 2022). Common projections include

- Mercator – preserves shapes but distorts sizes towards poles, often used for navigation.
- Equal Area – maintains accurate area representation but distorts shapes (e.g., Albers, Lambert).
- Conic – suitable for regions with east-west extents, preserving both shapes and areas.

Figure 2 provides different CRS examples, their EPSG codes along with brief descriptions of each CRS and its common usage.

Choosing the right CRS and projection is crucial. It affects accuracy and interpretation. For example, a local analysis might benefit from a projection that accurately preserves areas. While navigation might prefer preserving shapes. Transformations may be needed when working with data in different CRS. These conversions ensure data compatibility and accurate spatial analysis. Tools like Pyproj in Python help perform these transformations seamlessly.

Geometric Objects in Geospatial Analysis

Geometric objects serve as fundamental building blocks in geospatial analysis. They represent different features on maps. Three primary geometric objects include Points, Lines, and Polygons.

- Points – represent individual locations on the Earth's surface. These are defined by their coordinates (latitude and longitude). They indicate specific positions such as landmarks, cities, sampling sites etc.
- Lines – connect multiple points to form linear features on maps. These could represent roads, rivers, boundaries, pathways etc. They consist of connected vertices, each with its own set of coordinates.
- Polygons – enclose areas and consist of a sequence of connected lines forming closed shapes. They have multiple vertices that outline their boundaries. They represent regions such as countries, lakes, parks etc.

Points are crucial for marking specific locations, aiding in navigation, and pinpointing features. Lines help represent linear features like roads or rivers, essential for route planning or network analysis. Polygons define and enclose areas, facilitating analysis of regions or spatial extents. Python libraries like GeoPandas provide functionalities to create, manipulate, and analyze these geometric objects. Various operations like buffering around points, finding intersections between lines, calculating areas of polygons etc can be performed. Understanding these geometric objects is essential for accurate representation and analysis of spatial data. They allow for detailed spatial modeling, analysis, and visualization in various fields, from urban planning to environmental studies.

Topological Relationships and Spatial Operations

Topological relationships refer to the spatial connections and arrangements between geometric objects in geospatial analysis (Theobald, 2001). Understanding these relationships is crucial for performing spatial operations accurately. Common topological relationships are

- Containment – determines if one object completely contains another.
- Intersection – identifies where two or more objects overlap.
- Adjacency – establishes whether objects share a common boundary.
- Connectivity – describes how objects are connected.

Spatial operations involve manipulating geometric objects based on their topological relationships. These operations are essential for conducting analyses and deriving insights from spatial data. Examples of spatial operations include

- Union – combines multiple geometries to create a single geometry representing their collective extent.
- Intersection – determines the common area shared by two or more geometries.
- Buffering – creates a zone or area around a specific geometry, often used for proximity analysis.
- Dissolve – merges adjacent geometries with shared attributes to create a single geometry.

Topological relationships and spatial operations are employed extensively in geospatial analysis. They facilitate various tasks such as finding neighboring features, measuring distances, performing overlay analysis, and conducting geographic queries. Python libraries like GeoPandas, Shapely, and Pyproj offer functionalities to perform these spatial operations. Complex analyses can be executed by applying these operations to geometric objects easily. Understanding topological relationships and performing spatial operations accurately is critical for precise analysis in fields like urban planning, environmental modeling, transportation etc. These operations help derive valuable insights from spatial data.

POPULAR PYTHON LIBRARIES FOR GEOSPATIAL ANALYSIS

Python is a versatile programming language. It has established itself as a go-to platform for geospatial analysis (Toms et al., 2018). It owns a rich ecosystem of specialized libraries. These libraries cater to diverse geospatial tasks. Thus, enabling efficient management, analysis, visualization and manipulation of spatial data. From GeoPandas facilitating seamless handling of geospatial datasets to Shapely offering powerful geometric operations (McClain, 2022). These libraries empower users to delve into intricate spatial analysis tasks (Lawhead, 2015). This section explores the functionalities and applications of the most widely-used Python libraries in the realm of geospatial analysis.

GEOPANDAS

GeoPandas is a Python library that provides convenient data structures and functionalities to work with geospatial data (Hassan & Vijayaraghavan, 2019). It extends the capabilities of a popular data manipulation library "Pandas". It adds support for spatial operations and geometric data types. The Key Features of GeoPandas are as follows.

- GeoDataFrames: GeoPandas introduces "GeoDataFrame" as an efficient data structure for geospatial data management. GeoDataFrame is an extension of Pandas' DataFrame. It is capable of handling tabular data as well as geometric objects. Its structure simplifies storage and manipulation of geospatial datasets. Thereby, enabling efficient integration of spatial and attribute data.
- Handling Geometric Objects: GeoPandas allows the creation and manipulation of geometric objects. Geometric operations such as merging, buffering, finding intersections between geometries etc can be performed.
- Spatial Operations: GeoPandas offers a wide range of spatial operations, such as overlay analysis, spatial joins, proximity analysis etc. These operations facilitate the analysis of relationships between spatial datasets. Thereby, it aids in deriving valuable insights.
- Interoperability: GeoPandas supports multiple geospatial file formats such as Shapefiles, GeoJSON etc. It allows to read, write, and manipulate various types of geospatial data.

A few applications of GeoPandas are as follows.

- Data Exploration: GeoPandas simplifies exploration and visualization of geospatial data. Users can easily plot maps; can visualize spatial patterns; can gain insights into geographic datasets.
- Geospatial Analysis: GeoPandas enables users to perform complex geospatial analysis tasks. Examples include spatial aggregations, overlay operations, spatial joins etc. These functionalities are crucial for various domains.
- Integration with Other Libraries: GeoPandas integrates seamlessly with other Python libraries like Matplotlib, Seaborn, Folium etc. Thereby, enhancing visualization capabilities and enabling the creation of interactive maps.

Handling Geometric Data Structures

GeoDataFrames is a core component of GeoPandas. GeoDataFrames are specialized data structures designed for managing geospatial data in Python. These structures seamlessly integrate spatial information and tabular data. They combine the functionalities of Pandas DataFrames with geospatial capabilities. Few important characteristics of GeoDataFrames are as follows.

- Tabular and Spatial Data Integration: GeoDataFrames amalgamate tabular data (similar to a spreadsheet) with spatial data, allowing the simultaneous storage and manipulation of geometry-related information alongside attribute data.
- Geometric Representation: Each entry in a GeoDataFrame includes a geometry column representing different geometric objects like points, lines, or polygons. These geometries are linked to specific rows in the tabular data, enabling spatial analysis with associated attributes.
- Familiar Pandas-Like Interface: GeoDataFrames inherit the intuitive and user-friendly interface of Pandas. They support similar functionalities for data manipulation and analysis.

Few functionalities and operations supported by GeoDataFrames are as follows.

- Attribute Management: GeoDataFrames facilitate attribute management by allowing addition, deletion, and modification of attribute data associated with geometries.
- Spatial Operations: Spatial operations can be directly performed on GeoDataFrames. Examples include finding intersections between geometries, calculating distances, conducting spatial joins, creating buffers around geometries etc
- Data Visualization: GeoDataFrames enable easy visualization of spatial data using plotting functionalities. Users can generate maps, overlay different layers, and create insightful visualizations of geographic information.

A few applications supported by GeoDataFrames in geospatial analysis are as follows.

- Data Analysis: GeoDataFrames streamline the analysis of geospatial data. It allows to conduct exploratory data analysis, perform spatial queries, and to derive insights from spatial relationships within the data.
- Integration with Other Libraries: They integrate well with other Python libraries like Matplotlib. Thus, allowing for creation of maps and visualizations. Also, they allow to work in coordination with libraries like Scikit-learn for machine learning tasks involving spatial data.

Spatial Operations and Analysis Using GeoPandas

GeoPandas library extends the capabilities of Pandas DataFrames to handle geometric data. The idea is to enable efficient exploration, manipulation, and analysis of geospatial datasets (Korstanje, 2022). Spatial operations involve various processes that allow the comparison, combination, and manipulation of spatial data. Spatial analysis enables to derive insights and patterns from geographical information. The key aspects of spatial operations and analysis are as follows.

- Geometric Operations: GeoPandas facilitates operations like finding intersections between spatial objects, creating buffers around them, combining shapes, and performing geometric transformations.
- Data Integration: GeoPandas seamlessly integrates geometric and tabular data. Thereby, enabling the association of geographic features with attributes and allowing comprehensive analysis.
- Spatial Joins and Queries: The library enables spatial joins and queries, aiding in matching and retrieving data based on spatial relationships between geometries.
- Overlay Analysis: Overlay operations like union, intersection, and difference assist in analyzing areas where spatial data overlaps, helping to derive new information.
- Proximity Analysis: Functions like buffering aid in analyzing proximity, allowing the creation of zones around features.

Spatial operations and analysis are vital in various fields such as urban planning, environmental studies, transportation etc. They help in identifying patterns, assessing spatial relationships, and deriving insights critical for decision-making.

SHAPELY

Shapely is a Python library. It offers tools for handling geometric shapes and performing operations like intersections, buffering, and transformations (Westra, 2015). It allows precise manipulation and analysis of geometries which is crucial for spatial computations and analysis. Its functionalities aid in creating, examining, and modifying geometric objects. This makes it an essential tool for handling and processing spatial data within Python environments.

Working With Points, Lines, and Polygons

Shapely facilitates handling and manipulating points, lines, and polygons in geospatial analysis. Points represent individual locations, while lines connect points, forming linear features like roads or rivers. Polygons enclose areas, representing regions such as parks or countries. Shapely enables creation, modification, and analysis of these geometric objects. This allows to perform operations like finding intersections, measuring distances, determining containment between shapes etc. This capability is invaluable in spatial analysis. It aids in tasks like mapping, urban planning, environmental modeling etc.

Geometric Operations and Predicates

Shapely supports various geometric operations and predicates, enabling precise manipulations and analyses of geometric shapes. Geometric operations involve tasks like finding intersections, creating buffers, performing union and difference operations between shapes etc. Predicates help determine spatial relationships such as containment, adjacency, equality between geometries etc. Shapely's functionalities are crucial for spatial computations and analysis. Shapely empowers users to perform complex geometric tasks within their Python-based geospatial workflows.

FIONA

Fiona is an yet another Python library for geospatial analysis (Garrard, 2016). It serves as a vital tool for managing geospatial data within Python-based workflows. It specializes in reading and writing various geospatial file formats. Thus, providing essential functionalities to interact with geographic datasets. Fiona's primary purpose revolves around facilitating seamless access, manipulation, and transformation of geospatial data in Python. Fiona acts as a bridge between Python and different geospatial file formats. It enables to read, write, and manipulate geographic datasets effortlessly. Fiona simplifies geospatial data handling in Python. Its primary purpose lies in empowering Python developers, data scientists, and GIS professionals. It enables them to handle diverse geospatial data formats efficiently within their coding environments.

Functionalities of Fiona in the Geospatial Domain

- Reading and Writing Geospatial Data: Fiona supports various file formats commonly used in the geospatial domain. It allows to read and write datasets such as Shapefiles, GeoJSON etc. It offers methods to access spatial data and metadata stored in these formats.
- Geospatial Data Handling: Both geometries and attribute data stored within geospatial datasets can be accessed, explored, and manipulated. Fiona provides functionalities to retrieve information about spatial objects and their associated attributes.

- Data Transformation and Conversion: Fiona facilitates conversion and transformation of geospatial data between different formats. It allows to convert data structures into Fiona-compatible formats and vice versa. This provides interoperability across various geospatial tools.
- Attributes and Schema Handling: Fiona enables to handle attribute data and schema information within geospatial datasets. Attribute fields can be modified, added, and deleted. This provides flexibility in managing associated information.
- Integration and Interoperability: Fiona integrates well with other Python geospatial libraries like GeoPandas, Shapely etc. Fiona can be integrated into other tools to improve existing geospatial workflows seamlessly. Fiona in combination with other tools enhances their capabilities.

Supported File Formats

Fiona's supports various file formats. This enables to seamlessly handle and process geospatial data across a wide range of file types. This makes Fiona a versatile tool in working with different geospatial datasets.

- Shapefile (.shp): It is a widely used format for storing vector geospatial data. It consists of multiple files (.shp, .shx, .dbf) storing geometry, index, and attribute information. It is suitable for representing points, lines, and polygons.
- GeoJSON (.geojson): This file format is based on JavaScript Object Notation (JSON) for encoding geographic data structures. It allows storage of various geometries and their associated attributes in a human-readable format. This format is ideal for web-based mapping and data interchange because of its simplicity and compatibility with web services.
- GeoPackage (.gpkg): This is a modern, open-source format. it stores geospatial information in a single SQLite database file. It supports multiple layers containing various geometries and attributes. It provides a compact and portable format suitable for mobile applications and offline use.
- KML (Keyhole Markup Language) (.kml/.kmz): This file format is developed by Keyhole, Inc. (now Google) for expressing geographic annotation and visualization. It is commonly used in Google Earth and other geospatial software for displaying points, lines, and polygons along with their attributes.
- GML (Geography Markup Language) (.gml): This is an XML-based format for encoding geographical features and their attributes. It offers a standard way to represent and exchange geospatial data. It is widely used in web services and interoperability between different GIS software.

- WKT (Well-Known Text) and WKB (Well-Known Binary): Textual and binary formats are used to represent geometries in a human/machine readable form. WKT provides a textual representation. WKB represents geometries in binary format which is more efficient for storage and transmission.

Reading and Writing Geospatial Data

Fiona provides efficient methods and techniques for reading and writing geospatial datasets.

Reading Geospatial Data

- Reading Methods: Fiona offers functions like *fiona.open()* to read geospatial datasets from different file formats. Geometries, attributes, and metadata stored within these datasets can be accessed by using Fiona's read functionalities.
- File Format Handling: Fiona supports multiple file formats such as Shapefile, GeoJSON, GeoPackage, etc. Desired file format can be specified when reading. This gives flexibility in accessing different types of geospatial data.
- Data Structure Access: Geometries and other associated information stored within geospatial files can be accessed by using Fiona's methods. It provides easy-to-use functions to iterate through features, extract geometries, and retrieve attribute data.

Writing Geospatial Data

- Writing Methods: Fiona facilitates writing geospatial data to various formats using methods like *fiona.open()* in write mode. Users can create new datasets, modify existing ones, or append new features to geospatial files.
- Format Handling and Schema: Fiona allows specification of the file format, schema, and geometry types while writing geospatial data to a file. Users can define attribute fields, geometry types, and Coordinate Reference Systems (CRS) for creating new datasets.
- Data Structure Modification: Users can modify and update geometries, attributes, and metadata while writing geospatial data using Fiona's functionalities.

Handling Different File Formats and Data Structures

- Versatility in File Formats: Fiona's support for multiple file formats enables users to read and write data seamlessly across various file types.

- Flexibility in Data Structures: Fiona's capabilities extend to managing different data structures along with their attribute data. Users can manipulate, transform, and organize geospatial information across diverse data structures using Fiona.

Geospatial Data Handling

Fiona facilitates exploration and manipulation of geospatial data. It offers techniques to access geometries, attributes, and metadata stored within these datasets.

Exploring Geospatial Data

- Accessing Geometries: Fiona enables users to access different types of geometries within geospatial datasets. Methods like iteration and feature access allow users to retrieve and work with these geometries.
- Attribute Access: Users can retrieve attribute information associated with geometries using Fiona's functionalities. Accessing attribute data allows for analyzing and processing additional information linked to spatial features.

Manipulating Geospatial Data

- Geometry Modification: Fiona allows users to modify and manipulate geometries. Thereby, enabling operations such as transformation, simplification, conversion etc between geometry types.
- Attribute Manipulation: Users can modify attribute data associated with geometries, including adding, editing, and deleting attribute fields. These manipulations enable users to customize and organize attribute information within geospatial datasets.

Techniques for Accessing Metadata

- Metadata Retrieval: Fiona provides access to metadata information stored within geospatial datasets. Examples include CRS, data creation details, feature counts etc. Retrieving metadata aids in understanding and utilizing the characteristics and properties of the geospatial data.
- Metadata Utilization: Utilizing metadata information helps in setting up coordinate systems, understanding data sources, and determining the structure of the geospatial dataset.

FOLIUM

Folium is a Python library designed for creating interactive and visually appealing maps (Korstanje, 2022). It serves as a bridge between data manipulation in Python and visualization on interactive web-based maps. The primary purpose of Folium is to simplify the process of visualizing geospatial data by allowing users to create maps directly from Python code effortlessly. Folium yokes the capabilities of *Leaflet.js*. *Leaflet.js* is an open-source JavaScript library for interactive maps (Crickard, 2014). It seamlessly integrates interactive maps with Python. This integration is useful for users especially data scientists, GIS professionals, and researchers. They can generate maps with ease. They can combine Python's data manipulation prowess with dynamic mapping functionalities. Folium's simplicity, coupled with its ability to create interactive and feature-rich maps directly from Python code. This makes it an asset in the domain of geospatial visualization and analysis.

The purpose and applications of Folium are discussed below.

- Geospatial Visualization: Folium excels in geospatial visualization. It enables users to represent geographical data on interactive maps. It helps visualize data patterns and spatial relationships effectively.
- Exploratory Data Analysis (EDA): It serves as a valuable tool for performing exploratory data analysis. It provides support for plotting geospatial data, markers, heatmaps, choropleth maps, overlays etc. Thus, allowing users to gain insights from the data.
- Data Presentation and Communication: Folium facilitates creation of interactive maps suitable for presentations, web applications etc. It supports enhances data communication and storytelling.
- Geospatial Analysis and Decision-making: Folium aids in making informed decisions based on geospatial analysis. Users can overlay various data layers, analyze patterns, and draw conclusions from visualized information (Ozmen, 2021).
- Educational and Research Purposes: Folium is beneficial in educational settings. It assists students and researchers to visualize geographical data for study, analysis, and presentations.
- Web-based Mapping Applications: It finds applications in developing web-based mapping applications where interactive maps play a crucial role (Pohanka et al., 2016). It offers features like markers, pop-ups, tooltips for user engagement etc.

Creating Maps With Folium

- Initializing a Map: To create a map, users can start by initializing a map object using *folium.Map()* function. Specifying parameters like location coordinates (latitude and longitude) and initial zoom level sets up the map view.
- Adding Markers: Markers pinpoint specific locations on the map. They indicate points of interest and/or data points. Users can use *folium.Marker()* function to add markers. Values for the parameters such as coordinates, popup text can be provided for enhancing map interactivity.
- Controlling Zoom Level: Zoom level determines the map's level of detail, ranging from a broader view (lower zoom) to a more detailed view (higher zoom). Adjusting zoom level using the *zoom_start* parameter in *folium.Map()* allows users to control the initial map view.

Figure 3 shows an example geospatial map created using Folium. The map is created with markers, customized tiles, and controlled zoom levels.

Techniques to Customize Maps With Folium

- Map Tiles Customization: Customizing Map Tiles: Folium offers various map tile options like OpenStreetMap, Stamen Terrain etc. Users can select desired map tiles using the tiles parameter in *folium.Map()* function to customize map's appearance.
- Controlling Map Colors: Users can modify map colors by adjusting parameters like *fill_color*, *fill_opacity,* and *color* when creating geometrical shapes, choropleth maps etc. Changing these parameters allows customization of shape and boundary colors for better visualization.
- Icon Customization: Folium allows the usage of customized icons for markers instead of default symbols. Users can create personalized icons using *folium. Icon()* function by specifying custom icon images and their sizes.

Layer Control: Folium supports layer control. It allows to toggle between different overlays or data layers. The *folium.LayerControl()* function allows to manage layers. Thus user can control visibility

Figure 3 shows an example geospatial map created using Folium. The map is created with markers, customized tiles, and controlled zoom levels.

Techniques to Customize Maps With Folium

- Map Tiles Customization: Customizing Map Tiles: Folium offers various map tile options like OpenStreetMap, Stamen Terrain etc. Users can select desired

Figure 3. Example geospatial map with markers and zoom level—Created using folium

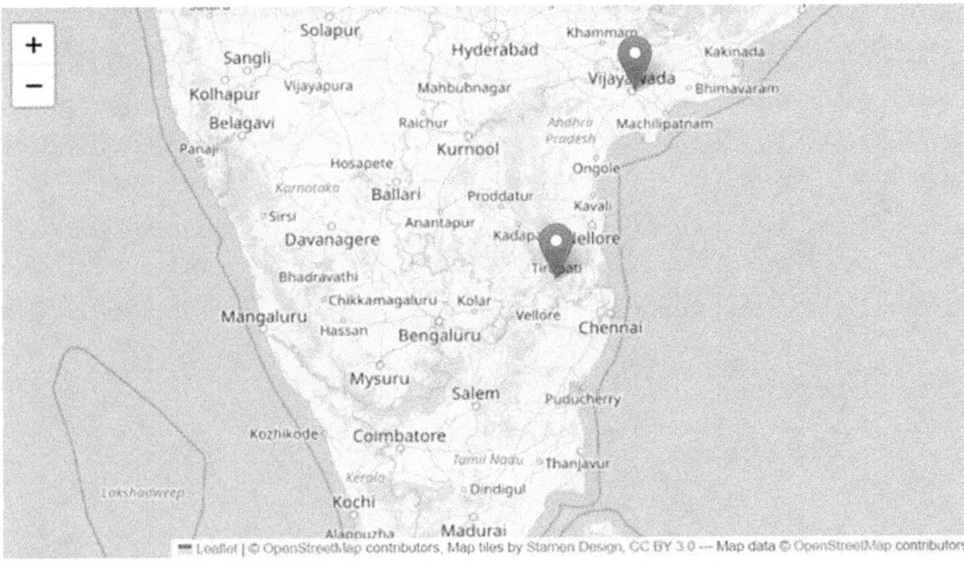

map tiles using the tiles parameter in *folium.Map()* function to customize map's appearance.

- Controlling Map Colors: Users can modify map colors by adjusting parameters like *fill_color*, *fill_opacity,* and *color* when creating geometrical shapes, choropleth maps etc. Changing these parameters allows customization of shape and boundary colors for better visualization.
- Icon Customization: Folium allows the usage of customized icons for markers instead of default symbols. Users can create personalized icons using *folium.Icon()* function by specifying custom icon images and their sizes.
- Layer Control: Folium supports layer control. It allows to toggle between different overlays or data layers. The *folium.LayerControl()* function allows to manage layers. Thus user can control visibility.

Interactive Features in Folium

- Tooltips: Tooltips can be incorporated to maps to enhance data context. Tooltips provide supplementary information when users hover over map elements. Folium enables addition of tooltips by using the *tooltip* parameter when adding markers or shapes.
- Pop-ups: Pop-ups display additional information upon clicking on map elements. Pop-ups contain detailed data, images, or links. They provide

Figure 4. Example geospatial map with map tiles customization, icon customization and tooltips—Created using Folium

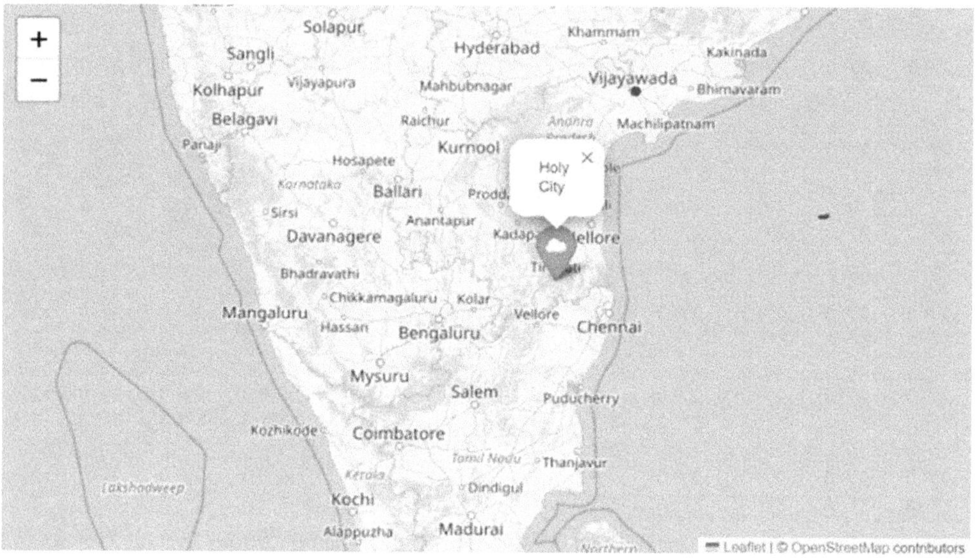

enhanced user interaction. They can be created by utilizing the *popup* parameter.

- Click Events: Folium supports click events. Thereby it enables actions triggered by user clicks on map elements. Actions can be associated with user events like clicking on markers, shapes, and other interactive elements.

Figure 4 shows an example geospatial map created using Folium. The map is created with Map Tiles Customization, Icon Customization and Tooltips.

Visualizing Geospatial Datasets With Folium

- Shapefiles: Folium allows users to visualize Shapefiles by converting them into GeoJSON format, which can be directly displayed on the map. Users can use Python libraries like GeoPandas, Fiona etc to read Shapefiles and convert them into GeoJSON before visualizing them with Folium.
- GeoJSON Files: Folium displays GeoJSON files on maps without requiring additional conversions. Users can directly add GeoJSON layers to Folium maps using *folium.GeoJson()* function. This function enables visualization of complex geometries and attributes.
- Pandas DataFrames: Folium supports visualizing geospatial data stored in Pandas DataFrames containing geometries and associated attributes. Users

can create GeoJSON-like structures from Pandas DataFrames and then utilize Folium's *folium.GeoJson()* or *folium.Choropleth()* functions for visualization.

POWERFUL TOOLS, PRUDENT CHOICES: A BALANCED LOOK AT GEOPANDAS, SHAPELY, FIONA, AND FOLIUM

The four Python libraries – GeoPandas, Shapely, Fiona, and Folium – form the backbone of many geospatial workflows. However, even the most powerful tools have limitations. The strengths and weaknesses of each are explored in this section to help you make informed choices.

GeoPandas

Strengths: Integrates seamlessly with Pandas, making data manipulation and analysis a breeze. Offers powerful geospatial operations for vector data (points, lines, polygons). Efficient for working with large datasets.

Challenges: Primarily focused on vector data. Limited support for raster data (images with geospatial information). Requires some understanding of Pandas data structures.

Shapely

Strengths: Versatile library for creating and manipulating geometric objects (points, lines, polygons) in Python. Offers advanced functions for complex geometric calculations.

Challenges: Primarily a low-level library for geometric operations. Requires coding expertise for data analysis tasks. Less user-friendly for beginners compared to GeoPandas.

Fiona

Strengths: Acts as a bridge between Python and various geospatial file formats (e.g., Shapefiles, GeoJSON). Enables reading, writing, and editing spatial data stored in these formats.

Challenges: Limited functionalities on its own. Relies on other libraries like GeoPandas or Shapely for data manipulation and analysis. Requires understanding of specific file formats.

Folium

Strengths: Simple and user-friendly for creating interactive web maps. Integrates well with GeoPandas for visualizing geospatial data. Excellent for quick visualizations and communication purposes.

Challenges: Limited customization options compared to advanced web mapping libraries. Performance can suffer with very large datasets. Focuses on basic interactivity. Not ideal for highly customized map designs.

Finding the Right Balance

The best library for your project depends on your specific needs and data type. Here's a simplified breakdown:

- For data manipulation and analysis with vector data: GeoPandas is a great choice.
- For advanced geometric calculations: Shapely offers unmatched power.
- For working with geospatial file formats: Fiona is the go-to library.
- For quick and interactive web maps: Folium shines.

The true strength lies in combining these libraries. For instance, you can use GeoPandas for data analysis, Shapely for complex geometries, Fiona to read data, and Folium for visualization.

CONCLUSION

Geospatial analysis involves studying data associated with specific geographical locations on Earth's surface. It includes a range of techniques to interpret, analyze and visualize geographical information. The primary aim of geospatial analysis is to derive meaningful insights into spatial patterns and relationships in order to make informed decisions. The importance of geospatial data spans across various domains. It impacts application areas like urban planning, environmental science, transportation, agriculture, disaster management etc. This book chapter explores the field of geospatial analysis. Fundamental concepts, including vector and raster data, coordinate reference systems, and geometric operations, set the foundation. Python libraries are presented as powerful tools for handling and visualizing geographic data. Python emerged as a formidable tool for geospatial analysis due to its versatile and extensive libraries. These libraries empower users to handle, process, and visualize geospatial data efficiently. Key Python libraries such as GeoPandas, Shapely, Fiona, and Folium are explored, demonstrating their significance for geospatial analysis. The chapter serves as a valuable resource for computer science enthusiasts, researchers, and practitioners seeking an understanding of geospatial analysis with Python.

REFERENCES

Aksha, S. K., Resler, L. M., Juran, L., & Carstensen, L. W. Jr. (2020). A geospatial analysis of multi-hazard risk in Dharan, Nepal. *Geomatics, Natural Hazards & Risk*, *11*(1), 88–111. doi:10.1080/19475705.2019.1710580

Andrienko, G. L., Andrienko, N., Keim, D. A., MacEachren, A. M., & Wrobel, S. (2011). *Challenging problems of geospatial visual analytics*.

Breunig, M., Bradley, P. E., Jahn, M., Kuper, P., Mazroob, N., Rösch, N., Al-Doori, M., Stefanakis, E., & Jadidi, M. (2020). Geospatial data management research: Progress and future directions. *ISPRS International Journal of Geo-Information*, *9*(2), 95. doi:10.3390/ijgi9020095

Cobb, C. D. (2020). Geospatial analysis: A new window into educational equity, access, and opportunity. *Review of Research in Education*, *44*(1), 97–129. doi:10.3102/0091732X20907362

Coetzee, S., Ivánová, I., Mitasova, H., & Brovelli, M. A. (2020). Open geospatial software and data: A review of the current state and a perspective into the future. *ISPRS International Journal of Geo-Information*, *9*(2), 90. doi:10.3390/ijgi9020090

Crickard, P. III. (2014). *Leaflet. js essentials*. Packt Publishing Ltd.

Dano, U. L., Balogun, A. L., Abubakar, I. R., & Aina, Y. A. (2020). Transformative urban governance: Confronting urbanization challenges with geospatial technologies in Lagos, Nigeria. *GeoJournal*, *85*(4), 1039–1056. doi:10.1007/s10708-019-10009-1

Emam, W. W., & Soliman, K. M. (2022). Geospatial analysis, source identification, contamination status, ecological and health risk assessment of heavy metals in agricultural soils from Qallin city, Egypt. *Stochastic Environmental Research and Risk Assessment*, *36*(9), 2437–2459. doi:10.1007/s00477-021-02097-8

Garrard, C. (2016). *Geoprocessing with python*. Simon and Schuster.

Grekousis, G. (2020). *Spatial analysis methods and practice: describe–explore–explain through GIS*. Cambridge University Press. doi:10.1017/9781108614528

Hassan, A., & Vijayaraghavan, J. (2019). *Geospatial Data Science Quick Start Guide: Effective techniques for performing smarter geospatial analysis using location intelligence*. Packt Publishing Ltd.

Hu, N., Meng, X., Liu, H., & Li, Y. (2022, June). Exploring Geospatial Data Visualization Based on Python. In *Proceedings of the 2022 2nd International Conference on Control and Intelligent Robotics* (pp. 858-862). IEEE. 10.1145/3548608.3559322

Jiang, B. (2015). Geospatial analysis requires a different way of thinking: The problem of spatial heterogeneity. *GeoJournal*, *80*(1), 1–13. doi:10.1007/s10708-014-9537-y

Korstanje, J. (2022). Creating Maps. In *Machine Learning on Geographical Data Using Python: Introduction into Geodata with Applications and Use Cases* (pp. 79–101). Apress. doi:10.1007/978-1-4842-8287-8_4

Lawhead, J. (2015). *Learning geospatial analysis with Python*. Packt Publishing Ltd.

Lee, J. G., & Kang, M. (2015). Geospatial big data: Challenges and opportunities. *Big Data Research*, *2*(2), 74–81. doi:10.1016/j.bdr.2015.01.003

Liu, X., Chen, M., Claramunt, C., Batty, M., Kwan, M.-P., Senousi, A. M., Cheng, T., Strobl, J., Cöltekin, A., Wilson, J., Bandrova, T., Konecny, M., Torrens, P. M., Zhang, F., He, L., Wang, J., Ratti, C., Kolditz, O., Klippel, A., & Lü, G. (2022). Geographic information science in the era of geospatial big data: A cyberspace perspective. *Innovation (Cambridge (Mass.))*, *3*(5), 100279. doi:10.1016/j.xinn.2022.100279 PMID:35967836

McClain, B. P. (2022). *Python for Geospatial Data Analysis*. O'Reilly Media, Inc.

Ozmen, E. (2021). Event density visualization with python folium library for public complaint management: An application with 311 calls. *Research & Reviews in Science and Mathematics-I*.

Patroumpas, K., Giannopoulos, G., & Athanasiou, S. (2014). Towards geospatial semantic data management: strengths, weaknesses, and challenges ahead. In *Proceedings of the 22nd ACM SIGSPATIAL International Conference on Advances in Geographic Information Systems* (pp. 301-310). ACM. 10.1145/2666310.2666410

Pohanka, T., Pechanec, V., & Hejlova, V. (2016). Python Web Server for Sensor Data Visualization. *International Multidisciplinary Scientific GeoConference SGEM*, *1*, 803–810. doi:10.5593/SGEM2016/B21/S08.101

Rajamani, S. K., & Iyer, R. S. (2023). Use of Python Modules in Ecological Research. In *Perspectives on the Transition Toward Green and Climate Neutral Economies in Asia* (pp. 182–206). IGI Global. doi:10.4018/978-1-6684-8613-9.ch011

Reece, A. S., & Hulse, G. K. (2020). Canadian cannabis consumption and patterns of congenital anomalies: An ecological geospatial analysis. *Journal of Addiction Medicine, 14*(5), e195–e210. doi:10.1097/ADM.0000000000000638 PMID:32187114

Roy, S., Saha, M., Dhar, B., Pandit, S., & Nasrin, R. (2021). Geospatial analysis of COVID-19 lockdown effects on air quality in the South and Southeast Asian region. *The Science of the Total Environment, 756*, 144009. doi:10.1016/j.scitotenv.2020.144009 PMID:33250248

Sunitha, G., Abouali, A., Galety, M. G., & Sriharsha, A. V. (2023). Dynamic Programming with Python. In *Advanced Applications of Python Data Structures and Algorithms* (pp. 102–122). IGI Global. doi:10.4018/978-1-6684-7100-5.ch005

Sunitha, G., Sriharsha, A. V., Yalgashev, O., & Mamatov, I. (2023). Interactive Visualization with Plotly Express. In *Advanced Applications of Python Data Structures and Algorithms* (pp. 182–206). IGI Global. doi:10.4018/978-1-6684-7100-5.ch009

Theobald, D. M. (2001). Topology revisited: Representing spatial relations. *International Journal of Geographical Information Science, 15*(8), 689–705. doi:10.1080/13658810110074519

Toms, S., Crickard, P., & Van Rees, E. (2018). *Mastering geospatial analysis with Python: explore GIS processing and learn to work with GeoDjango, CARTOframes and MapboxGL-Jupyter*. Packt Publishing Ltd.

Westra, E. (2015). *Python Geospatial Analysis Essentials*. Packt Publishing Ltd.

Zhang, J., Wang, Y., Li, S., & Shi, S. (2020). An architecture for IoT-enabled smart transportation security system: A geospatial approach. *IEEE Internet of Things Journal, 8*(8), 6205–6213. doi:10.1109/JIOT.2020.3041386

Chapter 6
Mastering Geospatial Analysis With Python:
Understanding Geopandas, GDAL, Fiona, Matplotlib, Data Integration, and GIS Tools

M. Shreemathi
Coimbatore Institute of Technology, India

B. Senthilkumar
https://orcid.org/0000-0001-8117-3314
Kumaraguru College of Technology, India

Sujithra M. Sujithra
Coimbatore Institute of Technology, India

A. Praisoodi
Coimbatore Institute of Technology, India

S. Rithika
Coimbatore Institute of Technology, India

ABSTRACT

This chapter explores the intricate realm of geospatial analysis leveraging the power of Python. This chapter embarks on a journey through the fundamentals of geospatial data types, formats, and sources, laying a robust foundation for navigating the complexities of spatial analysis. Key Python libraries such as Geopandas, GDAL, and Fiona are meticulously dissected, elucidating their pivotal roles in processing, analyzing, and visualizing geospatial data. Matplotlib's contribution to geospatial data visualization adds insight, enhancing spatial information's communicative power. Furthermore, the chapter delves into data integration techniques, showcasing how Python seamlessly integrates with GIS tools to extend, customize, and streamline spatial analyses. By unraveling the functionalities of these essential libraries and tools, this chapter equips readers with the knowledge and skills necessary to master geospatial analysis in Python.
DOI: 10.4018/979-8-3693-6381-2.ch006

INTRODUCTION

What Is Geospatial Data?

Geospatial data refers to information inherently tied to a specific location on Earth. It encompasses diverse data, including geographic coordinates, satellite imagery, terrain elevation, etc. This data is pivotal for understanding spatial relationships, mapping landscapes, and deriving insights from the geographical context. In essence, geospatial data adds a crucial dimension to information, allowing us to analyze, visualize, and interpret data in the context of its geographical location. The significance of geospatial data extends across various fields, from urban planning and environmental monitoring to navigation and disaster response.

Why Use Python?

Python is chosen for geospatial data analysis due to its simplicity, versatility, and powerful libraries. Its clear syntax makes it beginner-friendly, while libraries like GDAL/OGR, NumPy, and Matplotlib offer robust tools for reading, processing, and visualizing geospatial data. Python's active community ensures continuous support and improvement. Its integration with GIS tools, cross-platform compatibility, and role in data science contribute to its popularity, making Python an ideal language for beginners and geospatial analysis professionals.

UNDERSTANDING GEOSPATIAL DATA

Types of Data

In the vast landscape of geospatial data, the intricacies of Earth's features are captured through various data types, each offering a unique perspective on our surroundings. Among these, Vector Data and Raster Data are pillars shaping the foundation of geospatial analysis and interpretation.

Vector Data

Vertices and paths make up vector data. Points, lines, and polygons (areas) are the three fundamental forms of vector data. A spatial reference frame exists for every point, line, and polygon, such as latitude and longitude. First, XY coordinates are a simple form of vector points. This point or vertex is connected in a specific sequence to create vector lines. In the end, groups of vector lines combine to form polygons.

Nonetheless, it forms a polygon region by enclosing the beginning and last vertices (Kiwelekar et al., 2020).

Raster Data

Raster data is made up of pixels or grid cells. They are typically square and evenly spaced. However, rectangular rasters are also possible. Every pixel has a value assigned by rasters. Continuous rasters, like elevation or temperature, have values that vary steadily. However, discrete rasters assign a particular class to every pixel. For instance, we use a set of variables to represent land cover classifications.

SOURCES OF GEOSPATIAL DATA

Geospatial data encompasses diverse information that directly or indirectly references specific geographical locations. These data are collected from various sources using different methods, contributing to a rich and comprehensive understanding of the Earth's surface and features (Chang et al., 2020). Some common sources of geospatial data include:

- *Census Data:* Census data provides demographic and socio-economic information about populations in specific geographic areas. It includes data on population counts, age distribution, income levels, education attainment, and other socio-demographic characteristics.
- *Satellite Images:* Satellite imagery captures orbiting satellites' high-resolution images of the Earth's surface. These images are used for land cover classification, vegetation monitoring, urban planning, and environmental assessment.
- *Weather Data:* Weather data includes information about atmospheric conditions such as temperature, precipitation, humidity, wind speed, and air pressure. It is collected from weather stations, radar systems, satellites, and other meteorological instruments.
- *Mobile Phones:* GPS-enabled devices generate location-based data through GPS signals and network connectivity. This data includes information about users' movements, travel patterns, and points of interest, which can be used for location-based services, navigation, and urban planning.

FORMATS

Geospatial data can be stored in many formats. Choosing the right format for geospatial data significantly impacts geospatial data processing. The right format ensures data can be read, shared, and analyzed accurately and efficiently across various platforms.

GeoJSON

GeoJSON is a lightweight format for encoding geospatial data in JavaScript Object Notation (JSON) format. It supports geometric types such as Point, Line String, Polygon, MultiPoint, Multiline String, and Multi Polygon, along with feature collections that combine these geometry types with associated properties (Psaila, 2021).

Features

- Versatility: GeoJSON is versatile and widely supported by mapping libraries and platforms such as Leaflet, Map Box, and Open Layers, making it easy to integrate with various mapping tools.
- Simplicity: GeoJSON files are easy to create and understand, making them popular for sharing and distributing geospatial data.
- Compatibility: GeoJSON can be easily parsed and manipulated using JavaScript, making it suitable for web-based applications and interactive mapping.
- Limitations:
- Performance: GeoJSON may not be the most efficient format for large or highly detailed datasets due to its text-based structure, which can result in larger file sizes and slower processing times compared to binary formats.
- Data Types: GeoJSON has limited support for advanced data types and geometries compared to other formats, which may restrict its suitability for certain spatial data types.

Shapefile

Shapefile is a widely used vector data storage format developed by Esri for storing geographic features, attributes, and shapes. It consists of multiple files, including a .shp file for geometric data, a .dbf file for attribute data, and a .shx file for indexing (Kantabutra, 2021).

Features

- Compatibility: Shapefile is supported by numerous GIS platforms and applications, making it a de facto standard for sharing and exchanging geospatial data.
- Flexibility: Shapefile can store various geometries, including points, lines, polygons, and associated attribute data, making it suitable for diverse spatial analysis tasks.
- Interoperability: Shapefile can be easily converted to other formats and vice versa, facilitating data interoperability and integration with different GIS software and systems.

Limitations

- File Size Limit: Shapefile has a file size limit of 2 gigabytes for each component file (.shp, .dbf), which may restrict its suitability for handling large or complex datasets.
- Topology: Shapefile does not support topological relationships between features, which may limit its usefulness for certain spatial analysis and modeling tasks.
- Complexity: Managing multiple files and ensuring data integrity can be challenging, especially for large or multi-layered datasets.

KML

Keyhole Markup Language (KML) is an XML-based format for expressing geographic annotations and visualizations in two-dimensional and three-dimensional Earth browsers and maps. It allows users to add map overlays with features such as points, lines, and polygons to Google Earth, providing a rich and interactive way to visualize geospatial data (Urquhart, 2022).

Features

- Visualization: KML enables users to visualize geospatial data in Google Earth and other compatible Earth browsers, allowing for interactive exploration and analysis of spatial information.
- Customization: KML supports the creation of custom map overlays with rich styling options, including colors, icons, labels, and pop-up information windows, enabling users to tailor the presentation of their data to specific needs.

- Temporal Data: KML supports the representation of temporal data, allowing users to visualize changes over time, such as historical trends, environmental shifts, and demographic changes.

Limitations

- Complexity: KML files can become complex and cumbersome, especially for large or multi-layered datasets with numerous features and attributes.
- Compatibility: While Google Earth and other Earth browsers support KML, its compatibility with other GIS software and platforms may be limited, hindering data interoperability and integration.
- Performance: Rendering large or highly detailed KML files in Earth browsers may require significant computational resources and result in performance issues, such as slow loading times and reduced responsiveness.

APPLICATIONS AND SIGNIFICANCE OF GEOSPATIAL DATA

The following are the applications of Geospatial data (Coetzee et al, 2020).

Urban Planning and Development

- Geospatial data aids urban planners in designing efficient infrastructures, zoning areas for residential, commercial, and industrial purposes, and determining suitable locations for public amenities like parks, schools, and hospitals.
- Through spatial analysis, planners can assess the impact of proposed developments on the environment, transportation networks, and communities, leading to more sustainable and resilient cities.

Environmental Management

- Geospatial data is crucial in monitoring and managing natural resources, including forests, water bodies, and agricultural land.
- Remote sensing technologies allow tracking changes in land cover, deforestation rates, and biodiversity loss, enabling policymakers to implement conservation measures and mitigate environmental degradation.

Disaster Response and Emergency Management

- During natural disasters such as earthquakes, floods, or wildfires, geospatial data facilitates rapid response efforts by providing real-time information on affected areas, population distribution, and infrastructure damage.
- Geospatial analysis helps emergency responders prioritize rescue operations, allocate resources effectively, and plan evacuation routes, minimizing casualties and property loss.

Public Health

- Geospatial data supports epidemiological research by mapping disease outbreaks, identifying high-risk areas for infectious diseases, and analyzing the spread of epidemics.
- Health agencies use geographic information systems (GIS) to track vaccination coverage, monitor environmental health hazards, and plan healthcare services based on population demographics and healthcare facility locations.

Natural Resource Management

- Geospatial data supports sustainable resource utilization by mapping mineral deposits, oil and gas reserves, and renewable energy potential.
- Land-use planning tools utilize geospatial information to balance competing resource demands, minimize land-use conflicts, and promote conservation practices.

KEY LIBRARIES AND MODULES FOR GEOSPATIAL ANALYSIS IN PYTHON

In the vast landscape of geospatial analysis, Python is empowered by a suite of key libraries and modules that elevate its capabilities. This section unravels the essence of these tools, offering insights into their functionalities and applications within geospatial data analysis.

INTRODUCTION TO GDAL LIBRARY

Geospatial Data Abstraction Library (GDAL) is a free and open-source library that works with raster and vector geospatial data. It provides a unified interface for reading and writing geospatial data in various formats and performing geospatial operations such as reprojection, resampling, and format conversion (Lemenkova et al., 2023).

Key Features of GDAL

- *Data format support*: GDAL supports many geospatial data formats, including popular ones such as GeoTIFF, Shapefile, JPEG, and PNG.
- *Data transformation*: GDAL allows users to perform various geospatial data transformations, such as reprojection, resampling, and format conversion. This is essential for ensuring data compatibility and consistency in geospatial analysis.
- *Data access*: GDAL provides functions for reading and writing geospatial data, making it a valuable tool for data extraction, manipulation, and export. This is especially useful for working with remote sensing data, satellite imagery, etc.
- *Geometric operations*: GDAL provides functionality for performing geometric operations on geospatial data, such as clipping and mosaicking. This allows users to extract specific regions of interest from or combine multiple datasets.

Here is a basic example of using GDAL to open a raster dataset:

```
from osgeo import gdal
dataset = gdal.Open("path/to/raster.tif")
print("Raster Size:", dataset.RasterXSize, "x", dataset.
RasterYSize)
```

UNDERSTANDING GEOPANDAS FUNCTIONALITY

GeoPandas is a Python library that extends the Pandas Data Frame to support geospatial data. It provides a few features for working with geospatial data, including (Rajamani & Iyer, 2023):

- *Data structures*: GeoPandas introduces two new data structures, GeoSeries and Geodata Frame, extensions of pandas Series and Data Frame. GeoSeries and Geodata Frame objects can store geometric data alongside attribute data, such as points, lines, and polygons.
- *Data I/O*: GeoPandas provides convenient methods for reading and writing geospatial data in various formats, including Shapefile, GeoJSON, and GeoPackage.
- *Data exploration and analysis*: GeoPandas provides methods for exploring and analyzing geospatial data, such as head(), tail(), sample(), and plot(). It

also supports geometric operations, such as buffering, dissolving, intersection, union, and difference.

- *Spatial indexing*: GeoPandas can leverage spatial indexing to improve the performance of spatial queries on large datasets.
- *Spatial joins*: GeoPandas supports spatial joins between Geodata Frames, which are useful for aggregating attribute data from one dataset to another.

Unique Features of GeoPandas:

- *Seamless integration with Pandas:* GeoPandas is built on top of Pandas, inheriting all of Pandas' powerful data manipulation and analysis capabilities. This makes combining geospatial data with other data types, such as tabular and time series data, easy.
- *Support for a wide range of geospatial data formats:* GeoPandas can read and write geospatial data in various formats, including Shapefile, GeoJSON, GeoPackage, and PostGIS. This makes it easy to work with data from different sources and share data with other users.
- *GeoPandas is designed for high performance:* Even when working with large datasets, it leverages spatial indexing and other optimization techniques to ensure that spatial queries and operations are performed efficiently.

PROCESSING GEOSPATIAL DATA WITH PYTHON

Navigating the intricacies of geospatial data processing using Python requires a deep dive into essential libraries that shape the landscape of spatial analysis. This section explores key Python libraries, unraveling their functionalities and showcasing their significance in efficiently processing and analyzing geospatial data.

GDAL Library

The Geospatial Data Abstraction Library (GDAL) is a cornerstone in geospatial data processing. GDAL is a robust open-source library that facilitates the reading and writing of various raster and vector geospatial data formats (Lemenkova & Debeir, 2023).

Key Components of GDAL

- Dataset: At the core of GDAL is the dataset, a container holding raster data in bands, each representing a specific aspect of the information. It serves as a comprehensive representation of the contents within a single file.
- Raster Bands: These represent individual channels within an image, such as red, green, and blue bands in an RGB image. Each band contains pixel values corresponding to a specific attribute.
- Georeferencing Transform: GDAL supports affine transformations and ground control points for georeferencing. This ensures that raster coordinates can be accurately translated into real-world geographic coordinates.
- Coordinate System: The coordinate system within GDAL includes information about the projection, datum, units, and scale, which is crucial for understanding the spatial context of the data.
- *Metadata*: Additional information about the dataset is stored in metadata, providing insights beyond the raw pixel values. This can include details about the acquisition date, sensor specifications, and more.
- *Raster Data Processing*: GDAL provides a range of drivers for reading and writing diverse raster geospatial data formats. GDAL automatically selects a suitable driver based on the data type when reading a file. Users can choose a specific driver to create a new dataset for writing.

Sample Code for GDAL Library

```
from osgeo import gdal
dataset = gdal.Open('path/to/raster.tif')
width = dataset.RasterXSize
height = dataset.RasterYSize
num_bands = dataset.RasterCount
band = dataset.GetRasterBand(1)
data = band.ReadAsArray()
dataset = None
```

GEOPANDAS

GeoPandas is a Python library that extends Pandas' capabilities to handle geospatial data. It introduces a new data structure called Geodata Frame, an extension of the Pandas Data Frame, which includes geometry columns for efficient spatial data manipulation. The Geodata Frame, a central component of GeoPandas, introduces a "geometry" column that supports the representation of complex spatial entities such as points, lines, and polygons (Jordan, 2023).

Key Features of GeoPandas

- *Integration with Pandas:* GeoPandas seamlessly integrates with Pandas, allowing users to perform spatial operations using familiar Pandas syntax. This makes combining tabular data with spatial data easier for comprehensive analysis.
- *Handling Geometries:* GeoPandas supports geometry types such as points, lines, and polygons. The Geodata Frame's geometry column allows for the representation and manipulation of geometric objects, making it suitable for spatial analysis.
- *Spatial Operations:* GeoPandas enables spatial operations like overlay analysis, spatial joins, and buffering. These operations are crucial for identifying spatial relationships between different datasets and creating new spatial datasets.
- *File I/O Support:* GeoPandas supports the reading and writing of various geospatial file formats, including Shapefiles, GeoJSON, and more. This makes it easy to work with data from different sources and share results in widely used formats.
- *Visualization:* The library integrates with Matplotlib for straightforward Visualization of Geodata Frames. Users can easily create maps and visualize spatial patterns without extensive code.

Sample Code for Geopandas Library:

```
import geopandas as gpd
gdf = gpd.read_file('path/to/shapefile.shp')
buffered_gdf = gdf.buffer(100)
buffered_gdf.plot()
```

FIONA

Fiona is a lightweight yet robust Python library meticulously crafted for the intricate tasks of reading and writing vector geospatial data formats. Fiona is pivotal in geospatial data processing, offering a streamlined interface that caters to vector datasets. Its design philosophy revolves around simplicity, efficiency, and compatibility, making it an essential component in the geospatial data scientist's toolkit (Žuraulis & Pečeliūnas, 2023).

Key Features of Fiona

- *Format Agnosticism:* Fiona supports diverse vector geospatial data formats, ensuring seamless compatibility for reading and writing datasets, including Shapefiles and GeoJSON, offering flexibility in data source integration.
- *Efficient Reading and Writing:* Fiona prioritizes efficient handling of geospatial data, employing optimized algorithms for swift processing. Its streamlined processes enable responsive performance, especially with large-scale datasets.
- *Simplicity in Interface:* Fiona emphasizes a user-friendly interface for interacting with vector datasets. Its functions are designed to be straightforward, enabling users to read and write geospatial data without unnecessary complexities easily.
- *Integration with Shapely:* Fiona seamlessly integrates with Shapely for powerful geometry handling. This collaboration enhances Fiona's capabilities, providing users with efficient tools for working with vector geometries.
- *Pythonic Approach:* Fiona follows a Pythonic approach, aligning with Python conventions and idioms. This ensures Python users' accessibility and smooth integration with other Python-based geospatial and data science tools, promoting code readability.

Sample Code for Using Fiona:

```
import fiona
# Open a Shapefile for reading
with fiona.open('path/to/shapefile.shp', 'r') as src:
# Access metadata
schema = src.schema
crs = src.crs
# Iterate over features
for feature in src:
geometry = feature['geometry']
properties = feature['properties']
```

DATA MANIPULATION USING GDAL

Reading and Manipulating Raster Data with GDAL

In this practical example, open a raster file, perform a simple manipulation, and save the result in a new file. The code snippet demonstrates the fundamental steps in leveraging GDAL for raster data operations. This code serves as a template for opening a raster file, manipulating it (which should be filled in), and then saving the manipulated raster to a new file. The GDAL library is a powerful tool for working with geospatial data, and this script showcases a basic workflow for raster data processing.Top of Form

Working with Vector Data

GDAL's capabilities extend beyond raster data to encompass vector data handling. A practical example where GDAL is utilized for reading, modifying, and writing vector data, such as a Shapefile. The code snippet highlights the seamless integration of GDAL into vector data workflows.

Sample Code for GDAL:

```
from osgeo import ogr
input_file = 'input.shp'
driver = ogr.GetDriverByName('ESRI Shapefile')
dataSource = driver.Open(input_file, 0) # 0 for read-only
output_file = 'output.shp'
outputDataSource = driver.CreateDataSource(output_file)
outputDataSource.CopyLayer(dataSource.GetLayer(), 'new_layer_
name')
dataSource = None
outputDataSource = None
```

DATA PROCESSING ILLUSTRATIONS WITH FIONA

Fiona, a streamlined library for handling geospatial vector data, facilitates seamless data processing and manipulation within the Python ecosystem. In this section, we delve into practical examples that demonstrate Fiona's capabilities in reading, modifying, and writing geospatial vector data.

Reading and Modifying Features With Fiona

Let us start by opening a Shapefile using Fiona and iterating through its features. This example showcases how Fiona simplifies the process of reading and modifying attribute data within geospatial vector files.

Sample Code for Using Features with Fiona:

```
import fiona
input_shapefile = 'input.shp'
with fiona.open(input_shapefile, 'r') as src:
for feature in src:
feature['properties']['population'] *= 2
src = None
```

Writing Modified Data With Fiona

After modifying the attributes, we can use Fiona to write the updated features to a new Shapefile. The following example illustrates creating a new Shapefile and copying the modified features.

Sample Code for Writing Data - Fiona

```
output_shapefile = 'output.shp'
with fiona.open(output_shapefile, 'w', **src.meta) as dst:
for feature in src:
dst.write(feature)
dst = None
```

PYTHON VISUALIZATION FOR GEOSPATIAL DATA

Geospatial data, or spatial data, refers to information containing a geographic component, typically represented by latitude and longitude coordinates. This data type has various applications in various fields, including urban planning, environmental research, agriculture, transportation, and more. Visualizing geospatial data effectively allows for insights into spatial patterns, trends, and relationships. Python offers a rich ecosystem of libraries specifically designed for handling and visualizing geospatial data. These libraries provide various tools for creating maps, charts, and other visualizations that can help communicate insights from geospatial data. (Lemenkova & Debeir, 2022)

Common Types of Geospatial Data Visualizations

There are many different geospatial data visualizations, each with strengths and weaknesses. Some of the most common types of geospatial data visualizations include:

- Choropleth maps: Choropleth maps use color variations to represent data associated with different geographic regions. They often show population density, income distribution, or other quantitative data.
- Point symbol maps: Point symbol maps use symbols to represent individual data points. They are often used to show the locations of businesses, crime incidents, or other point-based data.
- Line maps: Line maps use lines to connect data points. They often show transportation routes, migration patterns, or other linear data.
- Heatmaps: Heatmaps use color gradients to represent the intensity of data over an area. They often show environmental data, such as air pollution levels or temperature distribution.

MATPLOTLIB FOR GEOSPATIAL DATA VISUALIZATION

Matplotlib, a versatile plotting library for Python, can be employed to create a variety of visualizations, including maps. Despite not being specifically designed for geospatial data, Matplotlib's adaptability and flexibility make it popular for generating basic maps and simple geospatial visualizations (Sarkar, 2022).

Matplotlib's key features for geospatial data visualization include:

- ***Plotting points, lines, and polygons:*** Matplotlib can plot individual data points, connecting lines, and polygons representing geographic areas. This enables the Visualization of various geospatial data types, including point distribution, road networks, and administrative boundaries.
- ***Customizing map appearance:*** Matplotlib offers a range of options for customizing the appearance of maps, including color palettes, symbol styles, and line properties. This allows users to create visually appealing and informative maps tailored to their needs.
- ***Integration with GeoPandas:*** Matplotlib can be used with GeoPandas, a Python library that extends Pandas to support geospatial data. This integration

allows for seamless manipulation and Visualization of geospatial data within the Matplotlib framework.

Consider a CSV file named 'data.csv' containing geospatial data with latitude and longitude coordinates. The following code snippet demonstrates a simple map creation using Matplotlib:

Sample Code for Using Geopandas and Matplotlib:

```
import geopandas as gpd
import matplotlib.pyplot as plt
data = gpd.read_file('data.csv')
fig, ax = plt.subplots()
data.plot(ax=ax)
ax.set_title('Geospatial Data Map')
ax.set_xlabel('Longitude')
ax.set_ylabel('Latitude')
plt.show()
```

Enhancing Data Representation With Matplotlib

Matplotlib, a versatile plotting library for Python, goes beyond basic data visualization. It offers a range of features to enhance data representation, making it more informative, visually appealing, and tailored to specific needs.

Customizing Plot Appearance

Matplotlib provides a variety of color palettes to customize the appearance of plots. These palettes range from simple color gradients to perceptually uniform color schemes, ensuring that color variations effectively represent data values.

Output: A histogram with a smooth color gradient from blue to red represents the distribution of data values.

Sample Code:

```
import numpy as np
import matplotlib.pyplot as plt
import seaborn as sns
x = np.linspace(0, 10, 100)
y1 = np.sin(x)
y2 = np.cos(x)
```

```
colors = sns.color_palette("husl", 2)
plt.figure(figsize=(8, 6))
plt.plot(x, y1, label='Sin Function', color=colors[0],
linewidth=2)
plt.plot(x, y2, label='Cos Function', color=colors[1],
linewidth=2)
plt.title('Customizing Plot Appearance with Matplotlib and
Seaborn')
plt.xlabel('X-axis')
plt.ylabel('Y-axis')
plt.legend()
plt.grid(True)
plt.show()
```

Line Properties

Matplotlib provides various options for customizing line properties, including width, style, and color. This allows for differentiating between multiple data series or emphasizing specific trends.

Sample Code for Line Properties:

```
import numpy as np
import matplotlib.pyplot as plt
x = np.linspace(0, 10, 100)
y1 = np.sin(x)
y2 = np.cos(x)
plt.figure(figsize=(8, 6))
plt.plot(x, y1, label='Sin Function', linewidth=2,
linestyle='--', color='red')
```

Output: A line plot with two data series, each represented by a distinct line style (solid and dashed) and color (blue and green)

Sample Code

```
plt.plot(x, y2, label='Cos Function', linewidth=1,
linestyle=':', color='blue')
plt.title('Customizing Line Properties with Matplotlib')
```

```
plt.xlabel('X-axis')
plt.ylabel('Y-axis')
plt.legend()
plt.grid(True)
plt.show()
```

Symbol Styles

Matplotlib offers a variety of symbol styles for plotting individual data points. These symbols can range from simple markers (circles, squares, triangles) to more intricate shapes, effectively representing different data types.

Matplotlib to Mark the Lines:

```
import numpy as np
import matplotlib.pyplot as plt
x = np.linspace(0, 10, 20)
y = np.sin(x)
plt.figure(figsize=(8, 6))
plt.scatter(x, y, label='Circle Marker', marker='o',
color='blue', s=50)
plt.scatter(x, -y, label='Square Marker', marker='s',
color='green', s=50)
plt.scatter(x, 2*y, label='Triangle Marker', marker='^',
color='orange', s=50)
plt.title('Symbol Styles for Data Points in Matplotlib')
plt.xlabel('X-axis')
plt.ylabel('Y-axis')
plt.legend()
plt.grid(True)
plt.show()
```

Output: A scatter plot with data points represented by red circles (marker='o'), each with a slightly transparent appearance (alpha=0.7).

Creating Informative Geospatial Maps

Geospatial maps are a powerful tool for visualizing and understanding spatial data. They are used in various fields, including urban planning, environmental research, agriculture, transportation, etc. Creating informative geospatial maps requires careful consideration of several factors, including:

- ***Data selection:*** Choose data relevant to the map's purpose, providing insights into the spatial patterns or relationships of interest.
- ***Map projection:*** Select a map projection that is appropriate for the mapped area and minimizes distortions.
- ***Map elements***: Use symbols, lines, and polygons to represent different types of geospatial data. Choose clear, easy-to-distinguish symbols that are well-suited to the data being represented.
- ***Color schemes:*** Use color palettes that are perceptually uniform and that effectively communicate the data values.
- ***Labels:*** Add clear and concise labels for geographic features, data values, and legends.
- ***Balance and simplicity:*** Aim for a map that is both informative and easy to understand. Avoid overwhelming the viewer with too much information or complex color schemes.
- ***Tools and libraries***: Utilize Python's rich ecosystem of geospatial data visualization libraries, such as Matplotlib, Folium, GeoPandas, and Geoplot, to simplify map creation.

UTILIZING MATPLOTLIB FOR BASIC GEOSPATIAL MAPS

Matplotlib, a versatile plotting library for Python, offers basic geospatial visualization capabilities, enabling the creation of simple maps and straightforward visualizations of geospatial data. While it is not as specialized as dedicated geospatial plotting libraries like Folium or Geoplot, Matplotlib provides a user-friendly and accessible tool for beginners and those who need to incorporate basic maps into their data analysis workflow.

Essential Steps for Creating Basic Geospatial Maps with Matplotlib

- Data Preparation: Ensure the geospatial data, such as a CSV file containing latitude and longitude coordinates, is structured.
- Importing Libraries: Import the necessary libraries, including matplotlib.pyplot for plotting and geopandas for handling geospatial data.
- Data Loading: Load the geospatial data into a Pandas Data Frame or Geodata Frame using the appropriate read_csv or read_file function.

- Map Creation: Create a figure and an axis using plt.subplots() and specify the projection type if necessary.
- Data Plotting: Plot the geospatial data using the appropriate Matplotlib functions, such as plot() for point data or pcolor() for raster data.
- Customization: Customize the map appearance using Matplotlib's extensive customization options, including color palettes, symbol styles, line properties, labels, and titles.

ADDING ANNOTATIONS AND CONTEXT TO MAPS WITH MATPLOTLIB

Matplotlib's annotation capabilities extend the visual communication power of geospatial maps, allowing you to add labels, comments, explanations, and other visual elements to enhance the data's clarity, context, and interpretability.

Text Annotations for Highlighting Key Points

Text annotations provide a direct way to add labels, comments, or descriptions to specific locations on the map, drawing attention to important features or patterns (Figure 1).

Color bars for Interpreting Color-Mapped Plots

Color bars are essential for interpreting color-mapped plots, such as heat maps or choropleth maps, where color variations represent data values. They provide a visual guide to the color scale, enabling viewers to associate color variations with specific data ranges (Figure 2).

Legends for Differentiating Data Series

Legends are crucial for distinguishing between multiple data series in line plots, scatter plots, or other visualizations that combine different data elements. They provide the key to identifying the meaning of each line, symbol, or color (Figure 3).

INTEGRATING PYTHON WITH GIS TOOLS

Geographic Information Systems (GIS) are pivotal in spatial analysis, mapping, and decision-making. Integrating Python with GIS tools enhances the capabilities of

Figure 1. Highlighting key points for text annotations

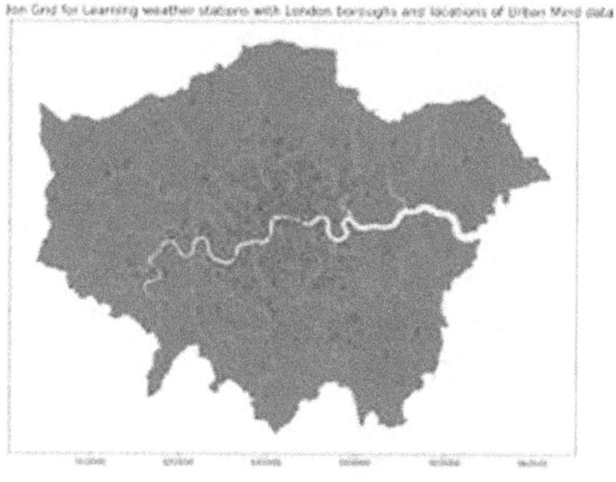

GIS platforms, enabling advanced geospatial analyses and workflow automation. This symbiotic relationship leverages Python's scripting power to extend, customize, and streamline GIS processes (Franch-Pardo et al., 2020).

Python With Arcgis for Advanced Analysis

In the realm of geospatial analysis, integrating Python scripts with GIS tools unveils a remarkable spectrum of possibilities, allowing for the execution of intricate analyses that transcend the conventional capabilities of standard GIS interfaces. The marriage

Figure 2. Color bars for color mapper plots

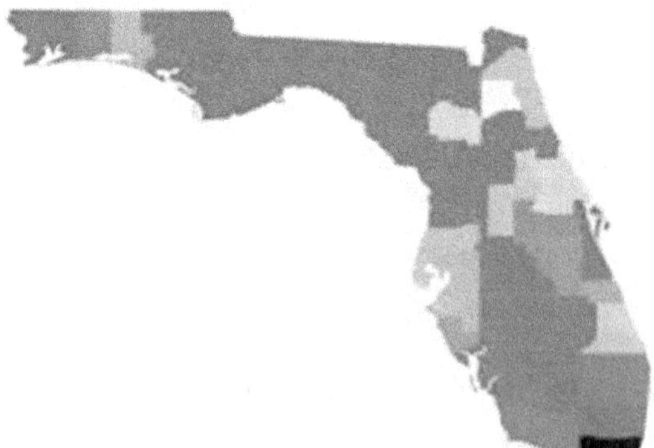

Figure 3. Legends for data series

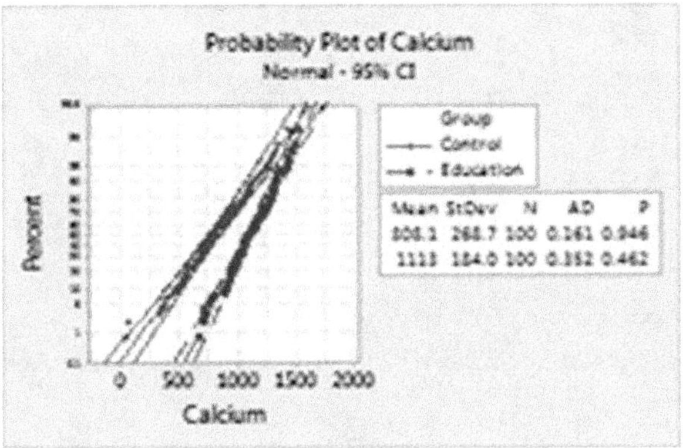

of Python's simplicity and versatility with the robust spatial functionalities of GIS tools introduces a new paradigm in spatial data exploration.

Initially, this script loads a spatial dataset from a shapefile, creating a Geodata Frame. The variable of interest is extracted from this dataset. The script then constructs a spatial weights matrix using the Queen contiguity criterion, defining spatial relationships between dataset observations. Subsequently, it utilizes PySAL's Moran's I analysis to quantify spatial autocorrelation, determining if similar attribute values cluster or disperse spatially. (Rey & Anselin, 2015)

QGIS and Python Integration

The integration of QGIS with Python marks a pivotal convergence, empowering programmers to augment the capabilities of this open-source Geographic Information System through the versatility of Python scripting. QGIS, renowned for its robust geospatial functionalities, becomes even more dynamic when coupled with Python's flexibility and automation prowess. This integration enables programmers to streamline geospatial workflows, automate repetitive tasks, and extend QGIS functionalities to meet specific spatial analysis requirements. Consider a scenario where a programmer needs to automate the generation of a series of thematic maps based on attribute values. When seamlessly integrated into QGIS, Python scripts become the driving force behind this automation. Through the PyQGIS library, which bridges Python and QGIS functionalities, the programmer can access and manipulate QGIS layers, apply symbology, and generate maps dynamically. (Lemenkova, 2020)

Sample Code: QGIS and Python Integration

```
from qgis.core import QgsProject, QgsVectorLayer, QgsSymbol,
QgsRendererCategory
project_path = 'path/to/your/project.qgs'
project = QgsProject.instance()
project.read(project_path)
layer_name = 'your_vector_layer_name'
vector_layer = QgsVectorLayer(f"{layer_name}?rendererV2=color",
layer_name, "ogr")
symbol = QgsSymbol.defaultSymbol(vector_layer.geometryType())
category = QgsRendererCategory('Category 1', symbol, 'label')
vector_layer.rendererV2().categories()[0] = category
0iface.layerTreeView().refreshLayerSymbology(vector_layer.id())
iface.mapCanvas().refresh()
```

ADVANCED GEOSPATIAL ANALYSIS AND MAPPING WITH PYTHON-GIS INTEGRATION

The fusion of Python and Geographic Information Systems (GIS) heralds a new era of advanced geospatial analysis and mapping. This integration empowers users to go beyond basic GIS functionalities, employing Python's versatility for complex analyses and customized mapping solutions.

Implementing GIS Analysis With Python Scripts

Fusing Python scripting with GIS tools unfolds a narrative of computational prowess. Python scripts, akin to the architects of a digital landscape, orchestrate intricate spatial analyses within the GIS environment. This synergy allows programmers to transcend the constraints of manual workflows, automating tasks and unleashing the potential for nuanced analyses that delve into the spatial intricacies of geospatial datasets. (Bunting et al., 2014)

The following steps provide a roadmap for implementing GIS analysis with Python scripts:

Acquire Geospatial Data: The first step involves obtaining geospatial data that serves as the canvas for analysis. This could be achieved by acquiring a shapefile, GeoJSON, or any other spatial data format that suits the analysis needs. Install

Required Libraries: Ensure the necessary Python libraries are installed. For basic geospatial operations, libraries like GeoPandas and PySAL are indispensable. Install them using the following command:

```
pip install geopandas pysal
```

Craft the Python Script: The heart of the GIS analysis lies in crafting a Python script that orchestrates spatial exploration. The script begins by importing the required libraries, such as GeoPandas and PySAL. Subsequently, it loads the acquired spatial dataset into a Geodata Frame using GeoPandas.

Interpret Results: Moran's I statistic, calculated through the spatial autocorrelation analysis, provides insights into spatial patterns within the dataset. A positive value indicates clustering of similar values, while a negative value suggests dispersion.

2. **Run the Script:** Execute the script in a Python environment. Ensure that the script has access to the required spatial dataset, and monitor the console for the output, especially Moran's I statistic.

Sample Code: Geopandas

```
import geopandas as gpd
from pysal.explore import esda
pip install geopandas pysal
file_path = 'path/to/your/spatial/data.shp'
gdf = gpd.read_file(file_path)
y - gdf['attribute_of_interest']
w = esda.weights.Queen.from_dataframe(gdf)
mi = esda.Moran(y, w)
```

Customized Maps and Visualizations With Integrated Tools

Python scripting with GIS tools enables insightful analyses and unveils a canvas for crafting customized maps and visualizations. Integrating Python libraries like Matplotlib and Folium with GIS tools allows programmers to extend beyond standard map representations, offering a palette of creative possibilities for tailored visualizations. Programmers can breathe life into static maps through the prism of Matplotlib, a versatile data visualization library. The script defines color gradients, line styles, and markers, allowing for the customization of map elements based on attribute values. This level of control empowers programmers to craft visual narratives that resonate with the complexity of the underlying geospatial data.

On a different note, integrating Folium, a Python library for creating interactive web maps, opens a portal to dynamic and engaging geospatial experiences. Programmers can embed interactive maps directly into their Python scripts, enriching the Visualization with tooltips, pop-ups, and layers that respond to user interactions. This level of interactivity transforms the map into a living entity, fostering a deeper understanding of the spatial nuances.

- ***Acquire Geospatial Data:*** As with any geospatial endeavor, the starting point is visualizing the spatial dataset. This dataset is the palette from which the map will draw its colors.
- ***Install Required Libraries:*** Ensure the necessary Python libraries are installed. For mapping and Visualization, commonly used libraries include Matplotlib and Folium. Install them using the following command:
- pip install matplotlib folium pip install matplotlib folium
- ***Craft the Python Script:*** The script, akin to a conductor's baton, guides the creation of customized maps. Import the essential libraries, such as GeoPandas for handling geospatial data and Matplotlib/Folium for Visualization. Load the spatial dataset into a Geodata Frame using GeoPandas.
- ***Design Customized Maps***: With the dataset, the script dives into crafting customized maps. For static maps, Matplotlib offers a versatile set of tools. The script can define colors, styles, and labels based on attribute values, creating a visual representation that aligns with the data's narrative.
- ***Interactive Maps with Folium:*** Folium takes a more dynamic approach. This library allows the creation of interactive web maps with ease. The script can add layers, markers, and tooltips, providing an engaging experience for the end-user.

Run the Script: Execute the Python script in an environment that supports graphical output. Watch as the script translates the spatial dataset into a visual masterpiece, conveying data points and a resonant narrative. This script, akin to an artisan's brushstroke, crafts static and interactive maps, transforming raw data into visual stories. (Rey & Anselin, 2009)

Sample Code:

```
import geopandas as gpd
import matplotlib.pyplot as plt
import folium
file_path = 'path/to/your/spatial/data.shp'
gdf = gpd.read_file(file_path)
fig, ax = plt.subplots(figsize=(10, 10))
```

```
gdf.plot(column='attribute_of_interest',ax=ax,legend=True,cmap=
'OrRd', legend_kwds={'label': "Legend"})
m=folium.Map(location=[gdf.geometry.centroid.y.mean(), gdf.
geometry.centroid.x.mean()], zoom_start=10)
for idx, row in gdf.iterrows():
folium.Marker([row.geometry.y,row.geometry.x],
popup=row['attribute_of_interest']).add_to(m)
m.save('customized_map.html')
```

FUTURE TRENDS AND APPLICATIONS IN GEOSPATIAL ANALYSIS

As technology continues to evolve, the field of geospatial analysis is poised for transformative advancements. Emerging trends and developments suggest a future where geospatial data becomes more accessible, analytical techniques become more sophisticated, and applications become increasingly diverse.

Increased Integration of AI and Machine Learning

Artificial intelligence (AI) and machine learning (ML) are revolutionizing geospatial analysis by automating previously time-consuming or infeasible tasks. AI algorithms can analyze large volumes of geospatial data, such as satellite imagery and sensor data, to extract valuable insights, identify patterns, and make predictions. Machine learning techniques, including deep learning, neural networks, and ensemble methods, enable the development of predictive models for various applications in fields like environmental monitoring, urban planning, and disaster management.

AI and ML can be applied to land cover classification, object detection, change detection, and predictive modeling tasks. For example, AI algorithms can analyze satellite imagery to detect land cover changes, monitor deforestation, or identify land use patterns. In disaster management, ML models can analyze historical data to predict the likelihood and severity of natural disasters, enabling proactive mitigation measures and emergency response planning. AI-driven analytics also play a crucial role in precision agriculture, optimizing resource allocation and increasing crop yields through data-driven decision-making.

Advancements in Remote Sensing Technologies

Remote sensing technologies, including satellites, drones, and aerial platforms, are evolving rapidly, leading to improved spatial, spectral, and temporal resolution. New sensor technologies, such as hyperspectral imaging and Light Detection and Ranging (LiDAR), enable the capture of rich and detailed information about the Earth's surface and atmosphere. These advancements enhance the quality and diversity of geospatial data available for analysis, enabling more accurate and comprehensive assessments of environmental, social, and economic phenomena. High-resolution satellite imagery, combined with advanced image processing techniques, can be used for land cover mapping, vegetation monitoring, urban growth analysis, and disaster response applications.

Emergence of Big Data Analytics

Big data analytics can be applied to diverse applications such as smart cities, precision agriculture, supply chain optimization, and disaster resilience. By analyzing large volumes of geospatial data, organizations can gain valuable insights into trends, patterns, and correlations that were previously inaccessible. Real-time analytics enable proactive decision-making and rapid response to dynamic events like natural disasters, disease outbreaks, or supply chain disruptions. Big data analytics also enable the integration of diverse data sources, including structured and unstructured data, to derive holistic insights and support data-driven decision-making across multiple domains. (Gholizadeh, 2022)

EXPLORING CLOUD-BASED GEOSPATIAL DATA ANALYSIS

In the ever-expanding domain of geospatial analysis, a profound transformation is underway, marked by the ascendancy of cloud-based methodologies. This paradigm shift signifies a fundamental change in the approach to harnessing the power of spatial data. In this visionary landscape, spatial data transcends the limitations of traditional infrastructure, finding a new abode in the ethereal realms of cloud computing. The cloud emerges as an expansive horizon where scalability takes center stage. Liberated from the constraints of fixed server capacities, geospatial analyses dynamically adapt to varying workloads and datasets of all sizes, offering unprecedented flexibility.

Collaboration takes on a new dimension in this cloud-driven environment. Regardless of their physical locations, geospatial analysts converge in a collaborative ecosystem where insights are shared seamlessly. The cloud becomes a global meeting ground, dissolving physical boundaries and making collaboration the cornerstone of geospatial endeavors.

Accessibility becomes omnipresent in this cloud-driven landscape. Geospatial tools and datasets are available anytime, anywhere, liberating analysts from the constraints of physical proximity and fostering a globally distributed approach to spatial analysis. (Ayano et al., 2023)

Sample Code:

```
import ee
ee.Authenticate()
ee.Initialize()
roi = ee.Geometry.Rectangle([-122.7, 36.8, -121.8, 37.5])
landsat_collection = ee.ImageCollection('LANDSAT/LC08/C01/T1_
SR') \
.filterDate('2020-01-01', '2020-12-31') \
.filterBounds(roi)
landsat_image = landsat_collection.first()
from IPython.display import Image
Image(url=landsat_image.getThumbUrl({'bands': ['B4', 'B3',
'B2'], 'min': 0, 'max': 3000}))
ndvi = landsat_image.normalizedDifference(['B5', 'B4'])
ndvi_vis_params = {'min': -1, 'max': 1, 'palette': ['blue',
'white', 'green']}
Image(url=ndvi.getThumbUrl(ndvi_vis_params))
task = ee.batch.Export.image.toDrive(image=ndvi,
description='ndvi_image',
folder='earth_engine_exports',
region=roi,
scale=30)
task.start()
```

Output: *This code snippet utilizes Google Earth Engine (GEE) to process satellite imagery and export an NDVI (Normalized Difference Vegetation Index) image to your Google Drive.*

CONCLUSION

In conclusion, this geospatial data and Python programming exploration has comprehensively understood the symbiotic relationship between technology and geography. Beginning with the fundamentals of geospatial data types, formats, and sources, the guide laid a strong foundation for navigating the intricate landscape of spatial analysis. The significance of Python in this context was underscored, demonstrating its prowess in handling diverse data types and facilitating efficient geospatial analysis.

Key Python libraries such as GDAL, GeoPandas, and Fiona were dissected, showcasing their integral roles in processing, analyzing, and visualizing geospatial data. Through practical examples, the guide illuminated the potential of Python scripts to automate tasks, manipulate data, and generate meaningful maps. Matplotlib's contribution to geospatial data visualization added insight, enhancing spatial information's communicative power.

REFERENCES

Chang, S., Wang, Z., Mao, D., Guan, K., Jia, M., & Chen, C. (2020). Mapping the essential urban land use in changchun by applying random forest and multi-source geospatial data. *Remote Sensing (Basel)*, *12*(15), 2488. doi:10.3390/rs12152488

Coetzee, S., Ivánová, I., Mitasova, H., & Brovelli, M. A. (2020). Open geospatial software and data: A review of the current state and a perspective into the future. *ISPRS International Journal of Geo-Information*, *9*(2), 90. doi:10.3390/ijgi9020090

Franch-Pardo, I., Napoletano, B. M., Rosete-Verges, F., & Billa, L. (2020). Spatial analysis and GIS in the study of COVID-19. A review. *The Science of the Total Environment*, *739*, 140033. doi:10.1016/j.scitotenv.2020.140033 PMID:32534320

Jordan, D. S. (2023). *Applied Geospatial Data Science with Python: Leverage geospatial data analysis and modeling to find unique solutions to environmental problems*. Packt Publishing Ltd.

Kantabutra, V. (2021). GeoDataLinks: A Suggestion for a Replacement for the ESRI Shapefile. In Advances in Data Science and Information Engineering: Proceedings from ICDATA 2020 and IKE 2020 (pp. 685-694). Springer International Publishing. doi:10.1007/978-3-030-71704-9_50

Kiwelekar, A. W., Mahamunkar, G. S., Netak, L. D., & Nikam, V. B. (2020). Deep learning techniques for geospatial data analysis. *Machine Learning Paradigms: Advances in Deep Learning-based Technological Applications,* 63-81.

Lemenkova, P., & Debeir, O. (2023). GDAL and PROJ libraries integrated with grass GIS for terrain modelling of the georeferenced raster image. *Technologies*, *11*(2), 46. doi:10.3390/technologies11020046

Psaila, G., & Fosci, P. (2021). J-CO: A platform-independent framework for managing geo-referenced JSON data sets. *Electronics (Basel), 10*(5), 621. doi:10.3390/electronics10050621

Rajamani, S. K., & Iyer, R. S. (2023). Use of Python Modules in Ecological Research. In *Perspectives on the Transition Toward Green and Climate Neutral Economies in Asia* (pp. 182–206). IGI Global. doi:10.4018/978-1-6684-8613-9.ch011

Sarkar, T. (2022). How to use python data science packages more productively. In *Productive and Efficient Data Science with Python: With Modularizing, Memory profiles, and Parallel/GPU Processing* (pp. 47–84). Apress. doi:10.1007/978-1-4842-8121-5_3

Urquhart, N. (2022). Linking to Real-world Data Sources and Services. In *Nature Inspired Optimisation for Delivery Problems: From Theory to the Real World* (pp. 191–207). Springer International Publishing. doi:10.1007/978-3-030-98108-2_9

Žuraulis, V., & Pečeliūnas, R. (2023). The Architecture of an Agricultural Data Aggregation and Conversion Model for Smart Farming. *Applied Sciences (Basel, Switzerland), 13*(20), 11216. doi:10.3390/app132011216

Chapter 7
Geospatial Data Storage and Management

Islom Mamatov
https://orcid.org/0000-0002-7029-5890
Samarkand International University of Technology, Uzbekistan

Mohammad Gouse Galety
https://orcid.org/0000-0003-1666-2001
Samarkand International University of Technology, Uzbekistan

Ruslan Alimov
Samarkand International University of Technology, Uzbekistan

A. V. Sriharsha
Mohan Babu University, India

Fanar Fareed Hanna Rofoo
Salahaddin University, Iraq

Gurram Sunitha
https://orcid.org/0000-0002-3305-8167
Mohan Babu University, India

ABSTRACT

This chapter comprehensively explores the evolving landscape of geospatial data storage and management, elucidating its critical role in unlocking the transformative potential of geospatial information across diverse applications. Focusing on recent advancements and future directions, this chapter delves into the intricacies of geospatial data management, encompassing vector data, raster data, and point-cloud data storage. Challenges such as data volume, variety, velocity, veracity, and standardization are meticulously addressed, highlighting the complexities of managing geospatial data effectively. Additionally, the chapter examines various storage options, including file formats, database systems, and cloud storage services, offering insights into their advantages and disadvantages. Moreover, it underscores the pivotal aspects of geospatial data management, such as data quality, metadata, and data integration.

DOI: 10.4018/979-8-3693-6381-2.ch007

INTRODUCTION

Efficient geospatial data management is necessary and a gateway to unlocking the transformative potential of geospatial information across diverse applications. Geospatial data's unique structure and functionality demand specialized management approaches to ensure its meaningful interpretation and practical application. This chapter takes you on a journey through the substantial advancements in geospatial data storage and management over the past decade, inspiring you with the evolving landscape of this field. It also outlines future research and application directions, offering insights into the exciting possibilities in geospatial data management.

Geospatial data management goes beyond mere storage and organization. It catalyzes earth observation and GIS applications, driving advancements in GIS/BIM integration, topology, 3D/4D geospatial data management, and the visualization of extensive geospatial features on web platforms. Integrating geosensor data sources has further broadened the scope of geospatial data applications, presenting new opportunities and challenges in data management (Cao et al., 2023).

Within the context of future developments, this chapter will delve into the burgeoning domain of geospatial data science, underscoring the significance of topology and the amalgamation of geospatial data management with data streaming libraries and "in-situ" geo-computing. Moreover, the discussion will extend to the possibilities of geospatial data visualization on web platforms and the creation of database support tailored to analyze substantial geospatial datasets, spotlighting the dynamic evolution of geospatial data management.

This chapter delves into the intricacies of geospatial data integration, including challenges such as data quality, inconsistent data, large file sizes, and processing time. The aim is to provide a comprehensive understanding of the complexities involved in managing geospatial data. The chapter also offers valuable insights into the best practices for geospatial data management, underlining the importance of creating a comprehensive catalog of all geospatial data. It suggests the establishment of a dedicated committee for catalog management and emphasizes the need to assess the implications before embarking on geospatial data integration efforts.

RELATED WORKS

Breunig et al. (2020) highlighted the indispensable role of geospatial data management in addressing contemporary challenges in big data applications, such as earth observation, geographic information system/building information modeling (GIS/BIM) integration, and 3D/4D city planning. Albrecht et al. (2020) examined the transformative impact of next-generation geospatial-temporal information technologies on disaster management, emphasizing the utilization of scalable technologies such as Apache Hadoop Distributed File Systems and NoSQL key-value stores to efficiently process and analyze large volumes of geocoded data for early warning, impact assessment, and response to natural disasters.

Li (2020) outlined the significance of high-performance computing (HPC) in addressing the challenges posed by geospatial big data, emphasizing its crucial role in efficiently collecting, managing, storing, and analyzing massive datasets collected from ubiquitous location-aware sensors. Dangermond and Goodchild (2020) discussed the evolution of visions for geospatial technology. They emphasized the necessity of a new vision centered around geospatial infrastructure to address challenges such as open access, sharing, engagement, and integrating emerging technologies like Big Data, artificial intelligence, and data science.

Rai, Mishra, and Singh (2022) edited a comprehensive volume titled "Geospatial Technology for Landscape and Environmental Management: Sustainable Assessment and Planning," which serves as an educational resource introducing the concepts, methods, and applications of geospatial technology to readers new to the field, featuring contributions from leading international practitioners. Gomes, Queiroz, and Ferreira (2020) provided a comprehensive overview of platforms for big Earth observation data management and analysis, highlighting novel technologies such as cloud computing and distributed systems and comparing seven prominent platforms, including Google Earth Engine, Sentinel Hub, and Open Data Cube, based on criteria relevant to the Earth observation community.

GEOSPATIAL DATA STORAGE

Geospatial data storage and management constitute fundamental components of geospatial information systems (GIS). Two main categories of geospatial data exist vector data and raster data. Each form of data possesses its distinct set of advantages and drawbacks, making them suitable for various applications (Breunig et al., 2020).

Vector Data

Vector data represents features on Earth's surface as points, lines, or polygons consisting of discrete geometric locations (x, y values) known as vertices (Leah Wasser et al., 2023). This data type is useful for representing discrete objects, such as cities, roads, or buildings, and can be easily scaled and manipulated in a GIS environment.

Examples of vector data are given in Figure 1.

Figure 1. Examples of vector data

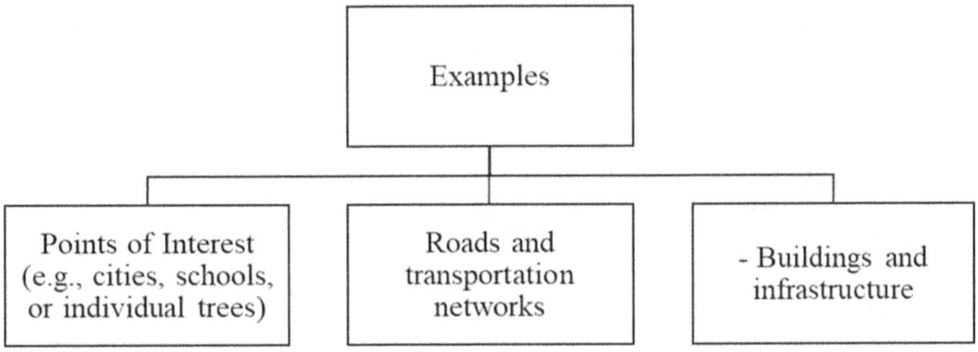

Raster Data

Raster data is stored as a grid of values, where each pixel represents a specific area on Earth's surface. Raster data can take on either a continuous form, such as elevation data, or a categorical form, like land use information. This data type is suitable for representing continuous phenomena, such as temperature, elevation, or spectral data from satellite images.

Examples of raster data are given in Figure 2.

Point-Cloud Data

Point cloud data constitutes a collection of points designed for dynamic information storage, often acquired from LiDAR (Light Detection and Ranging) systems. LiDAR, a sampling tool, emits over 160,000 pulses per second, and each pulse encompasses

Figure 2. Examples of raster data

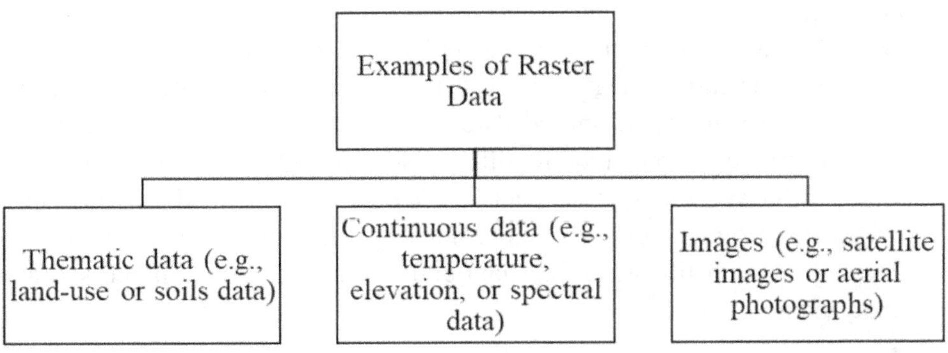

Figure 3. Challenges of geospatial data storage

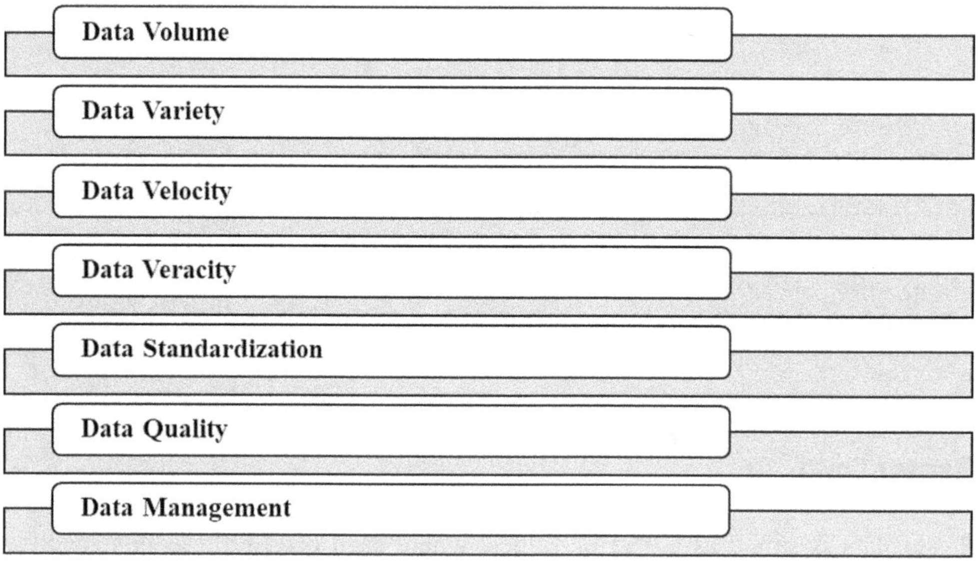

information components like X, Y, and Z coordinates, GPS timestamps, and inertial measurement units. Point cloud data finds application in diverse fields, including urban planning, infrastructure development, and environmental monitoring (Poux & Billen, 2019).

Storing geospatial data presents several challenges due to the characteristics of the data, such as volume, variety, and velocity. These challenges are discussed in more detail below (Figure 3):

- **Data Volume**: Geospatial data frequently possesses a substantial size, and the ongoing proliferation of high-resolution sensors and devices contributes to a continuous escalation in the volume of generated data. This can lead to data storage and processing difficulties as well as data transfer and analysis challenges.
- **Data Variety**: Geospatial data exhibits diverse formats and structures, encompassing raster images, vector data, and point cloud data. This variety can make developing a unified data management strategy challenging, as different data types may require different storage and processing methods.
- **Data Velocity:** Geospatial data is frequently generated in real-time, originating from sensors or satellite imagery. This can lead to challenges in keeping up with the streaming data and ensuring that it is processed and analyzed promptly.

- **Data Veracity**: Geospatial data may exhibit uncertainty, incompleteness, or inconsistency owing to various factors, including sensor errors, limitations in data processing, or human errors. This can make it difficult to rely on the data for decision-making and analysis.
- **Data Standardization**: The absence of standardization in geospatial data can pose challenges in integrating and analyzing data from diverse sources. This can lead to time-consuming and complex data cleaning and preprocessing steps, hindering the overall data management process.
- **Data Quality**: Safeguarding the quality and accuracy of geospatial data poses a substantial challenge. This can involve extensive data validation, cleaning, and standardization processes, which can be time-consuming and costly.
- **Data Management**: Effectively managing geospatial data to facilitate efficient access, analysis, and sharing is complex. This can involve developing robust data storage and processing systems and implementing effective data governance and collaboration strategies.

Various storage options are available for geospatial data, including file formats, database systems, and cloud storage services. Here are some examples of each:

FILE FORMATS

Understanding file formats is crucial in geospatial data storage and management. This chapter delves into the key file formats for storing and managing geospatial data effectively. Figure 4 depicts four prominent file formats that play integral roles in the geospatial domain: Esri Shapefile, GeoTIFF, Keyhole Markup Language (KML), and GeoJSON (Lawhead, 2015). Each file format serves distinct purposes and offers unique advantages in handling geospatial data. For instance, Esri Shapefile is a widely adopted format for storing vector data, providing essential geometry and attribute data files. On the other hand, GeoTIFF emerges as a prevalent raster format, embedding georeferencing details within TIFF files, and is commonly utilized for storing satellite imagery or elevation models.

Moreover, the Keyhole Markup Language (KML) presents itself as an XML-based open standard format, facilitating the exchange of GIS data. It is frequently used in web mapping applications to display geographic information (Li & Lu, 2018). Lastly, GeoJSON's ability to encode various geographic data structures is a favored format in web-based applications for representing spatial data (Portele, 2022) (Zouhar & Senner, 2020).

Figure 4. File formats

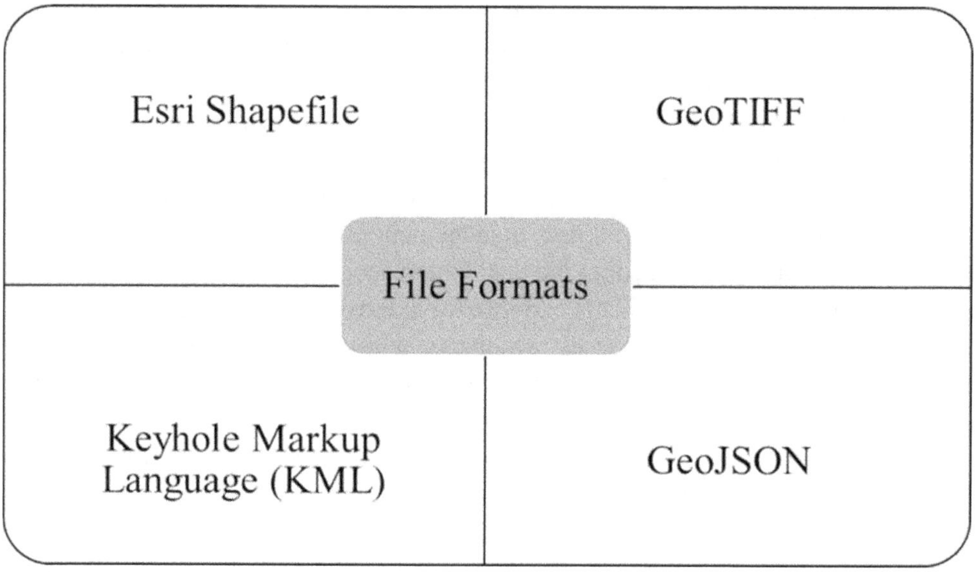

By comprehensively understanding these file formats, professionals in the field can optimize their workflows and make informed decisions when dealing with diverse geospatial datasets.

The file formats are given below and also in Figure 4:

1. Esri Shapefile
2. GeoTIFF
3. Keyhole Markup Language (KML)
4. GeoJSON

1. **Esri Shapefile:** A widely used file format for geospatial vector data. It includes three essential files: .shp for feature geometry, .dbf for attribute data, and .shx for the shape index position.
2. **GeoTIFF:** This raster file format embeds georeferencing details within a TIFF file, commonly utilized for storing raster data like satellite imagery or elevation models.
3. **Keyhole Markup Language (KML)**: An XML-based open standard format for exchanging GIS data, frequently employed in displaying geographic information in web mapping applications.

4. **GeoJSON**: This format encodes various geographic data structures and is commonly used to represent spatial data in web-based applications.

DATABASE SYSTEMS

In geospatial data storage and management, database systems are pivotal in efficiently organizing, accessing, and manipulating geographic datasets. This chapter delves into two significant database systems essential for managing geospatial data effectively: Esri Geodatabases and PostGIS. As depicted in Figure 5, these database systems offer comprehensive solutions for storing and managing diverse types of geospatial data, catering to the needs of various industries and applications.

Firstly, Esri Geodatabases stand out as enterprise storage solutions developed by Esri, capable of accommodating a wide range of geographic datasets, including vectors, rasters, point clouds, and 3D data. These geodatabases can be seamlessly integrated with enterprise-level database management systems such as Oracle, Microsoft SQL Server, IBM DB2, and PostgreSQL, providing scalability and robustness for handling large volumes of geospatial information. Secondly, PostGIS emerges as a powerful extension of the open-source database PostgreSQL, specifically designed to enhance its spatial database capabilities. PostGIS empowers users to store, query, and analyze geospatial data within the PostgreSQL environment, offering advanced spatial functions and indexing techniques for efficient data management.

By gaining a comprehensive understanding of Esri Geodatabases and PostGIS, professionals in the field can optimize their workflows, streamline data management processes, and unlock the full potential of geospatial data for various applications.

The various Database Systems are given below and also in Figure 5 (Dan Minney, 2021):

1. Esri Geodatabases (Arctur & Zeiler, 2004).
2. PostGIS (Corti et al., 2014)

- **Esri Geodatabases:** Esri's geodatabases are enterprise storage solutions that store various geographic datasets, including vectors, rasters, point clouds, and 3D data. They can be implemented on enterprise database management systems like Oracle, Microsoft SQL Server, IBM DB2, and PostgreSQL.
- **PostGIS**: PostGIS is an extension of the open-source database PostgreSQL, which provides spatial database capabilities for storing and managing geospatial data.

Figure 5. Database systems

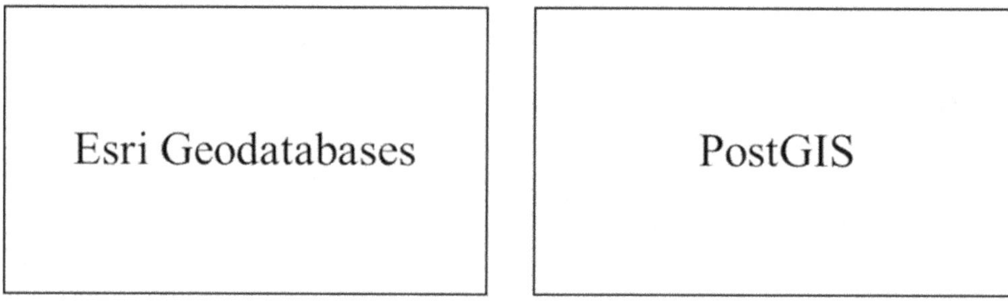

CLOUD STORAGE SERVICES

This section delves into cloud storage services tailored specifically for geospatial data, exploring the functionalities and advantages of prominent platforms such as Amazon Simple Storage Service (S3), Google Cloud Storage, and Microsoft Azure Blob Storage.

As illustrated in Figure 6, these cloud storage options provide organizations with flexible and robust solutions for storing, accessing, and managing geospatial data in the cloud environment.

Amazon Simple Storage Service (S3) stands out as a highly scalable cloud storage solution, renowned for its reliability and ease of use. It offers seamless storage and retrieval of substantial volumes of geospatial data, making it a preferred choice for organizations dealing with large datasets. Moreover, its compatibility with multiple GIS applications enhances its utility in the geospatial domain.

Google Cloud Storage, on the other hand, provides a comprehensive platform tailored specifically for geospatial workloads and applications. Its scalability, reliability, and high-performance capabilities cater to the diverse needs of organizations seeking efficient storage solutions for geospatial data.

Similarly, Microsoft Azure Blob Storage offers a cloud-based object storage service optimized for storing and managing geospatial data. Its scalability and support for scalable data transfer facilitate seamless integration with GIS applications, enabling organizations to leverage the power of the cloud for geospatial data management.

By understanding the capabilities and advantages offered by Amazon S3, Google Cloud Storage, and Microsoft Azure Blob Storage, organizations can make informed decisions and effectively leverage cloud technology to streamline their geospatial data workflows.

The various cloud storage services are as follows (Figure 6):

Figure 6. Cloud storage options

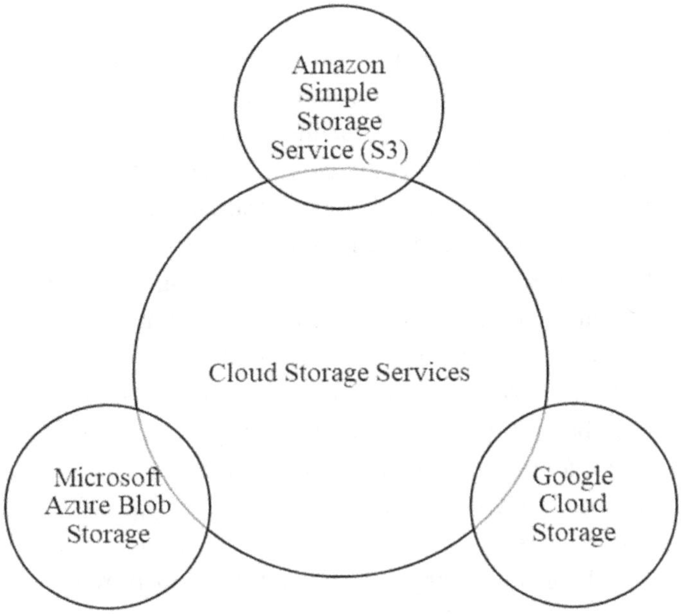

- Amazon Simple Storage Service (S3)
- Google Cloud Storage
- Microsoft Azure Blob Storage

- **Amazon Simple Storage Service (S3):** This scalable cloud storage solution facilitates storing and retrieving substantial data volumes. It is frequently used for geospatial data storage and can be integrated with multiple GIS applications.
- **Google Cloud Storage:** This platform offers a solution for storing and accessing geospatial data, delivering scalability, reliability, and high performance tailored for geospatial workloads and applications.
- **Microsoft Azure Blob Storage:** This cloud-based object storage service is suitable for storing geospatial data and supports scalable data transfer to and from the cloud.

These storage options cater to the diverse needs of geospatial data management, providing flexibility, scalability, and compatibility with various GIS applications and platforms.

ADVANTAGES AND DISADVANTAGES OF GEOSPATIAL DATA STORAGE OPTIONS

In geospatial data storage and management, selecting the appropriate data storage option is crucial for ensuring efficient organization, access, and manipulation of geographic datasets. This section delves into analyzing various geospatial data storage options, highlighting their advantages and disadvantages to aid in informed decision-making for data storage and management strategies. Figure 7 depicts several prominent data storage options, including Shapefile, File Geodatabase, GeoTIFF, KML, and GeoJSON. Each option offers distinct features and functionalities, catering to diverse requirements and preferences within the geospatial domain.

Firstly, we explore the advantages and disadvantages of Shapefile, a widely used format known for its compatibility and ease of sharing. Despite its widespread adoption, Shapefile has limitations, particularly regarding its functionality when dealing with complex geometrical or attributive information. Next, we delve into File Geodatabase, renowned for its comprehensive storage capabilities and capacity to accommodate various geographic datasets. However, its proprietary nature may pose interoperability challenges with applications not developed by Esri, necessitating careful consideration of compatibility requirements.

Furthermore, we analyze GeoTIFF, a raster file format known for its widespread use and georeferencing capabilities. While GeoTIFF offers versatility, it may not be universally suitable for all types of geospatial data, necessitating evaluation based on specific data requirements. Additionally, we examine KML, a straightforward markup language favored for ease of use and collaboration. Despite its advantages in sharing and distribution, KML's limited functionality for advanced geospatial analysis warrants consideration of alternative formats for complex data processing tasks.

Lastly, we explore GeoJSON, which is celebrated for its compatibility with web technologies and human-readable format. However, its potential for larger file sizes may impact storage and transmission efficiency, necessitating careful consideration of performance considerations. By analyzing the advantages and disadvantages of each storage option, this section aims to provide valuable insights into the suitability of different formats for diverse geospatial data management needs. With this knowledge, professionals can make informed decisions when crafting data storage and management strategies tailored to their specific requirements and objectives.

The various data storage options (Figure 7) are

- Shapefile
- Geodatabase
- GeoTIFF
- KML

Figure 7. Data storage options

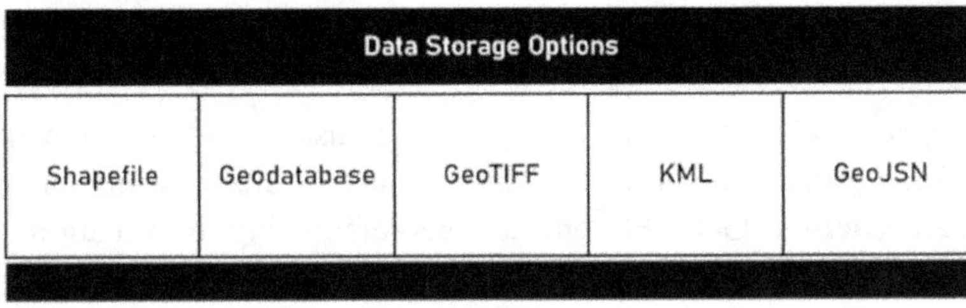

- GeoJSN

Advantages of Shapefile

- **Compatibility**: Shapefiles, not exclusive to Esri, are compatible with software from various providers, thus making them accessible to a broader audience.
- **Ease of Sharing**: Shapefiles offer ease of transmission, rendering them convenient for data sharing among civil engineers, surveyors, and other professionals.

Disadvantages of Shapefile

- **Limited Functionality**: Shapefiles come with certain constraints regarding geometry and attributes, potentially leading to data loss when dealing with or exchanging complex geometrical or attributive information.

Advantages of File Geodatabase

- **Comprehensive Storage:** File geodatabases can house various layers of points, polygons, and polylines, offering an all-encompassing storage option for geographic datasets.
- **Unified Storage:** File geodatabases can store diverse geographic datasets, such as vectors, rasters, point clouds, and 3D data.

Disadvantages of File Geodatabase:

- **Proprietary Nature:** File geodatabases, being proprietary to Esri, may face limitations in interoperability with applications not developed by Esri.

GeoTIFF

Advantages of GeoTIFF

- **Widespread Use:** GeoTIFF enjoys global usage and is compatible with open-source and commercial GIS and spatial data analysis software products.
- **Georeferencing:** GeoTIFF embeds georeferencing information within a TIFF file, making it well-suited for storing georeferenced raster imagery.

Disadvantages of GeoTIFF

- **Not Suitable for Every Data Type:** GeoTIFF might not be the most appropriate format for all kinds of geospatial data, as it is not universally suitable for every data type.

KML

Advantages of KML

- **Ease of Use:** KML, a straightforward and human-readable markup language, is relatively easy to learn and utilize for crafting geographic visualizations.
- **Collaboration and Sharing:** KML facilitates easy sharing and distribution among users across different devices, enhancing collaboration and interaction with geographic information.

Disadvantages of KML

- **Limited Functionality:** KML, primarily designed for display purposes, is not optimal for advanced geospatial analysis or complex data processing due to its limited analytical capabilities.
- **Performance Considerations:** KML files have the potential to grow large and complex, which can adversely affect performance, particularly on devices with lower-end specifications.

GeoJSON

Advantages of GeoJSON

- **Compatibility with Web Technologies:** GeoJSON's full compatibility and ease of processing with web technologies, notably JavaScript, contribute to its widespread use in web mapping applications.
- **Human-Readable Format:** GeoJSON is characterized by its simplicity and human-readable format, facilitating easy understanding and processing.

Disadvantages of GeoJSON

- **File Size:** Compared to other spatial vector layer formats, GeoJSON files can become relatively large, potentially affecting storage and transmission efficiency.

These advantages and disadvantages provide insights into the suitability of each storage option for different geospatial data management needs, helping in making informed decisions for data storage and management strategies.

GEOSPATIAL DATA MANAGEMENT

Geospatial data management represents a multifaceted discipline encompassing critical aspects such as data quality, metadata, and data integration. This section offers an in-depth exploration of geospatial data management, aiming to dissect each facet to illuminate its significance and implications for effectively handling and analyzing spatial data. Figure 8 illustrates this journey through the fundamental components of geospatial data management, which involves examining data quality. Precision, accuracy, and relevance are scrutinized within this domain, underlining the imperative nature of maintaining rigorous data quality standards. The repercussions of inadequate data quality, including errors, inconsistencies, and inefficiencies, underscore the necessity for unwavering attention and adherence to best practices to ensure the accuracy and reliability of spatial data and subsequent analyses.

Subsequently, attention is directed towards the pivotal role of metadata in GIS data management. As a vital repository of information about data resources, metadata offers insights into spatial data's source, quality, and structure. Adherence to metadata standards empowers organizations to establish a uniform framework for describing, managing, and sharing GIS dataset information, thereby promoting consistency and interoperability across many datasets. Organizations can surmount challenges associated with disparate data sources and formats through compatibility assessment, quality assurance, and harmonization methodologies, culminating in establishing a robust spatial database conducive to seamless querying and analysis.

The paramount importance of geospatial data management in ensuring the accuracy, reliability, and usability of spatial data and analyses cannot be overstated. Effective geospatial data management necessitates meticulous attention to detail and unwavering adherence to best practices, empowering organizations to leverage the full potential of spatial data for informed decision-making and strategic planning across diverse domains and industries.

Geospatial data management involves various aspects, including data quality, metadata, and data integration. Here are some details on each aspect (figure 8):

- Data Quality
- Metadata
- Data Integration

Data Quality

- Data quality encompasses various aspects, such as precision, accuracy, and relevance. In geospatial data management, maintaining data quality is crucial as it influences the accuracy, reliability, and usefulness of spatial data and its analysis. Inadequate data quality can result in mistakes, inconsistencies, inefficiencies, and the squandering of resources. For effective data quality management in GIS, rigorous attention and strict adherence to best practices are imperative.

Metadata

Metadata is the data about data details of a data resource, such as who, what, when, where, how, and why. It describes resources like maps, GIS files, imagery, and other location-based data in the geospatial context. Crucial for GIS data integration, metadata offers insights into the data source, quality, and structure, ensuring appropriate use and proper integration with other datasets. Metadata standards establish a uniform framework for describing, managing, and sharing GIS dataset information, promoting consistency and interoperability.

Data Integration

Integrating geospatial data entails merging spatial data from various sources and formats to form a comprehensive, unified dataset for analysis and decision-making.

Figure 8. Various aspects of geospatial data management

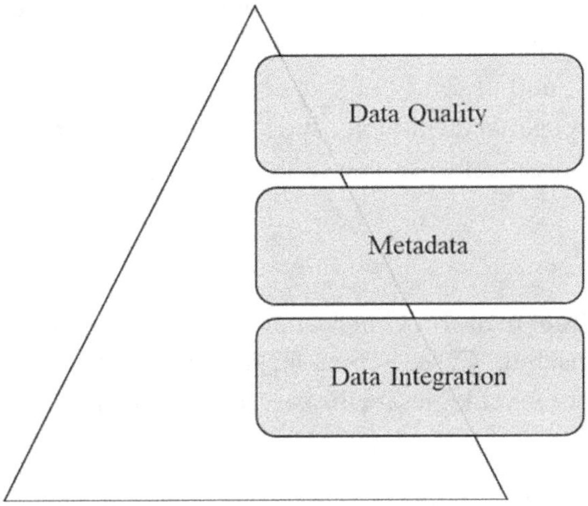

- This process involves aligning disparate data sources and formats, addressing data quality and accuracy variances, and establishing a cohesive spatial database for straightforward querying and analysis.
- GIS data integration facets include compatibility, quality, harmonization, and metadata.
- The techniques for data integration are continuously advancing, and applications are found in numerous fields, such as life sciences and engineering.

These aspects of geospatial data management are critical for ensuring the accuracy, reliability, and usability of spatial data and analysis. Effective geospatial data management requires meticulous attention to detail and adherence to best practices.

CONCLUSION

In conclusion, the chapter on "Geospatial Data Storage and Management" elucidates the critical role of efficient data management in unlocking the transformative potential of geospatial information across diverse applications. By exploring recent advancements and future directions in geospatial data storage and management, the chapter underscores the complexities of handling various data types, including vector, raster, and point-cloud data. Through a meticulous examination of challenges such as data volume, variety, velocity, veracity, and standardization, alongside an overview of storage options and best practices, the chapter equips professionals with the knowledge needed to optimize workflows and make informed decisions in geospatial data management. Emphasizing the significance of data quality, metadata, and data integration, the chapter highlights their role in ensuring the accuracy, reliability, and usability of spatial data and analysis, paving the way for advancements in geospatial technology and its applications across diverse domains.

REFERENCES

Albrecht, C. M., Elmegreen, B., Gunawan, O., Hamann, H. F., Klein, L. J., Lu, S., Mariano, F., Siebenschuh, C., & Schmude, J. (2020). Next-generation geospatial-temporal information technologies for disaster management. *IBM Journal of Research and Development*, *64*(1/2), 5–1. doi:10.1147/JRD.2020.2970903

Breunig, M., Bradley, P. E., Jahn, M., Kuper, P., Mazroob, N., Rösch, N., Al-Doori, M., Stefanakis, E., & Jadidi, M. (2020). Geospatial data management research: Progress and future directions. *ISPRS International Journal of Geo-Information*, *9*(2), 95. doi:10.3390/ijgi9020095

Breunig, M., Bradley, P. E., Jahn, M., Kuper, P., Mazroob, N., Rösch, N., Al-Doori, M., Stefanakis, E., & Jadidi, M. (2020). Geospatial data management research: Progress and future directions. *ISPRS International Journal of Geo-Information*, *9*(2), 95. doi:10.3390/ijgi9020095

Cao, Y., Xu, C., Aziz, N. M., & Kamaruzzaman, S. N. (2023). BIM–GIS integrated utilization in urban disaster management: The contributions, challenges, and future directions. *Remote Sensing (Basel)*, *15*(5), 1331. doi:10.3390/rs15051331

Dangermond, J., & Goodchild, M. F. (2020). Building geospatial infrastructure. *Geo-Spatial Information Science*, *23*(1), 1–9. doi:10.1080/10095020.2019.1698274

Gomes, V. C., Queiroz, G. R., & Ferreira, K. R. (2020). An overview of platforms for big earth observation data management and analysis. *Remote Sensing (Basel)*, *12*(8), 1253. doi:10.3390/rs12081253

Lawhead, J. (2015). *Learning geospatial analysis with Python*. Packt Publishing Ltd.

Li, D., & Lu, M. (2018). Integrating geometric models, site images and GIS based on Google Earth and Keyhole Markup Language. *Automation in Construction*, *89*, 317–331. doi:10.1016/j.autcon.2018.02.002

Li, Z. (2020). Geospatial big data handling with high performance computing: Current approaches and future directions. *High Performance Computing for Geospatial Applications*, 53-76.

Portele, C. (2022). Encoding of Geographic Information. In *Springer Handbook of Geographic Information* (pp. 95–106). Springer International Publishing. doi:10.1007/978-3-030-53125-6_4

Poux, F., & Billen, R. (2019). A smart point cloud infrastructure for intelligent environments. In *Laser Scanning* (pp. 127–149). CRC Press. doi:10.1201/9781351018869-9

Rai, P. K., Mishra, V. N., & Singh, P. (Eds.). (2022). *Geospatial technology for landscape and environmental management: sustainable assessment and planning*. Springer. doi:10.1007/978-981-16-7373-3

Zouhar, F., & Senner, I. (2020). Web-based visualization of big geospatial vector data. In Geospatial Technologies for Local and Regional Development: Proceedings of the 22nd AGILE Conference on Geographic Information Science 22 (pp. 59-74). Springer International Publishing. 10.1007/978-3-030-14745-7_4

Chapter 8
Machine Learning for Geospatial Analysis:
Enhancing Spatial Understanding and Decision-Making

Robin Rohit Vincent
https://orcid.org/0000-0003-1537-3902
Presidency University, India

E. Sakthivel
Presidency University, India

Meena Kumari
Presidency University, India

F. Nisha
https://orcid.org/0000-0001-9538-3048
CMR University, India

A. Rohini
Presidency University, India

ABSTRACT

Historically, statistical techniques have been a major component of geographic analysis, which is the study of geographical phenomena using geographic data. But the emergence of machine learning (ML) has completely changed the area, making it possible to analyse intricate spatial relationships and glean insightful information from enormous geographical datasets. The foundations, methods, and applications of machine learning for geographic analysis are all covered in detail in this chapter. The authors start with a brief overview of machine learning and then move on to talk about its application to geographic analysis. Subsequently, they explore particular machine learning approaches that are frequently employed in the domain, such as decision trees, support vector machines (SVMs), K-means clustering, convolutional neural networks (CNNs), geostatistics, reinforcement learning, time series analysis, and anomaly detection. The authors wrap up by outlining the possibilities of machine learning for geospatial analysis in the future and provide resources for more research.

DOI: 10.4018/979-8-3693-6381-2.ch008

INTRODUCTION

Combining geospatial analysis (H. Ouchra et al., 2023) with machine learning (M. Kumar et al., 2023) creates a novel synergy beyond conventional approaches to comprehending our spatial surroundings. The ability of machine learning to recognize intricate patterns and connections within data has made it an indispensable instrument for geospatial analysts. This integration makes complex analyses of satellite imagery, climatic data, and geographic data possible, providing previously complex insights to obtain via manual or rule-based methods.

Although machine learning holds great potential for geospatial analysis, there are still obstacles to overcome. A constant problem is guaranteeing machine learning models' interpretability and ethical usage, given the richness and volume of spatial data (J. Li et al., 2022). Despite this, the prospects brought about by this convergence of technology are revolutionary, opening up a whole new field for comprehending and handling the complexity of our globally interconnected environment. The future at the nexus of machine learning and geographic analysis presents intriguing opportunities for additional innovation, discovery, and refinement as we navigate this terrain.

Evolution of ML Within Geographic Analysis

The evolution is deeply intertwined with the advancements in both fields and the broader technological landscape. Here is a historical context of how ML has evolved within geographic analysis:

Early Years (1950s-1980s)

- In the early years of ML, the focus was primarily on developing foundational concepts and algorithms. This period saw the emergence of key ideas such as perceptrons, neural networks, and decision trees.
- Geographic analysis primarily relied on traditional statistical methods and manual interpretation of spatial data. GIS (Geographic Information Systems) emerged during this period, enabling spatial data storage, manipulation, and analysis.

Foundational Research (1980s-1990s)

- During the 1980s and 1990s, ML research experienced a resurgence with the development more sophisticated algorithms and approaches. Among notable advancements are backpropagation for training neural networks, support vector machines (SVM), and ensemble methods like random forests.

- Geographic analysis started incorporating computational techniques for spatial modeling and analysis. Remote sensing technologies became more prevalent, providing large volumes of spatial data for analysis.

Integration of ML and Geographic Analysis (2000s-2010s)

- The 2000s marked a significant shift as ML techniques began to be integrated into geographic analysis. Researchers explored the application of ML algorithms for tasks such as land cover classification, object detection in satellite imagery, and spatial prediction.
- Advances in computing power, data availability, and algorithmic sophistication enabled the scalability of ML approaches to handle large-scale geospatial datasets.
- This period saw the emergence of interdisciplinary research efforts, with collaborations between geospatial scientists, computer scientists, and statisticians.

Rise of Deep Learning and Big Data (2010s-Present)

- The past decade witnessed a surge in interest and adoption of deep learning techniques, particularly convolutional neural networks (CNNs), for geospatial analysis. Deep learning models demonstrated remarkable performance in tasks such as image classification, object detection, and semantic segmentation.
- The proliferation of big data, fueled by advancements in remote sensing, GPS, and social media, provided a wealth of geospatial data for analysis.
- Combined with big data analytics and cloud computing infrastructure, ML algorithms enabled real-time monitoring, predictive modeling, and decision support systems for geographic applications.

Current Trends and Future Directions

- ML plays a pivotal role in geographic analysis, with ongoing research focusing on addressing challenges such as model interpretability, uncertainty quantification, and ethical considerations.
- Emerging trends include the integration of ML with spatial-temporal modeling, explainable AI (XAI) for geospatial applications, and applying reinforcement learning in dynamic spatial decision-making.

- Future directions may involve harnessing ML techniques for addressing complex spatial challenges such as climate change mitigation, disaster response, urban planning, and sustainable development.

LITERATURE SURVEY

The authors of the paper (H. Ouchra et al., 2023) suggested a procedure for assessing and contrasting the capacity of the GEE platform and Landsat 8 OLI data to classify (Sand, Cropped, Liquid, Buildings, and Jungle regions) on Moroccan territory, as well as the producer, user, and classification accuracy of various algorithms. Finding a rapid technique to collect the training and validation points in addition to physically acquiring them is the study's main problem. Two other disadvantages are the requirement for an extremely reliable connection to the internet and the restricted availability of all the satellite information in GEE in the appropriate resolutions. There is a lack of relevant research on unsupervised algorithms or models based on deep learning that perform better at picture classification to improve the categorization and recognition of metropolitan areas in satellite imagery.

The scoping review in (Ylenia Casali et al., 2022) aims to: a) map the essential subjects, data sources, machine learning algorithms, and parameter selection techniques; b) pinpoint the most significant issues and trends in the application of machine learning; and c) pinpoint knowledge gaps to assist direct subsequent studies. They looked over and displayed the information in three different ways. As machine learning and AI gain traction, we anticipate that their applications will help address unresolved difficulties related to urgent sustainability concerns, including resilience and circularity. Standardizing the choices of data, algorithms, and settings is still necessary. Although meticulously disclosing parameters makes the work more transparent and repeatable, systematic comparisons of data and algorithm selection might help explore the applicability of these methods and their consequences. It will be advantageous for spatial machine learning to focus on the production and kinds of datasets. Creating spatial machine learning studies in data-poor regions (like the Global South) has constraints.

Additionally, research frequently employs heterogeneous data from several data sources. Spatial data-driven studies can benefit from learning about data integration and merging. The drawback is that it has brought to light several intriguing subjects for more research in all domains, ranging from more comparison studies to a deeper understanding of the impact of choosing appropriate algorithms, opt parameters, and pertinent datasets.

In (Negar Shabanpour et al., 2022), using three machine learning algorithms decision tree (DT) (Y. Lu et al., 2022), support vector regression (SVR) (Q. Wang et al., 2022), and linear regression (LR) (Kalva et al., 2021) In Iran's Isfahan province, the authors looked at how environmental factors affected leishmaniasis prevalence, geographical self-correlation spatial simulation, and the creation of a leishmaniasis risk map. Using three algorithms based on machine learning to create a map that forecasts the risk of leishmaniasis and enables a spatiotemporal examination of the disease was an inventive aspect of the current work. Its main limitations are the province's varying heights above sea level, its hilly and plain regions, and the difficulty of predicting leishmaniasis due to its reliance on numerous environmental factors. The majority of managers and decision-makers place the highest focus on identifying locations where leishmaniasis is a significant threat.

In a paper (Alkhatib, Ramez, et al., 2023), in their publication, the scientists describe how to incorporate machine learning methods into investigating forest fires. The articles we looked at suggested that creating sophisticated AI systems would open the door to foreseeing such significant environmental issues and assisting with legislative actions in the fight against forest fires. The main drawbacks are that during forest fires, machine learning (ML) systems, which require a large amount of data for training, are typically unavailable; The high processing power requirements of ML learning techniques can make them expensive and challenging to scale; and it could be difficult to assess the predictive efficacy of ML systems for learning in real-world scenarios.

The authors of (Tanzeela et al., 2023) proposed two methods for observation matrices that deal specifically with the geographical characteristics of the data prior to the start of the learning process without changing the learning algorithm. For any spatial machine learning application, spatial properties in missing data processing and sampling strategies are crucial. One of the main techniques for adding more spatial attributes of the data to the observed matrix is the generation of new spatial features. Due to their integration of LSTM (S. Güney and Ç. B. Erdaş, 2019) and CNN (Taye & Mohammad Mustafa, 2023), deep neural networks have certain constraints, such as the inability to undertake simultaneous learning across size, hierarchy, time, and space; the enormous amount of training data required; additionally, the large number of parameters that make DNNs expensive to compute and complex exceeds all limits due to the arbitrary nature of their architectural design.

WHAT IS MACHINE LEARNING (ML), FOR STARTERS?

Machine learning (ML) is a cutting-edge artificial intelligence (AI) area that allows computers to detect patterns and form opinions without explicit programming. It is an exciting and active field of study with the power to revolutionize how we use technology and approach challenging issues. The fundamental tenet of machine learning is that data can teach computers new things. Instead of directly programming a computer to complete a task, machine learning algorithms (ML algorithms) employ statistical techniques to allow the machine to improve its performance on a particular task gradually. For the system to recognize patterns, correlations, and trends in the data, it must be exposed to big datasets.

The Basic Workflow of Machine Learning

Figure 1 shows the diagram of the ML workflow (A. Giannopoulos et al., 2022).

- **Data Collection:** The first step in any machine learning endeavor is to collect relevant data. The amount and caliber of data are essential determinants of the machine learning model's performance.
- **Data Preprocessing:** Preprocessing is necessary to address missing numbers, outliers, and other irregularities in raw data since it is frequently untidy. In order to train the ML model, this step ensures the data is in the correct format.
- **Feature Selection/Extraction:** The input variables that the model is trained with are called features. While feature extraction mixes and modifies features to improve model performance, feature selection focuses on selecting the most pertinent characteristics.
- **Model Selection:** The kind of challenge determines which machine learning models are chosen. Often used models include support vector machines, decision trees, and neural networks.
- **Training the Model:** After training on the labeled dataset, the chosen model learns to forecast by minimizing errors in its parameter settings.
- **Evaluation:** The performance and generalization capabilities of the trained model are evaluated using an alternative dataset.
- **Prediction:** The model can be used to forecast fresh, unseen data after it has been trained and assessed.

Key Concepts of Machine Learning

Machine learning can be broadly categorized into three areas based on learning styles and approaches. These consist of unsupervised learning, reinforcement learning, and supervised learning.

Figure 1. ML workflow

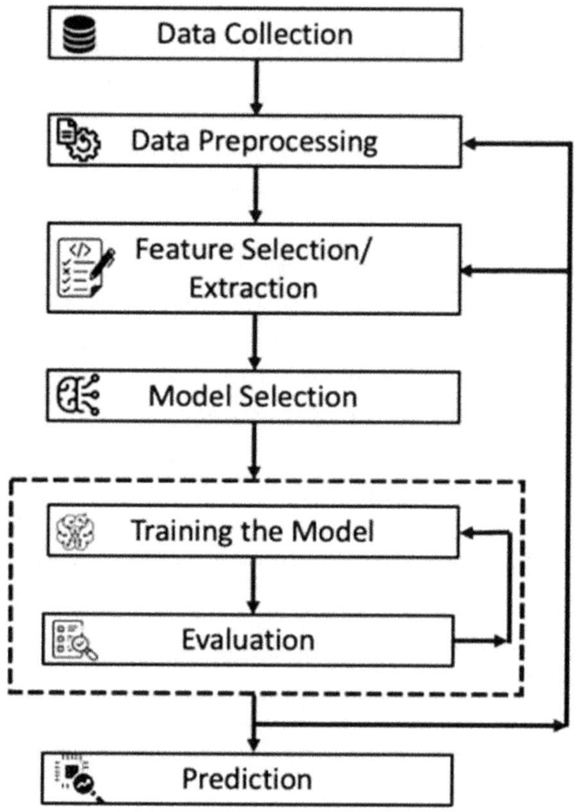

Supervised Learning

The fundamental artificial intelligence (AI) component, supervised learning (V. Gupta et al., 2022), unleashes data's amazing potential. Imagine providing an algorithm with a mountain of instances, each coupled with a desired result. The system gains the ability to link input data to predicted output through this potent training, which allows it to make predictions or judgments based on previously unknown data. This is what supervised learning is all about. There are two primary categories of supervised learning:

Classification: The algorithm gains the ability to classify data points into pre-established groups. Consider teaching an algorithm to distinguish between photos that feature dogs, cats, or neither. Based on its training on labeled images, the system can reliably classify fresh, unseen photos into these categories.

Regression: The main goal of this kind is to predict continuous values. For instance, based on a fresh wind speed measurement, an algorithm trained on past temperature data and associated wind speeds can forecast future temperatures.

Supervised Learning Algorithms

Various types of supervised learning algorithms can be distinguished according to the type of prediction task involved. Typical examples of supervised learning algorithms are as follows:

- **Linear Regression:** This traditional method uses a straight line to depict the relationship between the input and output variables. Consider utilizing linear regression to forecast the price of homes based on factors like location and size.
- **Decision Trees:** To arrive at a final forecast, these tree-like structures sequentially decide based on the qualities of the data. Consider a decision tree that considers variables like income and credit score to determine if a loan will be approved.
- **Support Vector Machines (SVMs):** These techniques identify the ideal hyperplane that divides data points from several classifications. Consider employing a Support Vector Machine to classify emails as either spam or non-spam.
- **Neural Networks:** These intricate networks (Montesinos López et al., 2022), which draw inspiration from the human brain, use interconnected layers to discover intricate links between data. They are quite good at things like translating languages and recognizing images.

Supervised Learning in Geospatial Analysis

Geospatial analysis uses supervised learning, a potent subset of machine learning that revolutionizes how we interpret and use Earth's geographical data. Supervised learning algorithms are essential for using labeled data to extract valuable insights, make predictions, and automate operations in the field of geospatial research.

Below are a few examples of how supervised learning is used in geographic analysis.

Land Cover Classification: Classifying land cover (H. Waghela et al., 2022) is one of the core uses of supervised learning in gcospatial analysis. While labeled, data describes land cover categories like forests, urban regions, and water bodies; satellite imagery with spectral information can be used as input features. By learning to distinguish between these distinct land cover types, the model makes mapping and monitoring more effective.

Predictive Modeling: Geospatial analysis uses supervised learning models' prowess in predicting a range of occurrences. For example, decision-makers in agriculture and environmental conservation are empowered when they can estimate crop production by analyzing soil and climate data or predict deforestation based on previous land cover changes.

Disaster Response and Management: Supervised learning facilitates quick response and management during natural disasters. The system can quickly analyze the effects of disasters by training models using pre-disaster geospatial data with matching damage labels. This helps with emergency response planning and resource allocation.

Unsupervised Learning

By training the algorithm on unlabeled data, unsupervised learning (K. Makkar et al., 2022) (M. Usama et al., 2019) enables the system to recognize patterns and structures without preset output labels. Unsupervised learning frequently involves tasks like clustering and dimensionality reduction, which help with data exploration and reveal links that may be hidden. Imagine discovering a new area by piecing together hints from the landscape, flora, and animal traces rather than using a map or guide. That is the fundamental idea behind unsupervised learning, a subfield of machine learning in which algorithms search through unlabeled data to find underlying structures and patterns without needing predetermined outputs or categories. There are two primary categories into which unsupervised learning falls:

- **Clustering:** Putting data pieces with comparable characteristics in groups, such as consumers based on past purchases.
- **Dimensionality reduction:** Reducing the complexity of complex data by highlighting its most crucial elements, which facilitates analysis and visualization.
- **Anomaly detection:** Spotting anomalous data points that differ from the norm and might be signs of fraud or system flaws.

Unsupervised Learning Algorithms

Clustering

- **K-means clustering:** Assembles data items according to how comparable their attributes are into a predetermined number (k) of clusters. It is easy to use and effective, but we have to decide in advance how many clusters to use.
- **Hierarchical clustering:** Creates a hierarchy of clusters by repeatedly dividing or merging them according to how related they are. It is flexible and does not require a predetermined cluster number, but it may incur high computing costs for large datasets.
- **Density-based spatial clustering of applications with noise (DBSCAN):** Detects clusters based on the density of data points in the feature space and is robust against noise and outliers. It needs exact parameter modification in order to get the best results.

Dimensionality Reduction

- **Principal component analysis (PCA):** Reduces redundancy and streamlines analysis by locating the data's most significant directions of variance and projecting it onto a lower-dimensional space.
- **Independent component analysis (ICA):** Removes statistically independent signals from mixed data, which helps reveal information sources and hidden patterns. Though more intricate than PCA, it can uncover unspoken connections.

Anomaly Detection

- **Local outlier factor (LOF):** Detects data points that differ noticeably from their neighbors and highlights any abnormalities or outliers (X. Yang et al., 2022).
- **Isolation forest:** Separates data points at random and makes efficient anomaly detection easier by needing, on average, fewer splits for outliers than for standard data.

Association Rule Learning

- **Apriori algorithm:** Finds common patterns and relationships among the items in an extensive dataset; this is useful for consumer segmentation and market basket analysis. It can be costly to compute for large datasets with intricate relationships.

Unsupervised Learning in Geospatial Analysis

Unsupervised learning, a subfield of machine learning, has become a potent instrument in geospatial analysis, offering a distinct viewpoint for identifying patterns and structures in the enormous amounts of geographical data on Earth. In contrast to supervised learning, unsupervised learning is not dependent on labeled datasets; instead, it investigates the innate structures present in geographical data, opening the door to meaningful data exploration, discovery, and clustering.

Below are a few examples of how unsupervised learning is used in geographic analysis.

Anomaly Detection: Unsupervised learning is quite useful when identifying irregularities in geospatial data. The algorithm can facilitate early intervention or investigation by identifying uncommon occurrences, such as unanticipated changes in land cover or abnormalities in environmental variables, by identifying patterns that depart from the norm.

Data Exploration and Pattern Discovery: Data exploration is encouraged by unsupervised learning, which makes hidden structures in geospatial datasets visible. This is especially helpful when working with intricate environmental or topographical data, where the correlations between variables might not be immediately evident. Novel patterns and links can be found by algorithms, which improves our comprehension of geographical phenomena.

Reinforcement Learning

An agent can be trained to make decisions in a particular environment to accomplish a specific objective through reinforcement learning (J. et al., 2022). Rewards or penalties are used as feedback to steer the agent towards the best action when making decisions. Applications of reinforcement learning include autonomous systems, robotics, and gaming.

Think of a hero from a video game traversing a perilous dungeon. They encounter challenges, gather rewards, and grow from every error they make until they finally grasp the game's nuances. This trip reflects the essence of Reinforcement Learning (RL), an intriguing branch of machine learning in which computers use trial and error to navigate uncertain settings to maximize rewards. There are three primary categories of reinforcement learning:

- **Model-based RL:** The agent creates an internal environment model to forecast future states and rewards. This allows for more strategic decision-making.
- **Model-free RL:** The agent gains knowledge directly from experience, refining its behavior through trial and error and reinforcement signals.
- **Deep Reinforcement Learning:** In conjunction with deep neural networks, reinforcement learning allows agents to carry out complex tasks in high-dimensional environments, including controlling robots or gaming consoles.

Reinforcement Learning Algorithms

- **Q-Learning:** Q-Learning is a model-free algorithm that calculates the predicted cumulative rewards for carrying out a specific action in a given state using a value function known as the Q-function.
- **Deep Q Network (DQN):** By approximating the Q-function with a deep neural network, DQN expands on Q-Learning. Because of this, the algorithm can work with high-dimensional state spaces, such as image data.
- **Policy Gradient Methods:** Policy gradient approaches maximize the number of activities that result in higher cumulative rewards by explicitly learning the policy that associates states with actions.
- **Actor-critic:** Aspects of value-based and policy-based methodologies are combined in the actor-critic approach. A performer suggests actions, while a critic assesses them and offers suggestions for improvement.
- **Proximal Policy Optimization (PPO):** PPO is an actor-critic algorithm designed to minimize significant policy modifications while optimizing policies. It assists in preventing oscillations in policy and stabilizing training.

Reinforcement Learning in Geospatial Analysis

The machine learning and reinforcement learning paradigm offers a dynamic and adaptive method for geospatial analysis, allowing systems to make successive decisions in ever-changing and complicated situations. This novel approach looks very promising for the interpretation of geospatial data and provides a fresh viewpoint for resource allocation and decision-making processes. Basic concepts in reinforcement learning:

- **Agent-Environment Interaction:** The foundation of reinforcement learning is the interaction between an agent and its surroundings. The environment represents the geospatial landscape in geospatial analysis, and the agent may be a robotic or algorithmic system. The agent eventually learns to take actions that maximize cumulative advantages as it acts and receives data in the form of awards or penalties.

- **Markov Decision Processes (MDPs):** Making decisions under ambiguity is standard in geospatial analysis. Reinforcement learning uses a structure of Markov Decision Procedures to mimic sequential decision-making challenges in which the present state and behavior dictate the forthcoming state alongside associated rewards (S. et al., 2021). For dynamic and ever-changing geographic contexts, this framework is especially well-suited.

The following lists a few uses for reinforcement learning in geographic analysis:

- **Autonomous Exploration:** Using reinforcement learning, geospatial environment exploration can be made more autonomous. Drones or autonomous systems can use reinforcement learning algorithms to navigate terrain, optimizing courses based on input received in real-time and the surrounding environment. This is particularly helpful in environmental monitoring or search and rescue missions.
- **Natural Resource Management:** Making choices about resource allocation, conservation, and land use is necessary for optimizing natural resource management. In order to guarantee sustainable practices, reinforcement learning models can dynamically adjust methods based on past data, tackling issues such as deforestation and water resource management.

ROLE OF MACHINE LEARNING IN GEOSPATIAL ANALYSIS

Applying machine learning algorithms has brought about a revolutionary transformation in geospatial analysis, which studies spatial data to derive meaningful insights. Through the lens of spatial information, this synergy provides a dynamic method for analyzing and comprehending our complex and linked reality.

The ability of ML algorithms to learn from large datasets and identify complex patterns within geographical data is at the core of this transformation. Machine learning plays a multifaceted role in geospatial analysis, with applications ranging from predictive modeling and land cover categorization to anomaly detection and spatial grouping.

Land Cover Classification

ML algorithms shine When automatically classifying different land cover forms in satellite data. By learning from labeled datasets, these models can precisely recognize and classify things like forests, cities, and water bodies. This capability improves environmental monitoring, urban planning, and conservation activities by giving precise and current land cover data.

Predictive Modeling

In predictive modeling for various geographical phenomena, machine learning is essential. Machine learning algorithms can predict events like disease spread, crop yields, and natural disasters by evaluating previous spatial data and linking it with environmental variables. Thanks to these predictive capabilities, decision-makers are empowered with information for proactive planning and resource allocation.

Anomaly Detection

Finding anomalies or odd patterns in geospatial data is essential for spotting possible problems before they become serious. Trained on standard patterns, machine learning algorithms can quickly detect deviations that could indicate altered environmental conditions, unpermitted land use, or other anomalies. Proactive anomaly detection facilitates prompt mitigation and response.

Spatial Clustering

Machine learning approaches enable the detection of spatial clusters based on innate patterns in geospatial datasets. This feature is useful for resource allocation, urban planning, and regionalization studies by putting similar places together. It offers a sophisticated comprehension of spatial interactions, assisting in better decision-making.

ML TECHNIQUES FOR GEOSPATIAL ANALYSIS

K-Means Clustering

Geospatial analysis, or the study of spatial data to derive essential insights, is a powerful ally of K-Means clustering (T. Kanungo et al., 2002). When geospatial datasets are divided into discrete groups or clusters according to innate patterns, this unsupervised machine-learning technique exposes important details about spatial relationships, distributions, and trends.

Algorithm

The K-Means clustering algorithm simplifies but effectively groups data points into clusters. Assigning data points to the closest cluster centroid iteratively, the algorithm recalculates the centroids until convergence. The number of clusters, or "k," is a crucial quantity that users specify based on the characteristics of the data or specific analysis requirements.

Advantages

- K-means clustering is helpful for tasks like classifying land cover or identifying urban growth patterns because it is skilled at revealing spatial patterns within geospatial datasets.
- The technique can handle large-scale geographical datasets since it is scalable and computationally efficient.
- K-Means is flexible and can be used in various geospatial domains, such as urban planning and environmental monitoring.

Challenges

- The choice of centroids at the algorithm's beginning can impact its performance and result in less-than-ideal results.
- The spherical and equal-size assumptions made by K-Means may not always match the intricate shapes seen in geographical data.
- Determining the appropriate number of clusters, or "k," can be difficult and may need extra validation techniques or domain knowledge.

Applications

- **Land Cover Classification:** K-Means clustering groups pixels with similar spectral properties to help in land cover classification.
- **Urban Planning:** Recognizing trends in urban growth and grouping places according to shared infrastructure features.
- **Natural Resource Management:** To manage resources sustainably, analyzing the spatial distribution of resources and grouping areas with comparable ecological characteristics.
- **Environmental Monitoring:** Collecting environmental data for proactive monitoring to identify abnormalities, including pollution or deforestation.

Decision Trees

Consider a decision tree as a flowchart that leads you to a conclusion by asking you a succession of yes/no questions. These inquiries encompass data parameters such as elevation, precipitation, and land cover in geospatial analysis.

Algorithm

A single node at the top of the tree represents the dataset. A decision rule based on a particular data property is applied at each node. Depending on how each data point responds to the rule, they are divided into branches. Recursively going forward, this process stops when every leaf node indicates a different class or prediction.

Advantages

- Robust against outliers and missing data; robust to interpretation and visualization; offers insights into decision-making.
- Able to manage data that is both continuous and categorical.

Challenges

- Overfitting may happen if the tree grows too complicated, resulting in subpar performance on unobserved data.
- It takes skill to select the best decision rules and tree depth.

Applications

- **Flood risk assessment:** Estimating the chance of flooding in various locations based on topography, precipitation, and land cover variables.
- **Medical diagnosis:** Helping physicians diagnose illnesses by using the patient's medical history and symptoms as a guide.
- **Fraud detection:** Identifying questionable financial transactions using account features and transaction patterns.

Support Vector Machines (SVMs)

Imagine using a piece of glass that is perfectly angled to divide marbles of various colors on a table. This ideal "glass separator" is found by SVMs for data points that belong to distinct classes.

Algorithm

Moves data points into a higher-dimensional space to facilitate easier separation. Find a hyperplane in this space that maximizes the margin, the distance between the hyperplane, and the closest data points for each class or the support vectors. New data points are classified according to which side of the hyperplane they are on.

Advantages

- Robust against noise and outliers; useful for complex non-linear relationships and high-dimensional data.
- Effective with modest datasets.

Challenges

- Selecting the best kernel function for translating data into a higher-dimensional domain can be difficult.
- It can be challenging to understand the decision boundary.

Applications

- **Land cover classification:** SVMs can accurately classify various land cover categories from high-resolution satellite data.
- **Handwritten digit recognition:** Recognising handwritten numbers from scanned papers or photos.
- **Facial recognition:** Identifying faces in pictures and movies.

Convolutional Neural Networks (CNNs)

Consider a group of knowledgeable investigators dissecting a convoluted crime scene image. CNNs examine minute details in photos to find hidden patterns, much like a team of AI detectives.

Algorithm

It consists of three layers: fully connected layers generate predictions, pooling layers lower the dimensionality of the data, and convolutional layers extract features. Convolutional filters go over an image to identify distinct elements such as forms, textures, and edges. By condensing these qualities, pooling layers enable the processing of data at a faster pace. Fully connected layers can make predictions or classify data by understanding the correlations between retrieved properties.

Advantages

- Capable of automatically learning complex features from data without explicit feature engineering; excellent at picture recognition and analysis; perfect for analyzing aerial images, satellite imagery, and maps.
- Accurately handle big, multi-dimensional image datasets.

Challenges

- It needs a significant quantity of processing power and training data.
- It can be challenging to prevent overfitting and adjust hyperparameters.
- It can be challenging to interpret the acquired characteristics.

Applications

- **Object detection:** Recognizing and finding particular items, such as buildings, roads, or cars, in satellite photos.
- **Land cover classification:** Accurately classifying various land cover types, such as urban areas, water bodies, and woods.
- **Change detection:** Examining aerial photography to spot long-term landscape changes, such as urbanization or deforestation.

Geostatistics

The weather forecast for your weekend cookout depends on the geographical interactions between weather patterns. Similar concepts are used in geostatistics (Q. Cao et al., 2022) to analyze and forecast various phenomena over large geographic areas.

Algorithm

In general, nearby places tend to be more similar than far-off ones. Depending on the distance between points, the degree of spatial dependence between them is calculated. A method of interpolation that forecasts values at unsampled locations by considering the spatial relationships of surrounding data points.

Advantages

- Takes spatial autocorrelation into account for more precise analysis and forecasts.
- Offers prediction uncertainty estimates that show the degree of confidence.
- Relevant to a range of spatially dependent phenomena, such as resource distribution, pollution levels, and soil moisture.

Challenges

- Needs familiarity with geostatistical principles and instruments.
- Selecting the correct kriging technique and semivariogram model might be challenging.
- Strongly non-linear relationship datasets might not be a good fit for this method.

Applications

- **Precision agriculture:** Forecasting crop yields, nutrient levels, and soil moisture content for focused resource management.
- **Environmental monitoring:** mapping pollution levels, forecasting flood dangers, and determining the accessibility of natural resources.
- **Public health:** Use geographical data to analyze disease outbreaks and forecast the development of epidemics.

Reinforcement Learning

Consider teaching a robot how to get around a maze. With the help of reinforcement learning, your AI agents may continuously adjust and enhance their behavior in dynamic contexts by learning via trial and error.

Algorithm

Agents engage with their surroundings, earning incentives for good deeds and punishments for bad deeds. Through trial and error, the agent gains experience and gradually improves behavior to maximize rewards. The agent is assisted in learning and optimizing its behavior using policy gradient approaches and Q-learning (Yadav & Sharma, 2023).

Advantages

- Has the ability to make difficult decisions in ambiguous situations.
- Learns from experience and adjusts to changing circumstances.
- Suitable for a wide range of dynamic geographic challenges.

Challenges

- The environment and incentive system must be carefully designed for learning to occur.
- It can be costly to compute for intricate jobs.
- It can be challenging to interpret learned behavior.

Applications

- **Route optimization:** Drones or autonomous cars can be trained to discover the best routes for delivery or exploration by considering traffic, weather, and terrain.
- **Resource management:** Directing robots to gather resources in dynamic settings, such as mines or woods, in an efficient manner.
- **Disaster response:** Enhancing emergency response teams' plans for floods and wildfires.

Time Series Analysis

Consider evaluating the ebb and flow of ocean tides or the rise and fall of market prices. You can use time series analysis to uncover and forecast temporal trends hidden in your data.

Algorithm

Finds patterns, seasonality, and trends in data that vary over time. It makes it possible to forecast and foretell future events. Applicable to a wide range of geographical phenomena, including environmental changes, disease outbreaks, and weather patterns.

Advantages

- Examine dynamic trends in data, such as pollution, disease outbreaks, or land cover, to forecast future conditions and track current events.

- Find hidden relationships between space and time, identify irregularities, and classify areas according to typical temporal dynamics.
- Apply insights from dynamic geospatial data to public health, precision agriculture, urban planning, and resource allocation.

Challenges

- The desired results and the properties of the data will determine which time series model is best.
- Handling outliers or missing data can affect the accuracy of the model.
- RNNs are sophisticated models that can be difficult to interpret.

Applications

- **Weather forecasting:** Use past weather data to forecast extreme events and weather patterns.
- **Disease outbreak prediction:** Forecasting the spread of infectious illnesses through the use of epidemiological data.
- **Air quality monitoring:** Estimating the amount of air pollution using both historical and current data.

Anomaly Detection

Imagine discovering a questionable transaction in your bank account or a wayward piece in a game of chess. Algorithms for anomaly detection are highly skilled in spotting odd trends and outliers in your data.

Algorithm

Data points far from any cluster can be flagged using K-means clustering, which may indicate anomalies. Data points with noticeably lower densities than their neighbors are identified using the local outlier factor (LOF), which suggests anomalies. One-class Support Vector Machines (OCSVMs) (Seokho et al., 2019) identify any points that considerably depart from the data's expected behavior and learn to recognize it.

Advantages

- Finds anomalous trends and data outliers.
- Beneficial for future problem prevention and early warning systems.

- Relevant to several geographical occurrences, including fraud, oil spills, and earthquakes.

Challenges

- The kind of abnormalities and the data distribution determine which anomaly detection method is best.
- It might be tricky to strike the right balance when deciding what level of abnormality to flag.
- It is necessary to carefully evaluate and validate results as false positives and negatives can happen.

Applications

- **Fraud detection:** Spotting questionable financial transactions or unusual credit card usage habits.
- **Earthquake prediction:** Searching for odd trends in seismic data that might point to an approaching earthquake.
- **Oil spill detection:** Keeping an eye out for any unusual changes in the patterns of ocean reflectance that might point to an oil spill on satellite photography.

Geostatistics and Time Series Analysis

Geostatistics and time series analysis are the Foundational methods in spatial and temporal data analysis, respectively, each offering unique insights into the characteristics and dynamics of data. When integrated with traditional machine learning (ML) techniques, these methods contribute to a more comprehensive understanding of spatial-temporal patterns by providing context, addressing specific challenges, and enhancing predictive capabilities. Here is how the theoretical underpinnings of geostatistics and time series analysis complement traditional ML techniques:

Geostatistics

Spatial Autocorrelation

- Geostatistics, rooted in spatial statistics, considers the spatial autocorrelation of data points, i.e., the degree of similarity between observations at different locations. This concept captures the spatial dependence structure, which is often ignored in traditional ML techniques.

- By accounting for spatial autocorrelation, geostatistical methods like variogram analysis and spatial interpolation techniques provide insights into spatial patterns and variability, helping to identify clusters, hotspots, and spatial trends.

Spatial Interpolation

- Geostatistical interpolation methods like kriging leverage spatial relationships between data points to estimate values at unsampled locations. Unlike traditional ML interpolation techniques, kriging incorporates spatial autocorrelation and uncertainty estimation, resulting in more accurate predictions and confidence intervals.
- By integrating kriging with ML models, spatial interpolation can enhance predictive accuracy and capture spatial variability in complex spatial datasets.

Spatial Regression

- Spatial regression models extend traditional regression techniques to account for spatial dependencies among observations. These models incorporate spatial weights matrices to capture the influence of neighboring locations on the response variable.
- When combined with ML algorithms, spatial regression enhances predictive modeling by incorporating spatial context and addressing spatial heterogeneity.

Time Series Analysis

Temporal Autocorrelation

- Time series analysis focuses on understanding and modeling the temporal dependencies within sequential data points. Autocorrelation functions quantify the degree of similarity between observations at different time lags.
- Time series models capture seasonality, trends, and cyclic patterns by considering temporal autocorrelation, which is essential for forecasting and anomaly detection.

Temporal Decomposition

- Time series decomposition techniques, such as seasonal decomposition of time series (STL), decompose the time series into trend, seasonal, and residual components. This decomposition helps identify underlying patterns and anomalies.

- Integrating time series decomposition with ML algorithms enables extracting features that capture both short-term and long-term temporal dynamics, enhancing predictive modeling accuracy.

Forecasting and Prediction

- Time series forecasting models provide predictive capabilities for future temporal observations, including autoregressive integrated moving averages (ARIMA), exponential smoothing methods, and recurrent neural networks (RNNs).
- ML techniques and time series forecasting models offer improved accuracy by leveraging nonlinear relationships, feature engineering, and ensemble methods.

Integration With Traditional ML Techniques

Feature Engineering

- Geostatistics and time series analysis provide valuable insights into feature engineering by identifying relevant spatial and temporal variables, autocorrelation structures, and trend components.
- These insights inform the selection and creation of features for traditional ML algorithms, enhancing their predictive performance and interpretability.

Model Calibration and Validation

- Geostatistical and time series methods contribute to model calibration and validation by providing tools for assessing model assumptions, diagnosing spatial and temporal patterns, and evaluating predictive uncertainty.
- Integrating these methods into model calibration and validation workflows, traditional ML models become more robust and reliable, particularly in spatial-temporal prediction tasks.

CONCLUSION

Machine learning and geographic analysis have ushered in a new era of insights, understanding, and decision-making in spatial data. Large-scale, dynamic geospatial datasets combined with state-of-the-art learning algorithms have revolutionized human perception, comprehension, and application of spatial data, opening up previously unthinkable possibilities. Across supervised, unsupervised, and reinforcement learning, machine learning algorithms have proven their worth in a variety of geographic analysis applications. Machine learning has become an essential technique for deriving actionable insight from spatial data, with applications ranging from autonomously navigating unmanned systems to accurately identifying land cover in satellite photos to predicting environmental occurrences. These algorithms' adaptability makes it possible to use them in various contexts, such as natural resource management, urban planning, environmental monitoring, and disaster response. Their capacity to detect irregularities, recognize intricate patterns, and forecast future trends has enabled businesses, scholars, and decision-makers to make well-informed judgments from a spatial perspective.

Notwithstanding the noteworthy advancements, several obstacles exist, such as the requirement for superior labeled datasets and the interpretability of intricate models. To overcome these obstacles and realize the full potential of this synergy, specialists in geography, data science, and machine learning must continue to collaborate because geospatial analysis and machine learning are interdisciplinary fields. Exciting potential lies ahead for machine learning in geospatial analysis. Developments in deep learning will further enhance these algorithms' capabilities, real-time data stream integration, and explainable artificial intelligence research. Our capacity to decipher the complexities of our globally interconnected world will advance along with machine learning, opening up new avenues for sustainable practices, well-informed decision-making, and a greater comprehension of the dynamic aspects of the Earth. The quest to fully utilize machine learning in geospatial analysis is evidence of our dedication to utilizing innovation for the benefit of our planet and its various ecosystems and the advancement of technology.

REFERENCES

Alkhatib, R., Sahwan, W., Alkhatieb, A., & Schütt, B. (2023). A Brief Review of Machine Learning Algorithms in Forest Fires Science. *Applied Sciences (Basel, Switzerland)*, *13*(14), 8275. doi:10.3390/app13148275

Cao, Q. (2022). Reshaping Geostatistical Modeling and Prediction for Extreme-Scale Environmental Applications. *SC22: International Conference for High Performance Computing, Networking, Storage and Analysis*, Dallas, TX, USA. 10.1109/SC41404.2022.00007

El Hamdani, S., Loudari, S., Novotny, S., Bouchner, P., & Benamar, N. (2021). *A Markov Decision Process Model for a Reinforcement Learning-based Autonomous Pedestrian Crossing Protocol.* 2021 3rd IEEE Middle East and North Africa COMMunications Conference (MENACOMM), Agadir, Morocco. 10.1109/MENACOMM50742.2021.9678310

Giannopoulos, A., Spantideas, S., Kapsalis, N., Gkonis, P., Sarakis, L., Capsalis, C., Vecchio, M., & Trakadas, P. (2022). Supporting Intelligence in Disaggregated Open Radio Access Networks: Architectural Principles, AI/ML Workflow, and Use Cases. *IEEE Access : Practical Innovations, Open Solutions, 10*, 39580–39595. doi:10.1109/ACCESS.2022.3166160

Güney, S., & Erdaş, Ç. B. (2019). *A Deep LSTM Approach for Activity Recognition.* 2019 42nd International Conference on Telecommunications and Signal Processing (TSP), Budapest, Hungary. 10.1109/TSP.2019.8768815

Gupta, V., Mishra, V. K., Singhal, P., & Kumar, A. (2022). *An Overview of Supervised Machine Learning Algorithm.* 2022 11th International Conference on System Modeling & Advancement in Research Trends (SMART), Moradabad, India. 10.1109/SMART55829.2022.10047618

Kang, S., Kim, D., & Cho, S. (2019). Approximate training of one-class support vector machines using expected margin. *Computers & Industrial Engineering, 130.* doi:10.1016/j.cie.2019.03.029

Kanungo, T., Mount, D. M., Netanyahu, N. S., Piatko, C. D., Silverman, R., & Wu, A. Y. (2002, July). An efficient k-means clustering algorithm: Analysis and implementation. *IEEE Transactions on Pattern Analysis and Machine Intelligence, 24*(7), 881–892. doi:10.1109/TPAMI.2002.1017616

Kumar, M., Ali Khan, S., Bhatia, A., Sharma, V., & Jain, P. (2023). *A Conceptual introduction of Machine Learning Algorithms.* 2023 1st International Conference on Intelligent Computing and Research Trends (ICRT), Roorkee, India. 10.1109/ICRT57042.2023.10146676

Kumar, T. S., Jothilakshmi, S., James, B. C., Prakash, M., Arulkumar, N., & Rekha, C. (2023). HHO-based vector quantization technique for biomedical image compression in cloud computing. *International Journal of Image and Graphics, 23*(03), 2240008. doi:10.1142/S0219467822400083

Li, J., Sun, F., & Chen, Y. (2022). *Research on spatial data management and application.* 2022 2nd International Conference on Electronic Information Engineering and Computer Technology (EIECT), Yan'an, China. 10.1109/EIECT58010.2022.00018

Lu, J., Qin, M., & Tong, Y. (2022). Research on Reinforcement Learning algorithms in Computer Vision. *2022 International Symposium on Advances in Informatics, Electronics and Education (ISAIEE)*, Frankfurt, Germany. 10.1109/ISAIEE57420.2022.00057

Lu, Y., Ye, T., & Zheng, J. (2022). Decision Tree Algorithm in Machine Learning. *2022 IEEE International Conference on Advances in Electrical Engineering and Computer Applications (AEECA)*. IEEE. 10.1109/AEECA55500.2022.9918857

Makkar, K., Kumar, P., Poriye, M., & Aggarwal, S. (2022). A Comparative Study of Supervised and Unsupervised Machine Learning Algorithms on Consumer Reviews. *2022 IEEE World Conference on Applied Intelligence and Computing (AIC)*, (pp. 598-603). IEEE. 10.1109/AIC55036.2022.9848880

Mohan, P., Neelakandan, S., Mardani, A., Maurya, S., Arulkumar, N., & Thangaraj, K. (2023). Eagle strategy arithmetic optimisation algorithm with optimal deep convolutional forest based fintech application for hyper-automation. *Enterprise Information Systems*, *17*(10), 2188123. doi:10.1080/17517575.2023.2188123

Montesinos López, O. A., Montesinos López, A., & Crossa, J. (2022). Fundamentals of Artificial Neural Networks and Deep Learning. In *Multivariate Statistical Machine Learning Methods for Genomic Prediction*. Springer. doi:10.1007/978-3-030-89010-0_10

Ouchra, H., Belangour, A., & Erraissi, A. (2023). An overview of GeoSpatial Artificial Intelligence technologies for city planning and development. *2023 Fifth International Conference on Electrical, Computer and Communication Technologies (ICECCT)*, Erode, India. 10.1109/ICECCT56650.2023.10179796

Ouchra, H., Belangour, A., & Erraissi, A. (2023). Machine Learning Algorithms for Satellite Image Classification Using Google Earth Engine and Landsat Satellite Data: Morocco Case Study. *IEEE Access : Practical Innovations, Open Solutions*, *11*, 71127–71142. doi:10.1109/ACCESS.2023.3293828

Shabanpour, N., Razavi-Termeh, S. V., Sadeghi-Niaraki, A., Choi, S.-M., & Abuhmed, T. (2022). Integration of machine learning algorithms and GIS-based approaches to cutaneous leishmaniasis prevalence risk mapping. *International Journal of Applied Earth Observation and Geoinformation, 112*. doi:10.1016/j.jag.2022.102854

Taye, M. M. (2023). Theoretical Understanding of Convolutional Neural Network: Concepts, Architectures, Applications, Future Directions. *Computation (Basel, Switzerland)*, *11*(3), 52. doi:10.3390/computation11030052

Usama, M., Qadir, J., Raza, A., Arif, H., Yau, K. A., Elkhatib, Y., Hussain, A., & Al-Fuqaha, A. (2019). Unsupervised Machine Learning for Networking: Techniques, Applications and Research Challenges. *IEEE Access : Practical Innovations, Open Solutions*, *7*, 65579–65615. doi:10.1109/ACCESS.2019.2916648

Waghela, H., Patel, S., Sudesan, P., Raorane, S., & Borgalli, R. (2022). *Land Use Land Cover Classification using Machine Learning. 2022 International Conference on Automation, Computing and Renewable Systems (ICACRS)*, Pudukkottai, India. 10.1109/ICACRS55517.2022.10029176

Wang, Q. (2022). Support Vector Machine Algorithm in Machine Learning. *2022 IEEE International Conference on Artificial Intelligence and Computer Applications (ICAICA)*, Dalian, China. 10.1109/ICAICA54878.2022.9844516

Yadav, P., & Sharma, S. C. (2023). *Q-Learning Based Optimized Localization in WSN*. 2023 6th International Conference on Information Systems and Computer Networks (ISCON), Mathura, India. 10.1109/ISCON57294.2023.10112130

Yang, X., Peng, K., An, L., Huang, P., & Feng, P. (2022). *GMBLOF: A Machine Learning Algorithm of Novelty Detection Based on Local Outlier Factor.* 2022 5th International Conference on Pattern Recognition and Artificial Intelligence (PRAI), Chengdu, China. 10.1109/PRAI55851.2022.9904135

Chapter 9
From Geospatial Data to Insight:
A Practical Guide to Machine Learning in Python for Real-World Problem-Solving

Assefa Senbato Genale
https://orcid.org/0000-0002-7411-9456
Bule Hora University, Ethiopia

Desalegn Aweke Wako
https://orcid.org/0000-0003-2803-7959
Bule Hora University, Ethiopia

Tsion Ayalew Dessalegn
Bule Hora University, Ethiopia

ABSTRACT

As technology advances, the potential applications for geospatial data will only continue to grow. However, conventional techniques for evaluating geographic data frequently involve manual interpretation or rule-based strategies, which take a long time and have a limited capacity to handle big datasets. Current technology has significantly enhanced geospatial analysis by providing powerful data collection, processing, and interpretation tools. This study used machine learning to analyze geospatial data and extract insights that would be difficult or impossible to obtain using traditional methods. Literature review, various Python libraries for geospatial data, building and evaluating machine learning models for algorithms like random forest, decision tree, linear regression, and K-means clustering using freely available geospatial data were presented. Machine learning makes analyzing geospatial data more effective for deriving deep understandings and extracting insights.

DOI: 10.4018/979-8-3693-6381-2.ch009

INTRODUCTION

Geospatial data contains useful information that can be used for various tasks, such as classifying land cover, detecting objects, doing geographical analyses, and other tasks(Abujayyab, 2019). Data analysis, modelling, and visualization with a geographic or geographical component are called geospatial analysis. Urban planning, environmental science, and public health are just a few industries that could benefit from this fast-expanding subject. However, conventional techniques for evaluating geographic data frequently involve manual interpretation or rule-based strategies, which take a long time and have a limited capacity to handle big datasets. Machine learning has revolutionized various fields, and geospatial analysis is no exception. With the increasing availability of geospatial data and advancements in computational power, machine learning algorithms have become powerful tools for extracting valuable insights from spatial data (Srivastava, 2023).

Machine learning is a powerful technique that can be used to analyze geospatial data and extract insights that would be difficult or impossible to obtain using traditional methods. A promising alternative is provided by machine learning, which automates the process of identifying patterns and making predictions from geospatial data (Zhalehdoost & Taleai, 2022). This chapter explores the use of Python, a popular programming language, for implementing machine learning algorithms specifically tailored for geospatial data analysis, and it will cover geospatial data preprocessing, perform geospatial analysis using different machine learning algorithms like supervised learning, unsupervised learning, and deep learning for classification, clustering, and object detection tasks. This chapter will apply machine learning algorithms to solve current real-world geospatial problems. We will use freely available geospatial datasets from open-source databases like Kaggle as a data source.

The main objective of this chapter is to provide a comprehensive understanding of how to frame geospatial problems, acquire and preprocess data, and fit a model using machine learning. Specifically, this chapter aims to address the following points: (1) To explore the application of machine learning techniques in geospatial analysis using the Python programming language; (2) To review and analyze existing literature on machine learning algorithms and their suitability for geospatial analysis; (3) To implement and demonstrate machine learning models in Python for geospatial analysis using real-world datasets; (4) To assess the performance and accuracy of the implemented models through appropriate evaluation metrics, and (5) To discuss potential applications and future directions for further research in this field. The chapter will contribute to advancing geospatial analysis by exploring how machine learning algorithms can be applied to extract meaningful insights from spatial data using Python programming language.

LITERATURE REVIEW

Recently, in geospatial data analysis, machine learning techniques have become powerful methods for extracting valuable insights from geospatial data. In this section, we have reviewed the most recent and relevant literature.

Authors (Jena, Pradhan, Prasanajit, and Alamri, 2021) have proposed a deep learning approach and geospatial analysis for assessing earthquake risks in NE India. This paper addressed a significant issue in the region by providing an innovative approach to evaluate earthquake risk prone to seismic activities. They used a convolutional neural network model and compared it with a conventional machine learning model. This study used DEM and shapefile data, spatial analysis, and a complete earthquake catalog to assess the earthquake risks in the region. Finally, they concluded that the deep learning model was the best choice for evaluating earthquake risks in the region, using geospatial data collected from the chosen area in NE India. In addition, the authors analyze the challenges and limitations of the findings for the future direction.

The authors (Srivastava, 2023) reviewed an application of artificial intelligence and different machine learning algorithms, including deep learning algorithms in geospatial data. This paper was intended to explore the various applications of artificial intelligence, machine learning, and deep learning in geospatial data analysis. The authors performed a comprehensive review of existing literature and case studies to identify the potential uses of these technologies in geospatial data. The authors have addressed various datasets from different sources, including satellite imagery, GPS data, and remote sensing data. They also investigated the effectiveness of using current technologies in geospatial data to extract insight into different data formats. Thus, their finding highlights the effectiveness of artificial intelligence, machine learning, and deep learning algorithms in land cover classification, object detection, and spatial pattern recognition.

Authors (Andronie & Iatagan, 2022) reviewed around 346 research papers published in different databases like Scopus, Web Science, and others to leverage deep learning techniques for smart process planning, robotic wireless sensor networks, and geospatial big data management in the context of the Internet of Manufacturing Things (IoMT). They have performed systematic reviews on papers related to Robotic Wireless Sensor Networks and Deep Learning-Assisted Smart Process Planning Geospatial Big Data Management Algorithms in the Internet of Manufacturing Things. The authors focused on developing deep learning models for process planning optimization, sensor data analysis, and geospatial big data management. The authors have used a combination of real-world manufacturing datasets and simulated sensor data for their experiments. The algorithms used include convolutional neural networks (CNNs), recurrent neural networks (RNNs), and long short-term memory (LSTM) networks. The results demonstrate significant improvements in process efficiency, sensor data analysis accuracy, and geospatial big data processing speed.

Authors (Abujayyab, 2019) presented geospatial dataset preparation and classification tools using machine learning algorithms for Mapping LULC From Rasat Satellite Images. They have proposed an integrated deep learning approach with machine learning to develop a tool for structuring geospatial machine learning datasets and performing land use/land cover (LULC) mapping from Rasat satellite images. This study shows the practical design of a dataset structuring tool that can preprocess satellite imagery into suitable input formats for machine learning models. The authors used Rasat satellite images as their primary dataset to develop and test their tool. They have employed various machine learning algorithms and deep learning models for LULC mapping. The finding indicated the effectiveness of the developed tool in accurately classifying different land cover types from Rasat satellite images.

A machine learning approach was presented by (Kanevski, Pozdnukhov, and Timonin, 2008) to explore the various applications of machine learning algorithms in analyzing geospatial data and to review existing knowledge of machine learning in geospatial data. Authors have done their experiments with different machine learning algorithms such as General Regression Neural Networks, K-NN cross-validation curve of the residuals, Support Vector Machines, Gaussian Mixture Model density estimator, and neural networks on diverse geospatial datasets covering areas like classification of environmental data, environmental monitoring, pollution data, and natural resource management. The authors have utilized publicly available geospatial datasets from government agencies and research institutions for their experiments. The results highlight the potential of machine learning algorithms in tasks such as spatial clustering, predictive modeling, anomaly detection, and pattern recognition within geospatial data.

In(Das, n.d.), the authors presented a significant deep learning technique for geospatial data analytics with various applications and software tools. The authors proposed deep learning using GIS technology tools to analyze geospatial data. They performed experiments with different machine learning algorithms such as K-means Clustering, CNN(Convolutional Neural Network), RNN(Recurrent Neural Network), and LSTM(Long Short-term Memory Network) for data geospatial data analysis on diverse geospatial datasets covering areas like classification of environmental data, environmental monitoring, pollution data, and natural resource management. They used deep learning packages like TensorFlow or PyTorch to perform tasks such as image classification from satellite imagery or object detection from LiDAR point clouds. The authors have utilized a combination of publicly available geospatial datasets and proprietary datasets specific to certain applications.

In(Reus, Kerbusch, & Schadd, n.d.), the authors presented a machine learning approach in tactical decision support using geospatial analysis. The study was intended to explore the use of geospatial analysis and machine learning in supporting tactical decision-making. The authors performed a demo using various machine learning algorithms that involved collecting geospatial data, preprocessing the data, optimizing abstract machine learning models, and evaluating the results. The dataset used in this study consists of geospatial information such as terrain, weather, and infrastructure data. The authors utilized machine learning algorithms to support tactical decision-making in analyzing geospatial data. The authors demonstrate the potential of using machine learning in tactical decision support, particularly in identifying patterns and making predictions based on geospatial information.

Geospatial analysis and machine learning methods were proposed (Zhalehdoost & Taleai, 2022) to predict air pollution. The authors reviewed various literature to examine the application of machine learning and geospatial analysis methods in predicting air pollution. The authors used statistical, machine learning, and hybrid methods to predict air pollution. The dataset used in this review paper comprises air quality monitoring data, meteorological data, and geographical information. The authors find that various machine learning algorithms, such as neural networks, support vector machines, and k-nearest neighbors, have been employed for air pollution prediction with promising results.

In(Kucklick, Müller, Beverungen, & Mueller, 2021), the authors proposed a deep learning approach with a geographic information system(GPS) to quantify the impact of location data on real estate appraisal. The authors used Convolutional Neural Network(CNN) with GPS technologies to appraise around 71,000 properties in Philadelphia. The authors utilized a dataset containing property attributes, location information, and historical sales data to train their deep learning model, and the result improved after they performed serious data preprocessing for geospatial data. They employed convolutional neural networks (CNNs) for feature extraction from spatial data and achieved better results in quantifying the impact of location on real estate appraisal. In addition, the authors assessed how location data can influence real estate valuation through deep learning techniques.

Authors (Dai et al., 2021) presented a deep-learning model for detecting diabetic retinopathy across the disease spectrum. The authors implemented this by collecting retinal images from patients with diabetic retinopathy, preprocessing the images, training a deep convolutional neural network (CNN), and evaluating its performance. The authors used a dataset of around 466,247 total images to train their models, and models were evaluated on a local dataset of around 200,136 fundus and three external datasets with a total of 209,322 images. The authors use a large dataset of retinal images labeled with different stages of diabetic retinopathy to train their deep-learning model. Their results demonstrate that the developed deep learning system can accurately detect diabetic retinopathy across different disease stages.

Authors (Acharya, Mondal, Bhadra, Abdelrahman, and Mishra, 2022) have presented the geospatial analysis of geo-ecotourism sites using GIS and AHP for resilient and sustainable tourism planning in West Bengal, India. This paper assessed the suitability of geo-ecotourism sites using the analytic hierarchy process (AHP) and geographic information systems (GIS) for sustainable regional tourism planning. The authors collect geospatial data on ecological features, accessibility, infrastructure, and cultural significance to evaluate site suitability for geo-ecotourism development.

Table 1. Literature review

Title	Authors	Method/S	Result/Gaps
Deep Learning-Assisted Smart Process Planning, Robotic Wireless Sensor Networks, and Geospatial Big Data Management Algorithms in the Internet of Manufacturing Things	Andronie & Iatagan	The algorithms used include convolutional neural networks (CNNs), recurrent neural networks (RNNs), and long short-term memory (LSTM) networks.	The authors focused on developing deep learning models for process planning optimization, sensor data analysis, and geospatial big data management. The results demonstrate significant improvements in process efficiency, sensor data analysis accuracy, and geospatial big data processing speed.
Geospatial Machine Learning Datasets Structuring And Classification Tool: Case Study For Mapping Lulc From Rasat Satellite Images	Abujayyab	They have employed various machine learning algorithms and deep learning models for LULC mapping.	The finding indicated the effectiveness of the developed tool in accurately classifying different land cover types from Rasat satellite images.
Machine Learning Algorithms for GeoSpatial Data	Kanevski, Pozdnukhov, & Timonin	General Regression Neural Networks, K-NN cross-validation curve of the residuals, Support Vector Machines, Gaussian Mixture Model density estimator, and neural networks	The results highlight the potential of machine learning algorithms in tasks such as spatial clustering, predictive modeling, anomaly detection, and pattern recognition within geospatial data.
Geospatial Data Analytics- A Deep Learning Perspective	Das	K-means Clustering, CNN, RNN, and LSTM	The authors have utilized a combination of publicly available geospatial datasets along with proprietary datasets specific to certain applications.
A deep-learning model for detecting diabetic retinopathy across the disease spectrum	Dai et al., 2021	Deep convolutional neural network (CNN),	Their results demonstrate that the developed deep learning system can accurately detect diabetic retinopathy across different disease stages.

They apply AHP to prioritize criteria based on stakeholder preferences and utilize GIS for spatial analysis. The results highlight suitable geo-ecotourism sites that can contribute to sustainable tourism planning in West Bengal. In addition, the authors analyze the challenges and limitations of the findings for the future direction.

GEOSPATIAL DATA

Geospatial data refers to information associated with a specific location on the Earth's surface. It is also known as geographic or spatial data, which is a general word that encompasses all kinds of information that are implicitly or explicitly associated with a specific location about objects, events, or phenomena on Earth's surface(Tamiminia, Salehi, Mahdianpari, Quackenbush, & Adeli, 2020). Geospatial data includes a wide range of information, including land use, topography, infrastructure, and environmental features. These data include coordinates, addresses, postal codes, and other location-based information. Geospatial data can be represented in various forms, such as maps, satellite imagery, and GPS coordinates. It is collected using various technologies such as GPS, satellite imagery, and remote sensing. It is used in many applications, including urban planning, environmental monitoring, natural resource management, and disaster response. One of the key characteristics of geospatial data is its ability to be analyzed and visualized using Geographic Information Systems (GIS)(Ighile, Shirakawa, & Tanikawa, 2022).

GIS technology allows users to overlay different layers of geospatial data to gain insights and make informed decisions. For example, urban planners can use GIS to analyze population density and land use patterns to determine the best locations for new infrastructure projects. Environmental scientists can use GIS to track changes in vegetation cover over time by analyzing satellite imagery(Gunawardena, 2014). Geospatial data is used in various industries or organizations, such as natural resource management, urban planning and development, disaster management, emergency management, etc. As technology advances, the potential applications for geospatial data will only continue to grow. Figure 1 illustrates an example of geospatial data.

Machine learning has become increasingly important in the analysis of geospatial data. Using machine learning algorithms, researchers and analysts can extract valuable insights from large volumes of geospatial data that would be difficult or impossible to uncover using traditional methods(Ighile et al., 2022). For example, machine learning algorithms can identify patterns in satellite imagery that indicate deforestation or illegal mining activities. Additionally, machine learning can predict future trends based on historical geospatial data, such as predicting the spread of diseases or identifying areas at risk for natural disasters.

Figure 1. An example of geospatial data
(Source: "Understanding-geospatial-data," 2016)

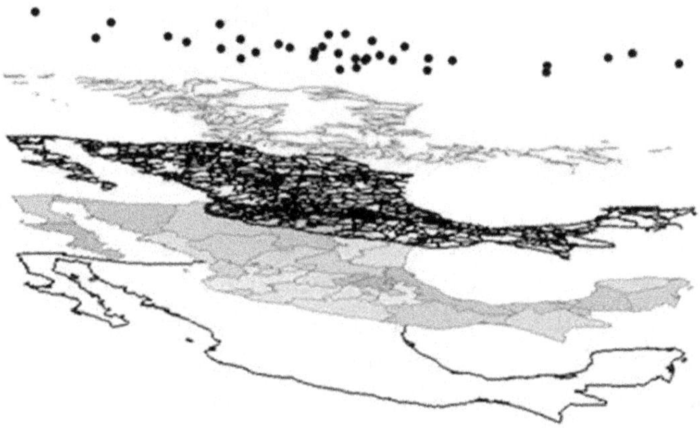

Geospatial Analysis

Geospatial analysis is analyzing and interpreting data with a geographic or spatial component. This analysis allows us to understand patterns, trends, and relationships within a specific location or area(Breunig et al., n.d.). With current technology, geospatial analysis has become more advanced and accessible, and spatial statistics are used to identify and quantify spatial patterns and relationships(Stewart, Robinson, Corley, & Ferres, n.d.). This can include analyzing the distribution of features within a specific area, identifying clusters or hotspots of activity, and measuring the proximity or connectivity between different geographic elements.

Current technology has significantly enhanced geospatial analysis by providing powerful data collection, processing, visualization, and interpretation tools. Integrating GIS with remote sensing, GPS technology, big data analytics, machine learning, and AI has expanded geospatial analysis capabilities across various domains, including environmental science, urban planning, public health, transportation management, etc. As technology continues to evolve rapidly, we will likely see even more advanced geospatial analysis applications in the future.

Machine learning has revolutionized geospatial analysis by enabling the development of predictive models and algorithms to process large volumes of geospatial data to extract meaningful insights(Kanevski et al., 2008; Zhalehdoost & Taleai, 2022). Python has emerged as a popular programming language for geospatial analysis due to its extensive data manipulation, visualization, and machine learning libraries. Python provides several libraries for working with spatial data, such as GeoPandas, Shapely, Fiona, and Pyproj. In addition, geospatial analysis based on machine learning with Python offers a powerful toolkit for extracting valuable insights from spatial data. By leveraging Python's rich ecosystem of libraries, developers can build sophisticated models that address complex challenges in fields ranging from urban planning to environmental conservation.

PYTHON LIBRARIES FOR GEOSPATIAL DATA ANALYSIS

Python has a wide range of libraries for geospatial data analysis using machine learning, which are used for various purposes such as mapping, spatial analysis, and visualization(Of, 2020). These libraries provide different functionalities for geospatial data, including data manipulation, visualization, and performing analysis using machine learning algorithms(Westra, n.d.). This section will discuss some popular Python libraries for geospatial data analysis and examples of their usage. Some of the most popular libraries for geospatial data analysis in Python include GeoPandas, Shapely, Fiona, Pyproj, Folium, and Scikit-learn.

GeoPandas

GeoPandas is an open-source library that provides Pandas with the capabilities to allow for easy manipulation and analysis of geospatial data. It provides a high-level interface for working with vector data such as points, lines, and polygons. It also integrates with other Python libraries, such as Matplotlib and Seaborn, for visualization. It allows users to read, write, and manipulate geospatial data in various formats, such as shapefiles, GeoJSON, and GeoPackage.

Installing Geopandas

We can install geopandas in two ways: Using conda commands: Make sure you have installed Anaconda on your computer, open up the Anaconda Powershell Prompt, and use the below command. Figure 2 demonstrates the installation of Geopandas using the conda command.

Using pip commands: open up the Anaconda Powershell Prompt and use the below command. Figure 3 depicts the installation of Geopandas using the pip command.

Importing geopandas: Open Jupyter Notebook from the Anaconda and run the below command on the Jupyter Notebook. Figure 4 showcases the process of importing Geopandas.

We can read various geospatial data types with the help of geopandas data frames. To read data from the computer, use the code below. Figure 5 demonstrates the procedure for reading a dataset using Geopandas.

Figure 2. Installing geopandas using conda command
(Source: Authors' Demo)

```
C:\WINDOWS\system32\cmd.

(tensorflow) C:\Users\Christ_In_Me>conda install geopandas
```

Figure 3. Installing geopandas using pip command
(Source: Authors' Demo)

```
C:\WINDOWS\system32\cmd.

(tensorflow) C:\Users\Christ_In_Me>pip install geopandas
```

Figure 4. Importing geopandas
(Source: Authors' Demo)

```
In [1]: import geopandas as gpd
```

Figure 5. Reading dataset using geopandas
(Source: Authors' Demo)

```
In [3]: # Load housing dataset
        df1 = gpd.read_file('C:/Users/Christ_In_Me/Downloads/housing.csv')
        df1.head()
```

Shapely

Shapely is a Python library for manipulating and analyzing planar geometric objects. It provides a set of geometric objects such as points, lines, and polygons and functions for performing geometric operations such as intersection, union, and buffering. Shapely is often used with GeoPandas to perform more advanced spatial analysis tasks. Figure 6 provides examples of the Shapely Python library in action.

Figure 6. Examples of Shapely Python library
(Source: Authors' Demo)

```
In [6]: from shapely.geometry import Polygon, LineString, Point
        # Create a polygon object
        polygon = Polygon([(0, 0), (0, 1), (1, 1), (1, 0)])
        # Create a line string object
        line = LineString([(0, 0), (1, 1), (2, 2)])
        # Create a point object
        point = Point(0, 0)
```

Figure 7. Examples of Fiona Python library
(Source: Authors' Demo)

```
In [7]: import fiona
        path = 'C:/Users/Christ_In_Me/Downloads/shapefile.shp'
        # Open a shapefile for reading
        with fiona.open(path) as src:
            # Print the schema of the shapefile
            print(src.schema)

            # Loop through each feature in the shapefile and print its geometry and properties
            for feature in src:
                print(feature['geometry'], feature['properties'])
```

Fiona

Fiona is a Python wrapper around the OGR library for reading and writing vector data formats such as Shapefiles and GeoJSON. It provides a simple interface for accessing geospatial data in Python and can be used with other libraries, such as GeoPandas or Shapely. Figure 7 illustrates examples of the Fiona Python library in practical use.

Folium

Folium is a Python library that makes visualizing geospatial data on interactive Leaflet maps easy. Using simple Python code, it allows users to create interactive maps with markers, polygons, popups, and other features.

Figure 8 showcases examples of the Folium Python library in action.

Figure 8. Examples of Folium Python library
(Source: Authors' Demo)

```
In [9]: import folium
        # Create a map object
        m = folium.Map(location=[51.5074,-0.1278], zoom_start=10)
        # Add a marker to the map
        folium.Marker([51.5074,-0.1278], popup='London').add_to(m)
        # Display the map
        m.save('map.html')
```

Figure 9. Examples of Pyproj Python library
(Source: Authors' Demo)

```
In [18]: import pyproj
         # Define input CRS
         input_crs = pyproj.CRS("EPSG:2100")
         # Define output CRS
         output_crs = pyproj.CRS("EPSG:2154")

         # Perform coordinate transformation
         transformer = pyproj.Transformer.from_crs(input_crs, output_crs)
         x_new, y_new = transformer.transform(0, 0)
         print(x_new, y_new)
```

Pyproj

Pyproj is a Python interface to the PROJ library that supports cartographic projections and coordinate transformations. It allows users to perform transformations between different coordinate reference systems (CRS) using various projection methods.

Scikit-Learn

scikit-learn is a popular machine-learning library in Python that includes several algorithms suitable for geospatial data analysis. For example, scikit-learn includes support vector machines (SVM), random forests, k-nearest neighbors (KNN), and other algorithms that can be applied to geospatial datasets to perform tasks such as classification or Regression.

MACHINE LEARNING FOR GEOSPATIAL ANALYSIS

Machine learning is a subfield of artificial intelligence that employs statistical techniques to automatically instruct a machine to learn and get better over time without explicit programming(Helm et al., 2020). Instead of utilizing computer programming that explicitly accomplishes a task specified by the programmer, like traditional programming, it uses data and output with a modular algorithm that can provide information about a group of data without the need to write any program that is particular to the situation. For instance, when creating a new program for a particular problem, we provide the generic algorithm with data to compute, and the algorithm then uses the data to construct its logical conclusion(Agrawal, 2020). Figure 10 illustrates the distinction between traditional programming and machine learning.

Machine learning techniques for geospatial analysis using Python have become increasingly popular due to the availability of large geospatial datasets and the

Figure 10. Difference between traditional programming and machine learning
(Source: Authors' elaboration)

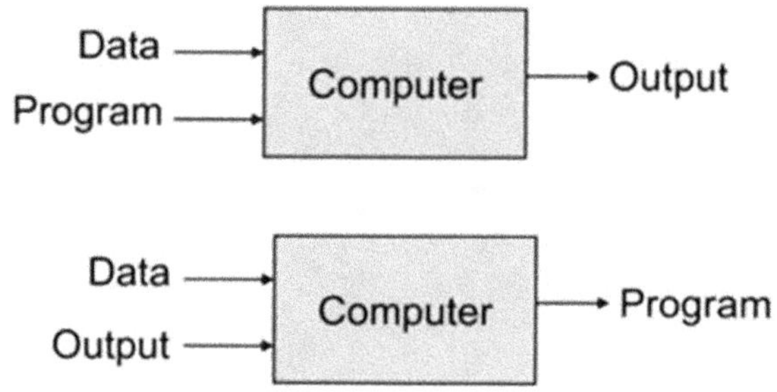

necessity of accurate and systematic analysis. Analyzing and interpreting data with a spatial component, which includes satellite data, Location information, or spatial databases, is known as geospatial analysis. These datasets can be processed using machine learning techniques to reveal important patterns and insights that have applications in various domains, including environmental control, emergency management, and urban planning. Several machine-learning techniques can be applied to geospatial analysis using Python. These consist of deep learning techniques, supervised learning, and unsupervised learning. We have covered a few of the most popular machine learning methods for geospatial analysis with Python to apply them to geospatial data.

Supervised Machine Learning

It is used to train machines using labeled datasets to predict known properties of data(takes labeled inputs and maps them to the known outputs)(KELLEHER, 2019). It can be applied in geospatial analysis to tasks including classifying land cover, identifying objects in satellite data, and forecasting environmental variables using geographic features. For instance, to perform land cover classification using supervised learning in Python, one can use algorithms such as Random Forest or Support Vector Machines from scikit-learn to train a model on labeled satellite imagery data. When compared to unsupervised learning, it is straightforward. Supervised learning is practiced when we have two variables and apply a mapping function to proceed from the input to the output.

In Figure 11. Supervised Learning Architecture, to predict the output variable(y) for the data, inputs(x) must map outputs(y) so precisely that a new input data set new input variable(new input x), which has never been seen before, can be employed. For example, this algorithm can detect objects from satellite imagery with the necessary information for further analysis. After learning from the data in the images, the

Figure 11. Supervised learning architecture
(Source: Authors' elaboration)

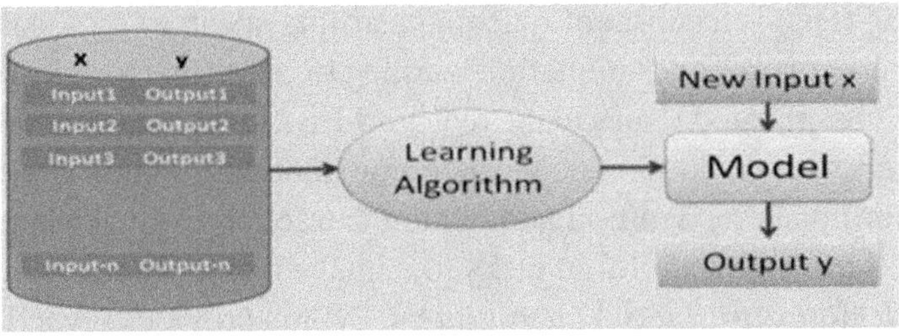

algorithm can predict using new images during testing. Typical examples include naive Bayes techniques, neural algorithms, support vector machine techniques, and decision tree techniques.

Unsupervised Machine Learning

It uses unlabeled data to train machines. It also understands patterns and trends in the output and learns about the data. Unlike supervised learning algorithms, unsupervised techniques do not need any trained data.

Figure 12 presents the architecture of unsupervised learning. Instead, they attempt to automatically uncover intriguing patterns in unlabelled data, for example, by clustering together similar examples (Nguyen, Nguyen, Tran, Pham, & Pestana, 2021). In geospatial analysis, unsupervised learning can cluster similar geographic features or identify patterns in spatial datasets. For instance, the scikit-learn library provides algorithms such as k-means clustering that can be applied to geographical datasets to discover clusters or trends. This algorithm is used when there is an input variable (x) but no corresponding output variable (y) (output variable).

Figure 12. Unsupervised learning architecture
(Source: Authors' elaboration)

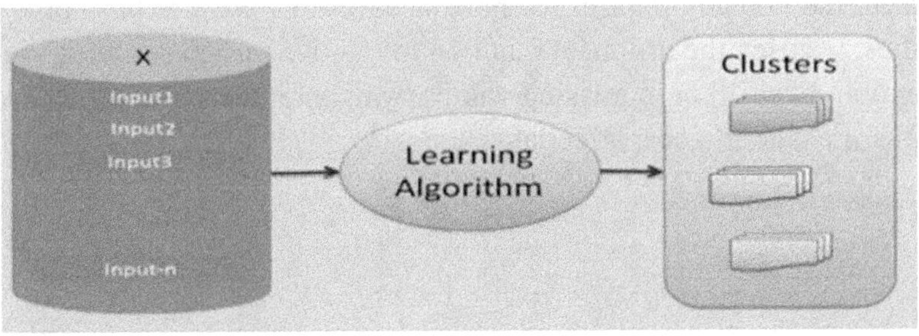

Deep Learning

Deep Learning (DL) is a subset of machine learning specifically focused on using artificial neural networks and simulation learning to create machine learning models that can solve complicated issues in the real world at a level comparable to cognitive abilities(Aggarwal et al., 2022). For instance, convolutional neural networks (CNNs) can be used for image recognition and feature extraction from geospatial imagery. For instance, this can be helpful in tasks like image recognition, satellite imagery, or extracting features from aerial photographs. Python provides several libraries for implementing deep learning algorithms for geospatial analysis. One of the most often used deep learning packages is TensorFlow, which provides tools for building and training neural networks on spatial data. Libraries like Keras assemble deep neural networks with minimal code by offering a high-level Application Programming Interface (API).

Preprocessing Geospatial Data for Machine Learning

Preprocessing geospatial data is an important stage in preparing the data for building machine learning models. Geospatial data refers to information that has a geographic component, such as coordinates, addresses, or boundaries. Preprocessing geospatial data for machine learning involves cleaning, transforming, and organizing the data to make it suitable for input into machine learning algorithms. In this chapter, the dataset that is freely available on https://www.kaggle.com/datasets/huyngohoang/housingcsv/data from the Kaggle database was used to analyze geospatial data with different Python libraries to extract insight from the dataset. This section has explored how to preprocess geospatial data using Python with examples.

Data Cleaning

Cleaning the data is the first stage in preparing geospatial data. This entails addressing missing values, fixing any data inaccuracies, and eliminating unnecessary or redundant information. The first option to clean our dataset is the dropna() function. For instance, the housing dataset we have used in this chapter has some missing values. Thus, we used the dropna() function to remove missing values. We can use another option, like filling in missing values by using means. Figure 13 showcases an example of cleaning geospatial data.

Figure 13. An example of cleaning geospatial data
(Source: Authors' Demo)

```
In [20]: df1= df1.dropna()
```

Feature Engineering

Through feature engineering, features can be transformed to improve their suitability for machine learning models, or new features can be created from existing ones. In geospatial data, feature engineering may involve calculating distances between points, extracting information from addresses, or encoding categorical variables. For instance, we used OneHotEncoder and StandardScaler libraries to perform feature engineering tasks. Figure 14 demonstrates an example of using the OneHotEncoder and StandardScaler libraries for performing feature engineering tasks.

Geocoding

Geocoding is translating addresses into geographical coordinates, such as longitude and latitude. This is often necessary when working with geospatial data, as many datasets contain address information rather than explicit geographic coordinates. For instance, in the housing dataset, we have used the following code to convert latitude and longitude columns to points as geometry columns. Figure 15 illustrates an example of the Geocoding Python library in action.

Figure 14. An example of OneHotEncoder and StandardScaler libraries to perform feature engineering tasks
(Source: Authors' Demo)

```
In [27]: # Encoding categorical variables
         encoder = OneHotEncoder(sparse=False)
         X_encoded = encoder.fit_transform(X)
```

```
In [28]: # Feature scaling
         scaler = StandardScaler()
         X_scaled = scaler.fit_transform(X_encoded)
```

Figure 15. An Example of Geocoding Python Library
(Source: Authors' Demo)

```
In [15]: #using geopandas to covert Long and Lat to points
         df_geo=gpd.GeoDataFrame(df,geometry=gpd.points_from_xy(df.longitude,df.latitude))
```

Spatial Join

Spatial join is a technique used to combine two datasets based on their spatial relationships. This is often useful when working with geospatial data such as containment, intersection, or proximity. For instance, we have two shapefile datasets: point locations and polygon boundary locations. For instance, we can perform a spatial join between these two datasets.

```
Importing geopandas as gpd
# loading point locations
points = gpd.read_file('point_locations.shp')
# Loading polygon boundaries
polygons = gpd.read_file('polygon_boundaries.shp')
result=gpd.sjoin(points, polygons, how='inner',
op='intersects')
```

Data Normalization

Normalization is essential for ensuring all variables are on a similar scale before feeding them into machine learning algorithms. This helps prevent certain features from dominating others during model training.

Handling Spatial Autocorrelation

Spatial autocorrelation refers to the tendency of nearby locations to have similar values due to underlying spatial processes. It is important to account for this phenomenon when analyzing geospatial data.

Figure 16. Feature selection using selectKBest Python library
(Source: Authors' Demo)

```
In [56]: # Feature selection
         selector = SelectKBest(score_func=f_classif, k=4)  # Select top k features
         X_selected = selector.fit_transform(X_scaled, y)
```

Case Studies: Predicting Median House Value Using Machine Learning

A city government wants to predict the median house value to minimize the need for houses for median-income residents and manage houses effectively. They have historical geospatial data with latitude, housing-median-age, longitude, total rooms, total bedrooms, median income, households, and population. The city wants to predict median house values using machine learning algorithms.

Building a Machine Learning Model

We are building a machine learning model for the housing dataset to predict median house value. The housing dataset contains geospatial data, so machine learning is used for geospatial analysis. Here are the important steps we have to follow up to build our model:

Data loading and Preprocessing: loading our dataset using the geopandas library from Python into Jupyter Notebook. Then, we must preprocess our data by performing activities like feature engineering, data cleansing, geocoding, transformation, and data normalization.

Feature Selection: Identify the relevant features from the dataset that will be used as input for the building model. It is one of the important tasks that enhance the performance of our model. Python has various libraries to perform feature selection tasks. In this chapter, we used the selectKBest Python library for feature selection. Figure 16 demonstrates feature selection using the selectKBest Python library.

Model Selection and Training

Selecting an appropriate model from several machine learning algorithms needs serious analysis. However, different researchers recommend selecting models from machine learning based on the nature of our problem and data. This chapter analyzes geospatial data using machine-learning techniques like decision trees, random forest,

Figure 17. Random forest algorithm code with geospatial data
(Source: Authors' Demo)

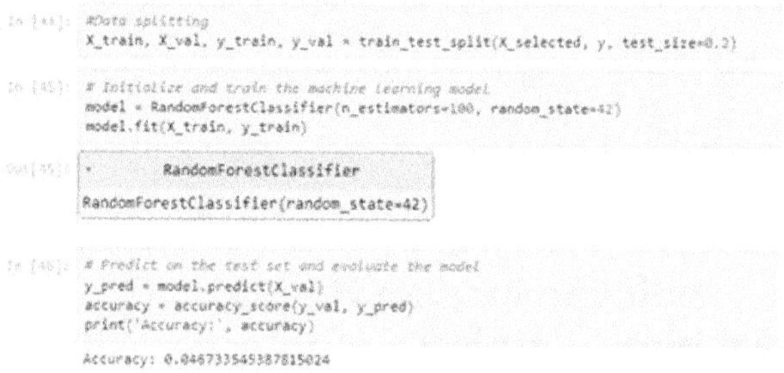

linear Regression, and k-means clustering, and it trains the selected model using the housing dataset.

Random Forest: is an ensemble learning technique that enhances prediction accuracy and manages over-fitting by combining several decision trees. It performs by constructing several decision trees and averaging the forecasts from each one. The equation for random forest can be represented as a combination of decision trees:

$$f(x)= 1/N*\sum(fi(x)) \quad (1)$$

Where f(x) is the predicted output, N is the number of trees in the forest, and f_i(x) are the predictions from individual trees. Figure 17 presents the code for implementing the Random Forest algorithm with Geospatial Data.

Decision Tree: is a supervised learning technique used in geospatial data analysis for regression and classification problems. It makes predictions based on the majority class or average value inside each subset after recursively splitting the data into subsets based on the value of selected features. The equation for the decision tree can be represented as:

$$f(x)= \sum(wi*xi)+b \quad (2)$$

Where f(x) is the predicted output, wi are the weights, xi is the input features, and b is the term used to express partiality. Figure 18 displays the code implementing the Decision Tree algorithm with Geospatial Data.

Linear Regression is a straightforward yet effective technique used in geospatial data analysis to forecast input variables. In order to perform, input features and output values must be fitted with the best possible linear connection. The equation for this algorithm can be represented as:

Figure 18. Decision tree algorithm code with geospatial data
(Source: Authors' Demo)

```
In [47]: clf=DecisionTreeClassifier()

In [48]: clf.fit(X_train, y_train)

Out[48]: ▸ DecisionTreeClassifier
         DecisionTreeClassifier()

In [49]: y_pred = clf.predict(X_val)

In [50]: accuracy = accuracy_score(y_val, y_pred)
         print("Accuracy:", accuracy)

         Accuracy: 0.046733545387815024
```

$$y = \beta 0 + \beta 1 * x1 + \beta 2 * x2 + ... + \beta n * xn \qquad (3)$$

Where y is the anticipated result, β0 is used for the intercept term, and βn is the input feature, xn coefficients. Figure 19 displays the code for implementing the Random Forest algorithm with Geospatial Data.

K-means Algorithms: is an unsupervised learning technique that groups or clusters geospatial information according to similarity. In order to reduce within-cluster variation, data points are iteratively assigned to clusters, and cluster centroids are updated. The equation for K-means involves figuring out the separations between cluster centroids and data points:

$$\arg\min\_c\Sigma\|x - \mu\_c\|\hat{}2 \qquad (4)$$

Figure 19. Random forest algorithm code with geospatial data
(Source: Authors' Demo)

```
In [58]: # Fit the linear regression model to the training data
         reg.fit(X_train, y_train)

Out[58]: ▸ LinearRegression
         LinearRegression()

In [59]: # Make predictions on the testing data
         y_pred = reg.predict(X_val)

In [60]: # Evaluate the model performance
         mse = mean_squared_error(y_val, y_pred)
         print("Mean Squared Error:", mse)

         Mean Squared Error: 13803715601.772472
```

Figure 20. Importing K-means package
(Source: Authors' Demo)

k-mean clustering for Geospatial data

```
In [61]: import pandas as pd

         from sklearn.cluster import KMeans
         import folium

         import matplotlib.pyplot as plt
         import seaborn as sns
         %matplotlib inline
```

where data points are represented by x, μ_c are cluster centroids, and ||x – μ_c||^2 represents the Euclidean distance between a data point and a centroid. Datasets for the K-means algorithm are available from https://www.kaggle.com/datasets/tekbahadurkshetri/uber-clust ering. Figure 20 illustrates the process of importing the K-means package. Figure 21 demonstrates the process of loading a dataset for K-means algorithms.

Figure 22 presents a visualization of longitude and latitude data. Figure 23 displays the determination of the number of clusters using the K-Means algorithm.

Figure 23 depicts the determination of the number of clusters using the K-Means algorithm. Figure 24 demonstrates the visualization of clusters using a Folium map. Figure 25 displays the visualization of three clusters using a Folium map.

Figure 25 displays the visualization of three clusters using a Folium map.

Visualization of geospatial data

In this section, we used Matplotlip and Folium geospatial visualization libraries to visualize the housing dataset of geospatial data. This visualization helps us to interpret more insightful information from geospatial data. You can find the full authors' Python code here https://github.com/Ase96/Geospatial_analysis_using_ML

Figure 21. Loading dataset for K-means algorithms
(Source: Authors' Demo)

```
In [62]: df = pd.read_csv('C:/Users/Christ_In_Me/Downloads/uber_clean.csv')
         df.head()
```

Out[62]:

	Date/Time	Lat	Lon	Base	Date
0	2014-07-01 0:03	40.7586	-73.9706	B02512	Tuesday
1	2014-07-01 0:05	40.7605	-73.9994	B02512	Tuesday
2	2014-07-01 0:06	40.7320	-73.9999	B02512	Tuesday
3	2014-07-01 0:09	40.7635	-73.9793	B02512	Tuesday
4	2014-07-01 0:20	40.7204	-74.0047	B02512	Tuesday

Figure 22. Visualization of longitude and latitude
(Source: Authors' Demo)

```
In [67]: plt.scatter(df['Lon'], df['Lat'], c=df['y'])
Out[67]: <matplotlib.collections.PathCollection at 0x1dfd1fbf190>
```

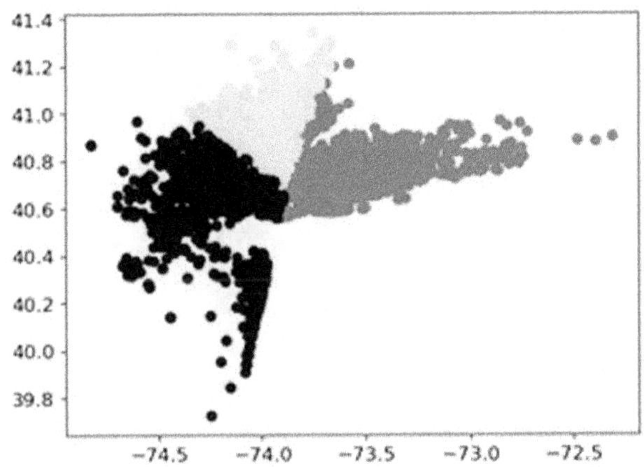

Figure 23. Numbers of cluster using K-means algorithm
(Source: Authors' Demo)

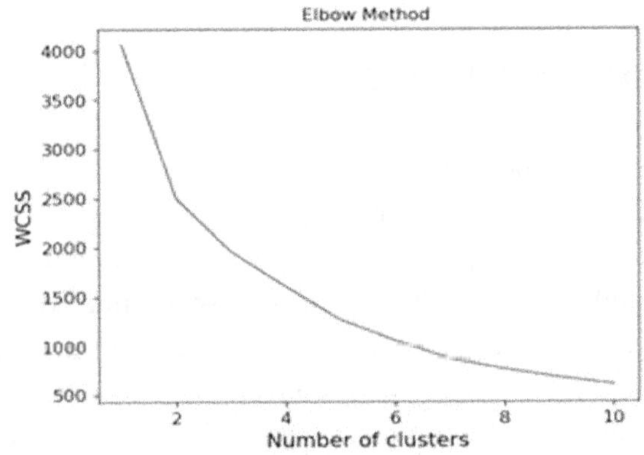

Figure 24. Visualization cluster using folium map
(Source: Authors' Demo)

```
In [52]: for i in cluster1:
             folium.CircleMarker(i, radius=2,color='blue',fill_color='lightblue').add_to(map)

         for i in cluster2:
             folium.CircleMarker(i, radius=2,color='red',fill_color='lightred').add_to(map)

         for i in cluster3:
             folium.CircleMarker(i, radius=2,color='green',fill_color='lightgreen').add_to(map)
```

Figure 25. Visualization of three clusters using folium map
(Source: Authors' Demo)

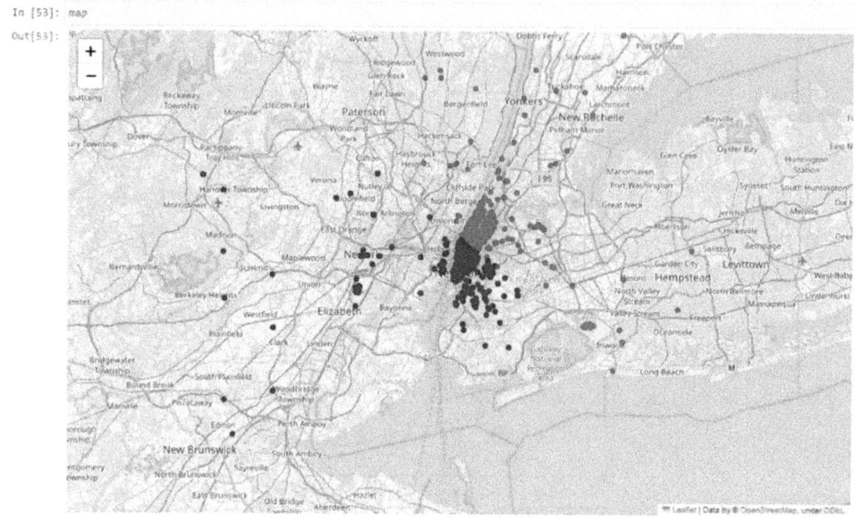

for this chapter. Figure 26 shows a scatter plot of housing longitude and latitude. Figure 27 displays a scatter plot of housing prices.

CONCLUSION AND FUTURE WORK

As technology advances, the potential applications for geospatial data will only continue to grow. However, conventional techniques for evaluating geographic data frequently involve manual interpretation or rule-based strategies, which take a long time and have a limited capacity to handle big datasets. This chapter used a machine learning algorithm to analyze geospatial data and extract insights that would be challenging or impossible to obtain with conventional techniques. In this chapter, we performed a literature review on the recent and relevant research for analyzing geospatial data using machine learning, python libraries for geospatial data like GeoPandas, Shapely, Fiona, Folium, and Pyroj with their examples.

Also, we discussed machine learning types like deep learning, unsupervised learning, and supervised learning techniques with their application in geospatial data. In addition, preprocessing geospatial data for machine learning using various techniques with their examples, Case study: Predicting median house value using Machine Learning, Machine learning algorithms selection, and building machine learning model using random forest, decision tree and linear regression algorithms and evaluation of model with demo were presented. Finally, from unsupervised

Figure 26. Housing longitude and latitude scatter plot
(Source: Authors' Demo)

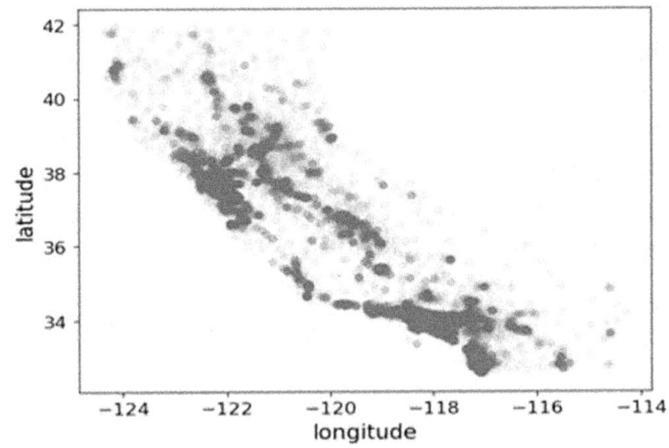

learning, we used a k-means clustering algorithm to cluster geospatial data, and geospatial data visualization was performed for two geospatial datasets taken from an open-source database(Kaggle). Thus, machine learning has revolutionized geospatial analysis by enabling the development of predictive models that can process large volumes of geospatial data to extract meaningful insights.

Figure 27. Housing price scatter plot
(Source: Authors' Demo)

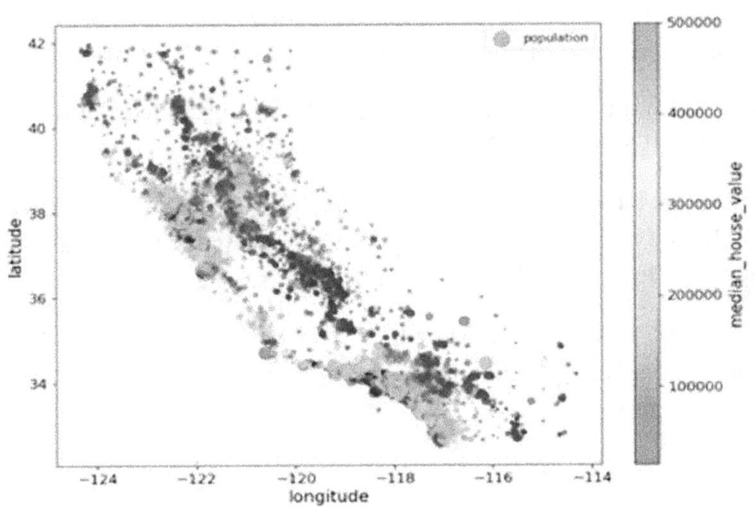

Future work in machine learning for geospatial analysis will focus on the following areas:

- Investigating the application of deep learning methods for geospatial research, such as recurrent and convolutional neural networks.
- Developing new machine learning models that can handle geospatial data more efficiently and accurately.
- Investigating the use of machine learning for real-time geospatial analysis, such as tracking natural disasters or monitoring traffic patterns.
- Exploring the use of machine learning for geospatial data visualization and examination, such as analyzing geospatial data to identify trends and anomalies.

REFERENCES

Abujayyab, S. K. M. (2019). Geospatial machine learning datasets structuring and classification tool. *Case Study For Mapping Lulc From Rasat Satellite Images, 42*(October), 1–3.

Acharya, A., Mondal, B. K., Bhadra, T., Abdelrahman, K., & Mishra, P. K. (2022). *Geospatial Analysis of Geo-Ecotourism Site Suitability Using AHP and GIS for Sustainable and Resilient Tourism Planning in West Bengal*. doi:10.3390/su14042422

Aggarwal, K., Mijwil, M. M., Alomari, S., Gök, M., Alaabdin, A. M. Z., & Abdulrhman, S. H. (2022). Has the Future Started? The Current Growth of Artificial Intelligence, Machine Learning, and Deep Learning. *Iraqi Journal for Computer Science and Mathematics*, 115–123. doi:10.52866/ijcsm.2022.01.01.013

Agrawal, R. (2020). Fundamentals of Machine Learning. In Machine Learning for Healthcare (pp. 1–16). doi:10.1201/9780429330131-1

Andronie, M., & Iatagan, M. (2022). *Deep Learning-Assisted Smart Process Planning*. Robotic Wireless Sensor Networks, and Geospatial Big Data Management Algorithms in the Internet of Manufacturing Things.

Breunig, M., Bradley, P. E., Jahn, M., Kuper, P., Mazroob, N., Rösch, N., & Jadidi, M. (n.d.). Geospatial Data Management Research : Progress and Future Directions.

Dai, L., Wu, L., Li, H., Cai, C., Wu, Q., Kong, H., Liu, R., Wang, X., Hou, X., Liu, Y., Long, X., Wen, Y., Lu, L., Shen, Y., Chen, Y., Shen, D., Yang, X., Zou, H., Sheng, B., & Jia, W. (2021). retinopathy across the disease spectrum. *Nature Communications*, *12*(1), 3242. doi:10.1038/s41467-021-23458-5 PMID:34050158

Das, S. K. (n.d.). *Geospatial Data Analytics- A Deep Learning Perspective.*

Gunawardena, N. K. (2014). *Introduction to geographic information system, (August).*

Helm, J. M., Swiergosz, A. M., Haeberle, H. S., Karnuta, J. M., Schaffer, J. L., Krebs, V. E., Spitzer, A. I., & Ramkumar, P. N. (2020). Machine Learning and Artificial Intelligence: Definitions, Applications, and Future Directions. *Current Reviews in Musculoskeletal Medicine*, *13*(1), 69–76. doi:10.1007/s12178-020-09600-8 PMID:31983042

Ighile, E. H., Shirakawa, H., & Tanikawa, H. (2022). *Application of GIS and Machine Learning to Predict Flood Areas in Nigeria.*

Jena, R., Pradhan, B., Prasanajit, S., & Alamri, A. M. (2021). Geoscience Frontiers Earthquake risk assessment in NE India using deep learning and geospatial analysis. *Geoscience Frontiers*, *12*(3), 101110. doi:10.1016/j.gsf.2020.11.007

Kanevski, M., Pozdnukhov, A., & Timonin, V. (2008). *Machine Learning Algorithms for GeoSpatial Data.* Applications and Software Tools Machine Learning Algorithms for GeoSpatial.

KELLEHER, J. D. (2019). *Deepl learning.*

Kucklick, J., Müller, J., Beverungen, D., & Mueller, O. (2021). *AIS Electronic Library (AISeL).* Quantifying The Impact Of Location Data For Real Estate Appraisal – A Gis-Based Deep Learning Approach Quantifying The Impact Of Location Data For Real Estate Appraisal – A Gis-Based Deep.

Nguyen, A. Q., Nguyen, H. T., Tran, V. C., Pham, H. X., & Pestana, J. (2021). A Visual Real-time Fire Detection using Single Shot MultiBox Detector for UAV-based Fire Surveillance. In *ICCE 2020 - 2020 IEEE 8th International Conference on Communications and Electronics* (pp. 338–343). IEEE. 10.1109/ICCE48956.2021.9352080

Of, S. A. (2020). Python libraries matplotlib, seaborn and pandas for visualization geospatial datasets generated by QGIS Polina LEMENKOVA 1, *LXIV*(1).

Reus, N. M. De, Kerbusch, P. J. M., & Schadd, M. P. D. (n.d.). *Geospatial analysis for Machine Learning in Tactical Decision Support*, 1–18.

Srivastava, N., & Saxena, N. (2023). Applications of Artificial Intelligence and Machine Learning in Geospatial Data. *Advances in Geospatial Technologies*, (May), 196–219. doi:10.4018/978-1-6684-7319-1.ch010

Stewart, A. J., Robinson, C., Corley, I. A., & Ferres, J. M. L. (n.d.). *TorchGeo : Deep Learning With Geospatial Data. The 30th International Conference on Advances in Geographic Information Systems (SIGSPATIAL' 22).* Association for Computing Machinery. 10.1145/3557915.3560953

Tamiminia, H., Salehi, B., Mahdianpari, M., Quackenbush, L., Adeli, S., & Brisco, B. (2020). Google Earth Engine for geo-big data applications : A meta-analysis and systematic review ISPRS Journal of Photogrammetry and Remote Sensing Google Earth Engine for geo-big data applications : A meta-analysis and systematic review. *ISPRS Journal of Photogrammetry and Remote Sensing*, *164*(May), 152–170. doi:10.1016/j.isprsjprs.2020.04.001

Westra, E. (n.d.). *Python Geospatial Development*.

Zhalehdoost, A., & Taleai, M. (2022). A Review of the Application of Machine Learning and Geospatial Analysis Methods in Air Pollution Prediction. *Pollution, 8*(3), 904–933. doi:10.22059/POLL.2022.336044.1300

Chapter 10
Geospatial Machine Learning and the Power of Python Programming:
Libraries, Tools, Applications, and Plugins

Munir Ahmad
https://orcid.org/0000-0003-4836-6151
Survey of Pakistan, Pakistan

Muhammad Awais Ali
https://orcid.org/0009-0007-3223-7140
Bahria University, Pakistan

Md. Rokibul Hasan
https://orcid.org/0000-0001-5220-6324
Gannon University, USA

Froilan D. Mobo
https://orcid.org/0000-0002-4531-8106
Philippine Merchant Marine Academy, Philippines

Shahid Iqbal Rai
https://orcid.org/0009-0006-1868-8407
University of Rome Tor Vergata, Italy

ABSTRACT

Machine learning can play a critical role in geospatial analysis, providing enhanced computing efficiency, flexibility, and scalability, improved predictive capabilities, complicated problem resolution, and information extraction from big datasets. Python has emerged as the predominant language for geospatial machine learning due to its user-friendly interface, extensive library support, and versatility. This chapter has explored a diverse ecosystem of Python libraries ranging from Geopandas, Fiona, Leafmap, Geemap, PySAL, and Shapely for geospatial data manipulation to Keras

DOI: 10.4018/979-8-3693-6381-2.ch010

Spatial, TorchGeo, Scikit-learn, and TensorFlow for deep learning applications. Complementing this, it also explored a variety of QGIS Python plugins that enhance geospatial machine learning capabilities, including smart-map, cluster analysis, PyQGIS, ClusterPoints, AI vectorizer, mapflow, deepness, and many more, offering functionalities for digital mapping, clustering, and map segmentation.

INTRODUCTION

Python is a broadly utilized general-purpose dynamic computer language. Python has been utilized in different disciplines due to its framework of modules, simple expansion, excellent graphics, object-oriented programming support, simple learning, fast development times, suitability for high-performance applications, and free download. For example, it is being used for computational physics (Borcherds, 2007), econometrics and statistics (Bilina & Lawford, 2012), and bioinformatics (Bassi, 2007).

In recent years, Python has become a powerhouse within the domain of geospatial data visualization and analysis visualization. Integrating Python with various open-source geospatial libraries and frameworks empowers experts and professionals to work with spatial information effectively for many tasks, such as point cloud analysis, real-time data processing, web-based mapping, clustering, and multivariate geophysical data analysis.

Furthermore, as highlighted by research findings, Python's dominance extends into machine learning applications. According to Anitha Elavarasi and Jayanthi (2022), the most commonly utilized programming languages for developing machine learning applications are Python by 57%, R by 31%, and Java by 17% data scientists. According to the Developer Nation survey, as shown in Figure 1, Python, with 16.9M developers, is the most popular language for Machine Learning (Dodd et al., 2023). This widespread adoption indicates Python`s suitability for high-performance applications and its appeal as a free and readily available language, cementing its position as a key player in modern data science and technology.

From the information mentioned above, it is clear that Python holds prominence as a popular programming language for machine-learning applications and is extensively employed in the development of geospatial machine-learning applications. Therefore, the primary aim of this chapter is to explore various avenues of geospatial machine-learning applications using the Python language. The chapter has been organized to address this objective: Section 2 provides the geospatial machine learning landscape. Section 3 outlines the key libraries, plugins, and cloud source solutions for spatial analysis using Python. The last section concludes the discussion.

Figure 1. Size of programming language communities
(Dodd et al., 2023)

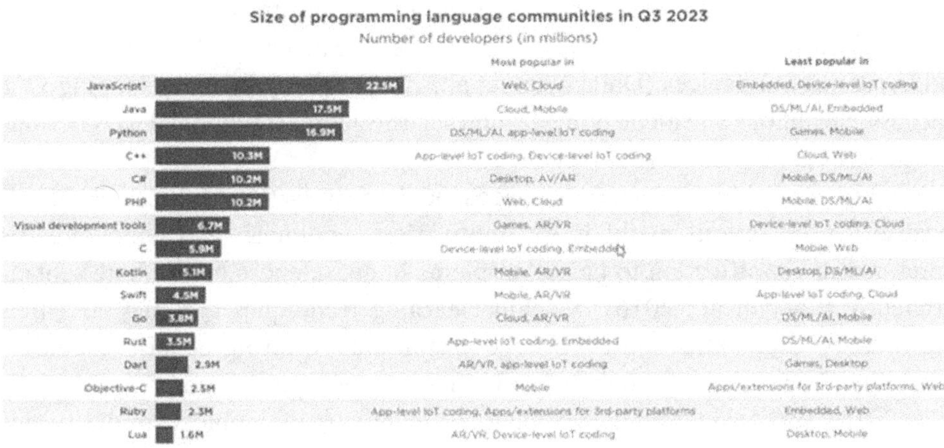

LITERATURE REVIEW

Evidence-based decision-making in a variety of fields, including marketing, manufacturing, healthcare, education, financial modeling, and law enforcement, can be facilitated by machine learning (Jordan & Mitchell, 2015). Integrating machine learning with domain-specific knowledge facilitates its application across diverse fields, leading to novel scientific insights and discoveries (Roscher et al., 2020). For example, many areas of physics, such as particle physics, cosmology, quantum many-body physics, quantum computing, and chemical and material physics, can benefit from machine-learning techniques (Carleo et al., 2019). In archaeology, machine learning techniques can be employed to accurately classify large archaeological datasets to support identifying and categorizing archaeological features and artifacts (Bickler, 2021). Even machine learning techniques can be applied to the estimation of certain types of poverty (Hall et al., 2022), understanding weather and climate extreme events (S. Jiang et al., 2022), exploring the interpretability of materials science models (Oviedo et al., 2022), improve tedious and challenging aspects of teaching and learning (Jalil et al., 2019) and accurate and efficient disease prevention and diagnosis in clinical settings (Zhang & Sejdić, 2019).

The field of geosciences bears substantial societal significance as it endeavors to address a myriad of urgent challenges faced by both the Earth and humanity (Karpatne et al., 2019). It is the most interdisciplinary in all the Science, Technology, Engineering, and Mathematics (STEM) fields, emphasizing Earth system studies and its many subsystems. Remote sensing is an emerging instrument within the field of geoscience, furnishing supplementary data for the examination of Earth and celestial bodies. It offers a comprehensive overview of large areas and environmental variables. It entails gathering data about the Earth and its surroundings using a variety of sensors, including optical, synthetic aperture radar, LiDAR, multispectral, and hyperspectral devices (Dalla Mura et al., 2015). It may involve mapping vast, inaccessible regions and identifying ground surface geomorphological, geothermal, and geostructural features (Saibi et al., 2018).

Machine learning is essential to geospatial analysis (Forghani et al., 2021) because it uses various methods for multivariate, nonlinear, and nonparametric regression or classification to tackle problems in geosciences and remote sensing efficiently (Lary et al., 2016). Machine learning techniques can play a critical role in knowledge extraction from large spatial data, offering benefits including scalability, flexibility, and computational efficiency for exposure modeling (Vopham et al., 2018). By considering user behavior, semantics, geographical similarity, and metadata properties, machine learning in geospatial data discovery enhances search ranking and produces intelligent data discovery (Y. Jiang et al., 2018). However, comprehending scientific logic and resilience using machine learning techniques necessitates a transdisciplinary approach combining computer science, statistics, and mathematics (Xie et al., 2017).

In the big data age, machine learning techniques such as support vector machine (SVM), semi-supervised and active learning, ensemble learning, and deep learning can handle spatial data well and offer various possible applications (Du et al., 2020). For example, supervised learning can facilitate the automated recognition of natural features and the deduction of intricate rules controlling spatial systems (Arundel et al., 2020). Similarly, as a special case of machine learning, active learning can automate the semantic interpretation of 3D point clouds, reducing the need for human input and enhancing output (Kölle et al., 2020). Moreover, the tree induction technique can help create generalizations from big data sets for classification and prediction (Whigham et al., 1992). Furthermore, deep learning can support seasonal forecasting and the modeling of spatial relationships across various timescales (Reichstein et al., 2019). Deep learning CNNs and other machine learning techniques can increase the precision of GIS analysis, supporting the social sciences and public policy formulation (M. Li et al., 2019).

Many programming languages are used to create machine learning applications, such as Python, R, Java, Julia, Matlab, JavaScript, Scala, Go, and C++. Python (with libraries such as Scikit-Learn, SciPy, NumPy, TensorFlow, Pandas, Seaborn, Theano, Matplotlib, and Keras) is a popular language. R is a widely adopted programming language for developing machine learning applications. It boasts proven software implementations for diverse tasks, including learning from tabular, behavioral, or textual data (Kliegr et al., 2020). Java is a sophisticated technology that amalgamates the Java language, AWT, and Java Virtual Machine. It provides diverse applications and solutions for various computing needs (Dubois, 1997). For example, MEKA is an open-source Java framework that facilitates machine learning applications, particularly those involving multi-label and multi-target data (Read et al., 2016).

The Julia language is a swift, user-friendly, and open-source programming language that balances efficiency and simplicity in developing machine learning applications (Gao et al., 2020). MATLAB's machine-learning methods can be used for the development of versatile applications in various industrial sectors, including but not limited to search engines, DNA sequencing, stock market analysis, and robot locomotion (Barber, 2012). Leveraging MATLAB Machine Learning Toolboxes, such as those for deep learning and instrument control, can result in robust systems for machine learning applications (Paluszek et al., 2022). MLitB, a framework based on JavaScript, can enable every internet-capable device to execute training algorithms and predictive models without requiring software installation. This approach makes machine learning ubiquitous, cost-effective, and collaborative (Meeds et al., 2015).

SciPipe is a workflow programming library crafted in Go that is designed to handle intricate and dynamic bioinformatics pipelines, particularly in machine learning. It incorporates extensive branching and dynamic scheduling to enhance its capabilities (Lampa et al., 2019). Scala is a fitting industrial development language for seamlessly integrating machine-learning models into various industrial applications. C++ serves as an extensively employed compiled language for crafting machine learning applications. For example, NEURObjects is a C++ library designed to facilitate experimental research in neural networks and enable rapid prototyping of inductive machine learning applications (Valentini & Masulli, 2002). Similarly, Yggdrasil Decision Forests is a C++ library specifically developed for training, serving, and interpreting decision forest models (Guillame-Bert et al., 2023).

Python is a versatile language widely used for developing geospatial applications to facilitate real-time observation, data processing, and analysis across various fields. Python can be applied for geospatial analysis, encompassing the handling of vector, raster, and 3D data. It is well-suited for extending existing tools and developing new ones in geospatial information (McInerney & Kempeneers, 2015). For example, the Python application DeepAutoMapping leverages deep learning technology to enable cost-effective and efficient data collection. It achieves this by generating geographical data and observing field data in real-time from video streams (Al-Azizi et al., 2022). Similarly, utilizing a Python script, the isolation forest method can be applied to identify anomalies in spatial data and predict the future potential trajectory of an invasion (Santosuosso et al., 2020). Meanwhile, Nansat, a Python toolbox, is specifically crafted to support studying and developing algorithms for self-governing processing systems. Its focus lies on processing two-dimensional geographical data (Korosov et al., 2016).

GEOSPATIAL MACHINE LEARNING

Geospatial machine learning means augmenting predictive capabilities and simulating complex Earth system issues by incorporating geographical and temporal information into machine learning models, especially those based on deep learning. It may encompass the application of data mining and machine learning approaches to address challenging tasks in the geospatial domain. These tasks may include hotspot detection, colocation detection, prediction, outlier detection, and teleconnection detection (Xie et al., 2017). It may also involve statistical learning within geospatial settings to tackle challenges such as spatial correlation and covariate shifts in the data (Hoffmann et al., 2021). It may encompass applying deep learning techniques, such as neural networks, to perform tasks like object recognition, image classification, and scene understanding within geospatial data (Kiwelekar et al., 2020).

The management of spatial data can be achieved by applying supervised and unsupervised machine learning techniques. Supervised learning generally involves training an algorithm on labeled data to predict the patterns. Unsupervised learning does not hang on labeled data but looks for patterns or relationships within the data. The utilization of these approaches for geospatial data analysis depends on the particular goals and the properties of the data in question. For example, a semi-supervised learning algorithm can enhance land cover classification accuracy significantly, achieving an improvement of 24%-36% compared to conventional methods by leveraging freely available unlabeled training samples and ancillary geospatial databases as underpinned by (Vatsavai & Bhaduri, 2011). Similarly, Chai et al., (2023) noted that a geophysics-steered self-supervised learning (SSL) approach can enhance seismic deconvolution accuracy and spatial continuity compared to traditional trace-by-trace methods.

Geospatial machine learning has vast applications. For example, by integrating machine learning techniques with multi-source semantic information, the accuracy of identifying geographic data-sharing websites can be enhanced (Cheng et al., 2021). Data-driven technologies, such as machine learning and artificial intelligence, can be used to develop a new generation of Geographic Information Systems (GIS) tailored for urban environments, as noted by (Golubev et al., 2016). Search ranking of geospatial data can be enhanced by utilizing real-time user clickstream data to dynamically update the ranking model (Y. Li et al., 2020). Cao et al., (2022) underscored that a machine learning-based approach can estimate and map economic development using multi-source open geospatial data by utilization of remote sensing imagery and OpenStreetMap road networks. Machine learning methods and geospatial knowledge can optimize transport and logistics processes and supply chains in logistics and transport (Kolesnikov et al., 2021).

The concept of geospatial artificial intelligence (GeoAI) amalgamates spatial science, machine learning, deep learning, data mining, and high-performance computing to extract knowledge from spatial big data (Vopham et al., 2018). Pierdicca, (2021) also underscored that GeoAI integrates AI, ML, and DL to analyze and interpret complex geomatics data, improving knowledge management and cost-effectiveness in various domains. GeoAI represents an emerging spatial analytical framework within data-intensive GIScience that harnesses machine learning and advanced computing for scalable processing and intelligent geospatial big data analysis (W. Li, 2020).

PYTHON FOR GEOSPATIAL MACHINE LEARNING

The popularity of Python in geospatial analysis originated from its user-friendly nature, extensive library support, and adaptability. Python is a proficient language for designing and implementing algorithms geared towards machine learning applications in geospatial contexts. These applications can encompass tasks such as feature extraction, prediction, and the categorization of spatial data. Integral to employing Python for geospatial machine learning are essential tools and frameworks like scikit-learn, TensorFlow, PyTorch, Rasterio, and many more. These resources can play a crucial role in manipulating spatial datasets, allowing the application of machine learning algorithms to extract and predict meaningful insights from spatial data.

Python Libraries for Spatial Data Handling

Several Python libraries are commonly used for spatial data handling to support implementing machine learning algorithms in the geospatial domain. The Python Package Index (PyPI) serves as a repository for software designed for the Python programming language, facilitating the discovery and installation of software created and shared by the Python community. 502,772 projects and 768,432 users have been listed on the website (https://pypi.org/). The search for different keywords on the website resulted as follows: 2,538 projects for "geo," 2,213 projects for "spatial," and 928 projects for "geospatial." Some notable libraries include:

Geopandas

Geopandas expands Pandas' functionality to facilitate the simple manipulation and analysis of geospatial data by adding spatial operations to Pandas' capabilities. It can help retrieve optimal image combinations from large-area remote sensing data, making it advantageous for processing large-area data (Yan et al., 2022). It is a versatile library and has been applied in fields for geospatial data analysis. For example, it can optimize the geocoding process using existing systems for cleaning and correcting addresses. It can help extract features from satellite imagery for water monitoring, land cover, and tree cover visualization (Yadav et al., 2022). It can provide array processing tools for geophysical data types like well logs, horizons, and seismic volumes (Pinte et al., 2012).

Key features of Geopandas (https://geopandas.org/en/stable/docs.html) include: -

i. GeoDataFrame and GeoSeries are the main data structures of Geopadas, which are an extension of pandas. DataFrame and pandas.Series respectively.
ii. It is fully open-source software, accessible to everyone for free use.
iii. It enables the execution of geospatial analysis through a wide range of spatial operations such as overlay, intersection, and buffering.
iv. It supports multiple formats for reading and writing spatial data, including Shapefiles, GeoJSON, and GeoPackage.
v. It easily interacts with Matplotlib to facilitate the visualizations of spatial data analysis.
vi. It facilitates spatial queries that allow for the filtering and analysis of data according to its spatial relationships.
vii. It supports reprojecting and manipulating spatial data and works with various CRS through Fiona and Rasterio libraries.

Fiona

Fiona is designed to read and write geospatial data formats. It works well in conjunction with Geopandas for data input and output tasks. It supports various GIS formats like GeoPackage, GeoJSON, and Shapefile. Fiona depends on GDAL but provides bindings designed for productivity and readability (Gillies, 2023a).

Shapely

Shapely is a Python package for manipulating and analyzing planar geometric objects using the GEOS open-source geometry library. It provides a feature-rich geometry interface for singular geometries. Shapely can collaborate with other Python GIS packages by utilizing GeoJSON-like dictionaries. Shapely does not read or write data files but can serialize and deserialize using various formats and protocols (Gillies, 2023b).

Keras Spatial

Keras, compatible with JAX, TensorFlow, and PyTorch, can streamline ML workflows with a user-friendly approach. It provides a clear API and natural abstractions and supports model movement across frameworks. A Keras model can be an effective and accurate tool for satellite image analysis. Keras Spatial, a Python package based on Keras, is a data generator designed to simplify the preprocessing of spatial data for deep learning applications. It reads samples directly from a raster data source, eliminating the need to create small, individual raster files before model execution. The raster data source may be a local or remote service, and any necessary reprojections and scaling are handled automatically. Keras Spatial is based on the rasterio package, which allows users to work in spatial coordinates rather than pixels and integrate data from different coordinate systems (Soliman & Terstriep, 2019; Terstriep, 2023).

PyTorch

PyTorch is a Python library tailored for deep learning applications, emphasizing its user-friendly design and convenience in constructing intricate neural networks (Imambi et al., 2021). Many branches have been developed based on Pytorch to deal with spatial data. A few are listed below.

i. TorchGeo is a PyTorch domain library that provides datasets, samplers, transforms, and pre-trained models specific to geospatial data. It aims to simplify the process for machine learning experts working with geospatial data and remote sensing experts exploring machine learning solutions (Stewart et al., 2022).
ii. Torch Spatiotemporal (tsl) is a Python library for neural spatiotemporal data processing, focusing on Graph Neural Networks. It is built upon the most used libraries of the Python scientific computing ecosystem, offering a straightforward process from data preprocessing to model prototyping (https://torch-spatiotemporal.readthedocs.io/en/latest/index.html accessed on 12 December 2023).
iii. PyG *(PyTorch Geometric)* is a library built upon PyTorch for deep learning that simplifies the creation and training of Graph Neural Networks (GNNs) for irregularly structured input data like graphs, and point clouds (https://pytorch-geometric.readthedocs.io/en/latest/ accessed on 12 December 2023).
iv. PyTorch Geometric Temporal is a deep learning framework for spatiotemporal signal processing, combining state-of-the-art machine learning algorithms for predictive performance in real-world problems (Rozemberczki et al., 2021).

Scikit-Learn

Scikit-learn is a machine learning package in the Python programming language that includes implementations of a comprehensive list of machine learning methods under unified data and modeling procedure conventions (Hao & Ho, 2019). Many offshoots have been developed based on Scikit-learn to deal with spatial data. A few are listed below.

i. Scikit-GStat is a SciPy-flavored geostatistical toolbox written in Python that uses classes, interfaces, and implementation rules from the SciPy package for scientific Python (Korosov et al., 2016).
ii. Scikit-network is a Python package for analyzing large graphs, providing state-of-the-art algorithms for ranking, clustering, classifying, embedding, and visualizing nodes (Bonald et al., 2020).

TensorFlow

TensorFlow is a powerful deep-learning library designed to simplify and expedite the research and implementation of neural network models. It is particularly instrumental in applications like data analysis and modeling within educational and behavioral sciences, offering robust support for various neural network architectures, including convolutional neural networks (Pang et al., 2020). TensorFlow, especially in conjunction with the Keras library, is a powerful framework for temporal and spatial deep learning tasks (Lee et al., 2021).

PySAL

PySAL is a comprehensive Python library focused on spatial data analysis, offering functionality for core spatial data structures, exploratory analysis, model estimation, and visualization (https://pysal.org/ accessed on 12 December 2023). It supports spatial analytical methods, including spatial autocorrelation, spatial clustering, and spatial filtering (Rey et al., 2015). PySAL Model has different modules to perform different tasks such as lib, explore, model, and viz as depicted in Figure 2.

Nansat

It is a Python toolbox designed with scientists in mind, tailored for the processing of 2D satellite Earth observation data to facilitate research and development of algorithms and autonomous processing systems (Korosov et al., 2016).

Figure 2. PySAL modules
(Source: https://pysal.org/)

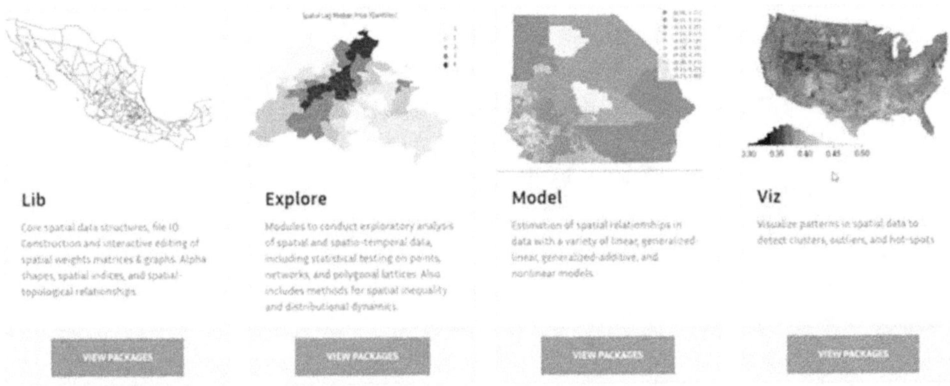

Geemap

It is a Python package for geospatial analysis and visualization with Google Earth Engine (GEE), facilitating interactive mapping (https://pypi.org/project/geemap/ accessed on 12 December 2023). It can facilitate diverse tasks such as generating Landsat timelapse animations, exploring locations and datasets through the Earth Engine Data Catalog, employing a time-series inspector for monitoring landscape changes, and exporting Earth Engine maps as HTML files and PNG images (Wu, 2020).

Leafmap

Leafmap is a Python package specifically crafted for interactive mapping and geospatial analysis within Jupyter environments (https://leafmap.org/ accessed on 12 December 2023) as depicted in Figure 3. This free and open-source tool ensures accessibility on platforms like Google Colab and Jupyter Notebook. With the capability to simplify the creation of interactive maps using just a single line of code, Leafmap supports multiple mapping backends, including ipyleaflet, folium, kepler.gl, pydeck, and bokeh. Users can dynamically alter base maps, incorporate XYZ, WMS, and vector tile services, and personalize legends and color bars. Leafmap facilitates the display of various data formats, encompassing Shapefile, GeoJSON, GeoPackage for vectors, and GeoTIFFs for raster data. Additionally, it empowers the creation of split-panel and linked maps, enabling users to conduct comparative analyses effortlessly (Wu, 2021b, 2021a).

Figure 3. Snap of Leafmap
(Source: https://leafmap.org/)

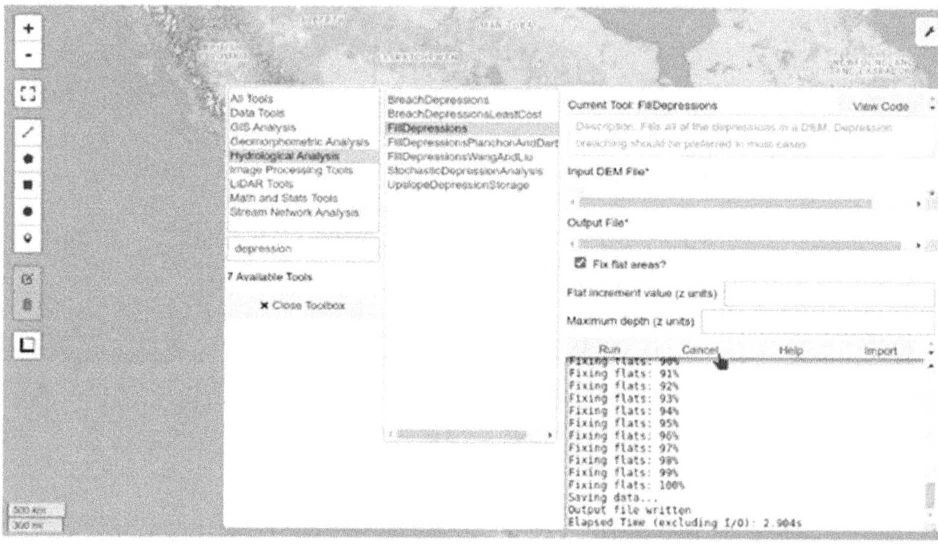

Cityseer

The Cityseer is a comprehensive suite of computational tools meticulously designed for in-depth network and land-use analysis at a fine-grained level. Tailored to assess the morphological elements that contribute to vibrant neighborhoods, it utilizes rigorous network-based methods developed from the ground up (https://pypi.org/project/cityseer/ accessed on 12 December 2023).

GeoUtils

GeoUtils is an open-source Python package specifically crafted for geospatial analysis, placing a strong emphasis on providing consistent Raster and Vector objects to ensure seamless interfacing. This package is geared towards facilitating user-friendly geospatial analysis methods. GeoUtils relies on essential dependencies such as Rasterio, GeoPandas, Pyproj, NumPy, and Xarray, enabling support for a diverse range of functionalities, including lazy loading, parallel computing, and object conversion (https://github.com/GlacioHack/geoutils accessed on 12 December 2023).

Figure 4. Workflow of GemGIS
(SoTurce: https://gemgis.readthedocs.io/)

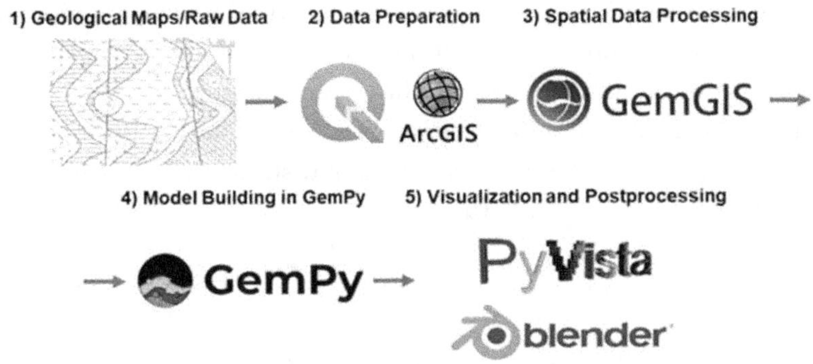

GemGIS

It is an open-source Python library for geographic information processing. It adeptly preprocesses a variety of spatial data types, encompassing vector data (shape files, geojson files, geopackages), raster data (tif, png), online service data (WCS, WMS, WFS), and XML/KML files. The preprocessed data can be efficiently stored in a dedicated Data Class, ready for utilization in the geomodeling package GemPy. GemGIS streamlines and automates workflows, leveraging GeoPandas, rasterio, OWSLib, Pandas, Shapely, PyVista, and NumPy to enhance functionality. Results derived from GemPy can be further processed and seamlessly exported to popular geoinformation systems such as QGIS, ArcGIS, or Google Earth (https://gemgis.readthedocs.io/en/latest/index.html accessed on 12 December 2023). Figure 4 depicts a workflow of GemGIS.

stco

This Python library offers a straightforward approach to analyzing spatial clusters over time, leveraging overlay analysis. Instead of directly applying clustering to spatiotemporal data, the approach involves running spatial clustering on periods, effectively creating "snapshots" of the data. This process generates clusters for each time step. Subsequently, a basic or weighted overlay method can be employed to investigate the temporal evolution of clusters (https://pypi.org/project/stco/ accessed on 12 December 2023).

Arches

Arches is a geospatial information system designed as a web-based solution for the inventory and management of cultural heritage. Tailored specifically for the international cultural heritage sector, Arches is crafted to capture a diverse range of immovable heritage, encompassing archaeological sites, buildings, historic structures, landscapes, and heritage ensembles or districts (https://pypi.org/project/arches/ accessed on 12 December 2023).

Geo-Dream

The Godream library is tailored for geospatial tools in Python, with a specific emphasis on remote sensing (RS) and Geographic Information System (GIS) analysis. Offering a suite of functions and tools, the library facilitates the manipulation, processing, and visualization of geospatial data, streamlining the execution of intricate analyses within these domains for users (https://github.com/DreamPTK/Godream accessed on 12 December 2023).

Damage Scanner

Damage Scanner is a specialized Python toolkit created to conduct direct damage assessments related to natural hazards. It is important to note that this package is presently in the development phase. Its functionality depends on essential libraries, including NumPy, pandas, geopandas, matplotlib, rasterio, tqdm, xarray, and pyproj (https://github.com/VU-IVM/DamageScanner accessed on 12 December 2023).

The geospatial landscape benefits from a diverse set of Python libraries that cater to different aspects of spatial data analysis, machine learning, and visualization. Geopandas stands out by expanding Pandas' capabilities with spatial operations, enabling efficient manipulation and analysis of geospatial data. It excels in optimal image retrieval from large remote sensing datasets and geocoding optimization. Fiona complements Geopandas for reading and writing geospatial data formats, while Shapely provides powerful geometry manipulation. Keras Spatial simplifies the preprocessing of spatial data for deep learning, leveraging Keras, TensorFlow, and PyTorch. Through its branches like TorchGeo, PyG, and PyTorch Geometric Temporal, PyTorch offers specialized geospatial machine learning tools. Scikit-learn extends its capabilities with Scikit-GStat and Scikit-network for geostatistics and large graph analysis. TensorFlow excels in temporal and spatial deep learning, especially in collaboration with Keras. PySAL is a comprehensive library for spatial data analysis, supporting various spatial analytical methods.

Nansat and Geemap cater to satellite Earth observation data processing and geospatial analysis with Google Earth Engine, respectively. Leafmap simplifies interactive mapping within Jupyter environments, while Cityseer and GeoUtils focus on fine-grained network and land-use analysis. GemGIS aids in geographic information processing, and stco facilitates spatial cluster analysis over time. Arches serves as a web-based solution for cultural heritage inventory and management. The Godream library specializes in remote sensing and GIS analysis, and DamageScanner is a toolkit for direct damage assessments related to natural hazards. This rich ecosystem of Python libraries empowers researchers and practitioners in diverse geospatial domains with a wide array of tools and functionalities. A summary of these libraries is provided in the following table for quick reference: Table 1 provides an overview of various Python libraries dedicated to handling geospatial data.

Plugins for Geospatial Machine Learning

Utilizing diverse Python libraries, numerous plugins have been created for a range of GIS software, encompassing both proprietary and open-source solutions. The repository of QGIS Python Plugins shows 2012 records as shown in Figure 5.

A few plugins that are relevant to geospatial machine learning are listed below.

Smart-Map

It is an open-source QGIS plugin that uses machine learning algorithms like Support Vector Machine (SVM) to digitally map soil attributes (Pereira et al., 2022).

Cluster Analysis

Cluster Analysis, a Python plug-in for QGIS, offers functionalities for the entire clustering process, supporting data exploration, pattern recognition, and prediction in geospatial data applications (Folini et al., 2022).

PyQGIS

The PyQGIS plugin can produce Land Surface Temperature (LST) maps from Landsat imagery using various algorithms, including the Planck function, Mono Window Algorithm, Single Channel Algorithm, and Radiative Transfer Equation (Ndossi & Avdan, 2016).

Table 1. Python libraries for geospatial data

Library	Purpose/Functionality	Key Features
Geopandas	Geospatial data manipulation and analysis	GeoDataFrame, spatial operations, open-source
Fiona	Reading and writing geospatial data formats	GeoPackage, GeoJSON, Shapefile support
Shapely	Manipulating and analyzing planar geometric objects	GEOS library integration, geometry interface
Keras Spatial	Preprocessing spatial data for deep learning	Raster data source integration, Keras-based
PyTorch	Deep learning for geospatial data	TorchGeo, Torch Spatiotemporal, PyG
Scikit-learn	Machine learning for spatial data	Scikit-GStat, Scikit-network
TensorFlow	Deep learning for temporal and spatial tasks	Keras integration, TensorFlow ecosystem
PySAL	Spatial data analysis	Various spatial analytical methods
Nansat	Processing 2D satellite Earth observation data	Experimental research support
Geemap	Geospatial analysis with Google Earth Engine	Interactive mapping, data exploration
Leafmap	Interactive mapping and geospatial analysis in Jupyter	Multiple mapping backends, data display
Cityseer	Computational tools for fine-grained network and land-use	Rigorous network-based methods
GeoUtils	Geospatial analysis with consistent Raster and Vector objects	Support for lazy loading, parallel computing
GemGIS	Geographic information processing	Workflow automation, integration with GIS
stco	Spatial cluster analysis over time	Overlay analysis for temporal cluster changes
Arches	Web-based cultural heritage inventory and management	Capture diverse immovable heritage
Godream	Remote sensing and GIS analysis	Suite of functions and tools
Damage Scanner	Toolkit for direct damage assessments related to hazards	Development phase, relies on essential libraries

Deep Learning Datasets Maker

It is a QGIS Python plugin for handling raster and vector data. The plugin provides tools to split data into small, equal-sized pieces suitable for machine-learning datasets. It specifically allows the division of remote-sensing images and their labels into datasets of specified sizes (https://plugins.qgis.org/plugins/deep-learning-datasets-maker/ accessed on 13 December 2023).

Figure 5. A snip of QGIS Python plugins

Map Segmentation

It provides pre-trained models to enhance workflow efficiency, although an upgraded license is necessary to achieve superior accuracy. The functionality relies on sci-kit-learn, opencv-python, and torch. There are two available models: K-Means (free) and CNN (licensed, delivering enhanced results) (https://github.com/sirebellum/qgis-segmentation accessed on 14 December 2023).

Deepness: Deep Neural Remote Sensing

It facilitates the utilization of ONNX Neural Network models for segmentation, detection, and regression on raster orthophotos. It supports processing diverse raster layers, such as custom orthophotos and those from online providers like Google Satellite. The plugin allows the restriction of processing to specific areas defined by vector layer polygons or visible parts of the raster. It accommodates common model types, including segmentation, regression, and detection (https://plugins.qgis.org/plugins/deepness/ accessed on 14 December 2023).

Mapflow is a QGIS Python plugin for extracting real-world objects from satellite imagery. It provides AI mapping pipelines for buildings, roads, fields, forests, and construction sites. Users can choose a web imagery provider or upload their images for feature extraction. Extractable features can encompass buildings, agricultural fields, forests (with optional height), roads, and construction sites (https://plugins.qgis.org/plugins/mapflow/ accessed on 14 December 2023).

AI Vectorizer

It employs machine learning to automatically vectorize lines and polygons from raster maps. It efficiently aids in digitizing old raster maps, accommodating users of varying GIS expertise (https://github.com/BuntingLabs/buntinglabs-qgis-plugin accessed on 12 December 2023).

ClusterPoints

This plugin is designed to cluster points on a map. Users can select from K-means, Fuzzy C-means, and hierarchical clustering algorithms. Clustering can be performed based on mutual distance or supplementary information from attributes, incorporating both geographical coordinates and numerical fields. The plugin offers various cluster algorithms, including SLINK and Lance-Williams distance updates. It provides explicit cluster IDs for individual points and membership probabilities for Fuzzy C-Means. Users can specify multiple numerical fields, controlling the balance between spatial and attribute-based clustering (https://plugins.qgis.org/plugins/clusterpoints/ accessed on 14 December 2023).

The QGIS Python plugins for machine learning provide diverse geospatial data analysis and mapping functionalities. Smart-Map utilizes machine learning algorithms like Support Vector Machine (SVM) to map soil attributes digitally. Cluster Analysis, a Python plug-in for QGIS, supports the entire clustering process, aiding in geospatial applications' data exploration, pattern recognition, and prediction. PyQGIS produces Land Surface Temperature (LST) maps from Landsat imagery. Deep Learning Datasets Maker facilitates handling raster and vector data for machine-learning datasets.

Map Segmentation offers pre-trained models for efficient workflow with enhanced accuracy. Deepness enables utilizing ONNX Neural Network models for segmentation, detection, and regression on raster orthophotos. Mapflow is designed for extracting real-world objects from satellite imagery, offering AI mapping pipelines for various features. AI Vectorizer automates the vectorization of lines and polygons from raster maps, aiding in digitizing old maps. ClusterPoints is designed to cluster points on a map, providing flexibility in algorithms and clustering based on both spatial and attribute information. These plugins are summarized in the following table for quick reference.

Table 2 outlines a collection of Python QGIS plugins designed specifically for machine learning tasks.

Table 2. Python QGIS plugins for machine learning

Plugin Name	Functionality	Key Features
Smart-Map	Digital mapping of soil attributes using SVM	Open-source, SVM algorithm
Cluster Analysis	Clustering process support for geospatial data applications	Data exploration, pattern recognition, prediction
PyQGIS	Production of LST maps from Landsat imagery	Various algorithms for LST maps
Deep Learning Datasets Maker	Handling raster and vector data for machine-learning datasets	Data splitting tools for remote-sensing images
Map Segmentation	Pre-trained models for enhanced workflow efficiency	K-Means (free) and CNN (licensed) models
Deepness	Utilization of ONNX Neural Network models for segmentation, detection, and regression	Processing diverse raster layers supports common model types
Mapflow	Extraction of real-world objects from satellite imagery	AI mapping pipelines for various features
AI Vectorizer	Automatic vectorization of lines and polygons from raster maps	Machine learning-based vectorization aids in digitizing old maps
ClusterPoints	Clustering points on a map with various algorithms	K-Means, Fuzzy C-Means, hierarchical clustering, attribute-based clustering

Cloud-Based Solutions for Spatial Data Handling

In addition to dedicated plugins for GIS software, numerous Python implementations specifically tailored for handling spatial data in cloud environments have been developed. Below, a selection of these implementations is highlighted.

PODPAC (Pipeline for Observational Data Processing Analysis and Collaboration)

PODPAC can perform geospatial analyses either locally or in the cloud. It can enable widespread exploitation of NASA earth science data by enabling multi-scale and multi-windowed access, exploration, and integration of datasets, standardized sharing of data and algorithms, and seamless transition to the cloud (Ueckermann et al., 2020).

CyberGIS-Compute Framework

CyberGIS-Compute framework provides an easy-to-use application interface and a Python SDK for geospatial applications like spatial accessibility and wildfire evacuation simulation (Padmanabhan et al., 2021).

Google Earth Engine (GEE)

Google Earth Engine (GEE) is a cloud-based geospatial processing platform for large-scale environmental monitoring and analysis. GEE is a free-to-use platform providing access to petabytes of publicly available remote sensing imagery and ready-to-use products via an explorer web app. GEE stands out for its high-speed parallel processing capabilities and integration of machine learning algorithms, harnessing Google's computational infrastructure. GEE features a library of Application Programming Interfaces (APIs) that support popular coding languages such as JavaScript and Python. Notably, GEE has made significant strides in addressing global challenges associated with analyzing large geospatial datasets (Tamiminia et al., 2020).

GeoEngine

GeoEngine is a reproducible and production-ready geospatial machine learning research platform, simplifying access to insights locked behind petabytes of imagery (Verma et al., 2022).

CHALLENGES IN USING GEOSPATIAL PYTHON TOOLS AND LIBRARIES

Navigating the landscape of geospatial data analysis and machine learning is challenging. Users may encounter challenges that span from learning complexities to integration intricacies. These kinds of problems require much consideration, expertise, and skillset. The following section outlines some potential challenges practitioners may face in harnessing these powerful tools.

i. A lot of this toolkit and library is not easy to understand, be it for beginners or those with a background in geospatial information technology and machine learning. Knowing how to use their APIs, the principles behind their functions, and their main benefits may take time and experience.

ii. The geospatial data can be variable, including different formats, coordinate variables, and data structures. Cleaning and processing data of such type may be hard, and it may be even more difficult if there are issues like data being very big or numerous. Thus, using geospatial data in various formats for comprehensive analysis using these tools, can be a changeling task.
iii. These libraries usually process other packages or software framework dependencies, so they may suffer version conflicts or testing issues. These dependencies' management, particularly in complex environments, is often troublesome because of the complexity of operations.
iv. When handling huge spatial datasets or when executing heavy machine learning algorithms that are computationally intensive, the computational resource can get strained. When it comes to performance and scalability, optimization is essential, but it can be tricky for beginners who have little or no experience in this field.
v. It is challenging to integrate various libraries and tools into a working environment that best suits the current environment, mainly because of the differences in functionality and APIs. Allowing for the smooth sharing of information between the various parts of the workflow necessitates careful considerations and planning for implementation.
vi. Several of these tools and libraries come with a lot of documentation and community support, but the documentation might be too narrow or inconsistent in some areas. This can create a hard-to-solve or troubleshooting issues. Whether it is limited or outdated support for a certain feature or function, the user may face unsolicited problems.
vii. Dealing with sensitive or confidential geospatial information could give rise to data security and privacy issues since data is often confidential in many applications. Ensuring compliance with regulations and adopting appropriate data security protocols may complicate the data analysis workflows of geospatial information processing.
viii. Machine learning algorithms, parameter settings, and domain expertise choice can be tricky due to the task specificity of geospatial analysis, which is often complicated. Running iterations that optimize models' performance and get the model's best results can be time-consuming and resource-intensive.

POTENTIAL GIS APPLICATIONS BASED ON GEOSPATIAL PYTHON TOOLS

Despite the difficulties mentioned above, geospatial data analysis tools and libraries find applications across various domains, including environmental monitoring, urban planning, and beyond. The following section highlights some of the common applications that can leverage the capabilities of these tools.

i. Geospatial Python tools and resources are quite common in remote sensing applications for studying satellite data to monitor environmental changes, track land and cover use evolution, and perform analyses.
ii. These tools can be used in urban planning by analyzing and assessing land suitability, in infrastructure development through spatial analysis, and in disaster management by mapping vulnerabilities.
iii. These tools can be employed to develop GIS applications that are very important in precision farming for crop monitoring, yield forecasting, soil mapping, and irrigation management using spatial analysis and machine learning.
iv. Use of these tools can be considered in transportation planning, route optimization, and logistics management for evaluating traffic patterns, computing travel times, and selecting ideal sites for infrastructural development.
v. These tools can be harnessed in cultural heritage management, archaeology, and historical preservation for site mapping, inventorying heritage, and spatial analysis of cultural land buildups.
vi. GIS based on these tools can be implemented in health research, epidemiology, and disease surveillance as a basis for modeling of the spread of diseases, healthcare resources provisioning, and planning of outbreak response.
vii. These tools can be handy in conducting spatial analysis related to climate change adaptation, disaster risk reduction, and resilience planning as they can assess vulnerability, predict hazards, and suggest mitigation measures.
viii. These tools can be applied in environmental monitoring, biodiversity conservation, and natural resource management for habitat mapping, species distribution modeling, and ecological risk assessment.

CONCLUSION

Python has become a dominant language for geospatial machine learning, driven by its user-friendly nature, extensive library support, and adaptability. Key tools and frameworks such as sci-kit-learn, TensorFlow, PyTorch, Rasterio, and more play a crucial role in manipulating spatial datasets, enabling the application of machine learning algorithms for tasks like feature extraction and prediction. The ecosystem of Python libraries for spatial data handling is diverse, ranging from Geopandas, Fiona, Leafmap, Geemap, PySAL, and Shapely for geospatial data manipulation to Keras Spatial, TorchGeo, Scikit-learn, and TensorFlow for deep learning applications.

Each library serves specific purposes, supporting tasks like preprocessing, model development, and spatial data analysis.

The rich landscape of QGIS Python plugins further enhances geospatial machine learning capabilities. Smart-Map, Cluster Analysis, PyQGIS, ClusterPoints, AI Vectorizer, Mapflow, Deepness, and others provide functionalities for digital mapping, clustering, and map segmentation. These plugins leverage machine learning algorithms, enhancing workflow efficiency and accuracy. Moreover, cloud-based solutions like PODPAC, CyberGIS-Compute framework, Google Earth Engine, and GeoEngine extend geospatial analysis to the cloud, enabling multi-scale access and collaboration.

For future work, continuous development and integration of machine learning techniques into geospatial workflows should be explored. This involves enhancing existing libraries and plugins and fostering collaboration among the geospatial and machine-learning communities. Additionally, the adoption of emerging technologies such as ONNX models and further integration with cloud-based solutions could contribute to more scalable and efficient geospatial machine learning applications. The ever-expanding Python ecosystem in this domain presents exciting opportunities for interdisciplinary research and innovation.

REFERENCES

Al-Azizi, J. I., Shafri, H. Z. M., Hashim, S. J., & Mansor, S. B. (2022). DeepAutoMapping: Low-cost and real-time geospatial map generation method using deep learning and video streams. *Earth Science Informatics*, *15*(3), 1481–1494. doi:10.1007/s12145-020-00529-7

Anitha Elavarasi, S., & Jayanthi, J. (2022). Programming language support for implementing machine learning algorithms. In *Research Anthology on Machine Learning Techniques*. Methods, and Applications. doi:10.4018/978-1-6684-6291-1.ch028

Arundel, S. T., Li, W., & Wang, S. (2020). GeoNat v1.0: A dataset for natural feature mapping with artificial intelligence and supervised learning. *Transactions in GIS*, *24*(3), 556–572. doi:10.1111/tgis.12633

Barber, D. (2012). Bayesian Reasoning and Machine Learning. In Bayesian Reasoning and Machine Learning. doi:10.1017/CBO9780511804779

Bassi, S. (2007). A primer on Python for life science researchers. In PLoS Computational Biology, 3(11). doi:10.1371/journal.pcbi.0030199

Bickler, S. H. (2021). Machine Learning Arrives in Archaeology. In Advances in Archaeological Practice, 9(2). doi:10.1017/aap.2021.6

Bilina, R., & Lawford, S. (2012). Python for Unified Research in Econometrics and Statistics. *Econometric Reviews*, *31*(5), 558–591. doi:10.1080/07474938.2011.553573

Bonald, T., De Lara, N., Lutz, Q., & Charpentier, B. (2020). Scikit-network: Graph analysis in python. *Journal of Machine Learning Research*, 21.

Borcherds, P. H. (2007). Python: a language for computational physics. *Computer Physics Communications, 177*(1-2 SPEC. ISS.). doi:10.1016/j.cpc.2007.02.019

Cao, R., Tu, W., Cai, J., Zhao, T., Xiao, J., Cao, J., Gao, Q., & Su, H. (2022). Machine learning-based economic development mapping from multi-source open geospatial data. *ISPRS Annals of the Photogrammetry, Remote Sensing and Spatial Information Sciences*, *5*(4), 259–266. doi:10.5194/isprs-annals-V-4-2022-259-2022

Carleo, G., Cirac, I., Cranmer, K., Daudet, L., Schuld, M., Tishby, N., Vogt-Maranto, L., & Zdeborová, L. (2019). Machine learning and the physical sciences. *Reviews of Modern Physics*, *91*(4), 045002. doi:10.1103/RevModPhys.91.045002

Chai, X., Yang, T., Gu, H., Tang, G., Cao, W., & Wang, Y. (2023). Geophysics-steered self-supervised learning for deconvolution. *Geophysical Journal International, 234*(1), 40–55. doi:10.1093/gji/ggad015

Cheng, Q., Zhu, Y., Zeng, H., Song, J., Wang, S., Zhang, J., Qian, L., & Qi, Y. (2021). A method for identifying geospatial data sharing websites by combining multi-source semantic information and machine learning. *Applied Sciences (Basel, Switzerland)*, *11*(18), 8705. doi:10.3390/app11188705

Dalla Mura, M., Prasad, S., Pacifici, F., Gamba, P., Chanussot, J., & Benediktsson, J. A. (2015). Challenges and Opportunities of Multimodality and Data Fusion in Remote Sensing. *Proceedings of the IEEE, 103*(9), 1585–1601. doi:10.1109/JPROC.2015.2462751

Dodd, L., Muir, R., Noll, B., Cubero, A. R., Solodkov, N., & Koraktis, K. (2023). State of the Developer Nation Q3 2023. In SlashData (Issue August).

Du, P., Bai, X., Tan, K., Xue, Z., Samat, A., Xia, J., Li, E., Su, H., & Liu, W. (2020). Advances of Four Machine Learning Methods for Spatial Data Handling: a Review. In Journal of Geovisualization and Spatial Analysis, 4(1). doi:10.1007/s41651-020-00048-5

Dubois, P. F. (1997). Is Java for Scientific Programming? *Computers in Physics*, *11*(6), 611–617. doi:10.1063/1.4822611

FoliniA.LenziE.BiraghiC. A. (2022). CLUSTER ANALYSIS: A COMPREHENSIVE AND VERSATILE QGIS PLUGIN FOR PATTERN RECOGNITION IN GEOSPATIAL DATA. *International Archives of the Photogrammetry, Remote Sensing and Spatial Information Sciences - ISPRS Archives, 48*(4/W1-2022). doi:10.5194/isprs-archives-XLVIII-4-W1-2022-151-2022

Forghani, A., Kazemi, S., & Bruce, D. (2021). A Machine-Learning Approach to Generalisation of GIS Data. *International Journal of Geoinformatics*, *17*(2), 41–59. doi:10.52939/ijg.v17i2.1757

Gao, K., Mei, G., Piccialli, F., Cuomo, S., Tu, J., & Huo, Z. (2020). Julia language in machine learning: Algorithms, applications, and open issues. In Computer Science Review (Vol. 37). doi:10.1016/j.cosrev.2020.100254

Gillies, S. (2023a). *The fiona user manual*. Read the Docs. https://fiona.readthedocs.io/en/stable/manual.html

Golubev, A., Chechetkin, I., Parygin, D., Sokolov, A., & Shcherbakov, M. (2016). Geospatial Data Generation and Preprocessing Tools for Urban Computing System Development. *Procedia Computer Science*, *101*, 217–226. doi:10.1016/j.procs.2016.11.026

Guillame-Bert, M., Bruch, S., Stotz, R., & Pfeifer, J. (2023). Yggdrasil Decision Forests: A Fast and Extensible Decision Forests Library. *Proceedings of the ACM SIGKDD International Conference on Knowledge Discovery and Data Mining*. ACM. 10.1145/3580305.3599933

Hall, O., Ohlsson, M., & Rögnvaldsson, T. (2022). A review of explainable AI in the satellite data, deep machine learning, and human poverty domain. In Patterns, 3(10). Cell Press. doi:10.1016/j.patter.2022.100600

Hao, J., & Ho, T. K. (2019). Machine Learning Made Easy: A Review of Scikit-learn Package in Python Programming Language. In Journal of Educational and Behavioral Statistics, 44(3). doi:10.3102/1076998619832248

Hoffimann, J., Zortea, M., de Carvalho, B., & Zadrozny, B. (2021). Geostatistical Learning: Challenges and Opportunities. *Frontiers in Applied Mathematics and Statistics*, *7*, 689393. doi:10.3389/fams.2021.689393

Imambi, S., Prakash, K. B., & Kanagachidambaresan, G. R. (2021). PyTorch. *Programming with TensorFlow: Solution for Edge Computing Applications*, 87–104.

Jalil, N. A., Hwang, H. J., & Dawi, N. M. (2019). Machines learning trends, perspectives and prospects in education sector. *ACM International Conference Proceeding Series*, (pp. 201–205). ACM. 10.1145/3345120.3345147

Jiang, S., Bevacqua, E., & Zscheischler, J. (2022). River flooding mechanisms and their changes in Europe revealed by explainable machine learning. *Hydrology and Earth System Sciences*, *26*(24), 6339–6359. doi:10.5194/hess-26-6339-2022

Jiang, Y., Li, Y., Yang, C., Hu, F., Armstrong, E. M., Huang, T., Moroni, D., McGibbney, L. J., & Finch, C. J. (2018). Towards intelligent geospatial data discovery: A machine learning framework for search ranking. *International Journal of Digital Earth*, *11*(9), 956–971. doi:10.1080/17538947.2017.1371255

Jordan, M. I., & Mitchell, T. M. (2015). Machine learning: Trends, perspectives, and prospects. In Science, 349(6245). doi:10.1126/science.aaa8415

Karpatne, A., Ebert-Uphoff, I., Ravela, S., Babaie, H. A., & Kumar, V. (2019). Machine Learning for the Geosciences: Challenges and Opportunities. *IEEE Transactions on Knowledge and Data Engineering*, *31*(8), 1544–1554. doi:10.1109/TKDE.2018.2861006

Kiwelekar, A. W., Mahamunkar, G. S., Netak, L. D., & Nikam, V. B. (2020). *Deep Learning Techniques for Geospatial Data Analysis*. doi:10.1007/978-3-030-49724-8_3

Kliegr, T., Bahník, Š., & Fürnkranz, J. (2020). Advances in Machine Learning for the Behavioral Sciences. *The American Behavioral Scientist*, *64*(2), 145–175. doi:10.1177/0002764219859639

Kolesnikov, A., Kikin, P., & Panidi, E. (2021). Modelling of traffic flows and supply chains based on geospatial knowledge. *Journal of Physics: Conference Series*, *2068*(1), 012042. doi:10.1088/1742-6596/2068/1/012042

Kölle, M., Walter, V., Schmohl, S., & Soergel, U. (2020). Hybrid Acquisition of High Quality Training Data for Semantic Segmentation of 3D Point Clouds Using Crowd-Based Active Learning. *ISPRS Annals of the Photogrammetry, Remote Sensing and Spatial Information Sciences*, *5*(2), 501–508. doi:10.5194/isprs-annals-V-2-2020-501-2020

Korosov, A. A., Hansen, M. W., Dagestad, K.-F., Yamakawa, A., Vines, A., & Riechert, M. (2016). Nansat: A Scientist-Orientated Python Package for Geospatial Data Processing. *Journal of Open Research Software*, *4*(1), 39. doi:10.5334/jors.120

Lampa, S., Dahlö, M., Alvarsson, J., & Spjuth, O. (2019). Scipipe: A workflow library for agile development of complex and dynamic bioinformatics pipelines. *GigaScience*, *8*(5). doi:10.1093/gigascience/giz044 PMID:31029061

Lary, D. J., Alavi, A. H., Gandomi, A. H., & Walker, A. L. (2016). Machine learning in geosciences and remote sensing. *Geoscience Frontiers*, *7*(1), 3–10. doi:10.1016/j.gsf.2015.07.003

Lee, T., Singh, V. P., & Cho, K. H. (2021). *Tensorflow and Keras Programming for Deep Learning*. doi:10.1007/978-3-030-64777-3_11

Li, M., Perrier, E., & Xu, C. (2019). Deep hierarchical graph convolution for election prediction from geospatial census data. *33rd AAAI Conference on Artificial Intelligence, AAAI 2019, 31st Innovative Applications of Artificial Intelligence Conference, IAAI 2019 and the 9th AAAI Symposium on Educational Advances in Artificial Intelligence, EAAI 2019*. IEEE. 10.1609/aaai.v33i01.3301647

Li, W. (2020). GeoAI: Where machine learning and big data converge in GIScience. *Journal of Spatial Information Science*, *20*(20). doi:10.5311/JOSIS.2020.20.658

Li, Y., Jiang, Y., Yang, C., Yu, M., Kamal, L., Armstrong, E. M., Huang, T., Moroni, D., & McGibbney, L. J. (2020). Improving search ranking of geospatial data based on deep learning using user behavior data. *Computers & Geosciences*, *142*, 104520. doi:10.1016/j.cageo.2020.104520

McInerney, D., & Kempeneers, P. (2015). Open source geospatial tools: Applications in Earth observation. In *Open Source Geospatial Tools*. Applications in Earth Observation., doi:10.1007/978-3-319-01824-9_20

Meeds, E., Hendriks, R., Al Faraby, S., Bruntink, M., & Welling, M. (2015). MLitB: Machine learning in the browser. *PeerJ Computer Science, 2015*(7). doi:10.7717/peerj-cs.11

Ndossi, M. I., & Avdan, U. (2016). Application of open source coding technologies in the production of Land Surface Temperature (LST) maps from Landsat: A PyQGIS plugin. *Remote Sensing (Basel)*, *8*(5), 413. doi:10.3390/rs8050413

Oviedo, F., Ferres, J. L., Buonassisi, T., & Butler, K. T. (2022). Interpretable and Explainable Machine Learning for Materials Science and Chemistry. *Accounts of Materials Research*, *3*(6), 597–607. doi:10.1021/accountsmr.1c00244

Padmanabhan, A., Ziao, X., Vandewalle, R. C., Baig, F., Michel, A., Li, Z., & Wang, S. (2021). CyberGIS-compute for enabling computationally intensive geospatial research. *Proceedings of the 3rd ACM SIGSPATIAL International Workshop on APIs and Libraries for Geospatial Data Science, SpatialAPI 2021 - In Conjunction with the 29th International Conference on Advances in Geographic Information Systems, ACM SIGSPATIAL 2021*. ACM. 10.1145/3486189.3490017

Paluszek, M., Thomas, S., & Ham, E. (2022). MATLAB Machine Learning Toolboxes. In Practical MATLAB Deep Learning. doi:10.1007/978-1-4842-7912-0_2

Pang, B., Nijkamp, E., & Wu, Y. N. (2020). Deep Learning With TensorFlow: A Review. In Journal of Educational and Behavioral Statistics, 45(2). doi:10.3102/1076998619872761

Pereira, G. W., Valente, D. S. M., de Queiroz, D. M., Coelho, A. L. de F., Costa, M. M., & Grift, T. (2022). Smart-Map: An Open-Source QGIS Plugin for Digital Mapping Using Machine Learning Techniques and Ordinary Kriging. *Agronomy (Basel)*, *12*(6), 1350. doi:10.3390/agronomy12061350

Pierdicca, R. (2021). *GeoAI:artificial intelligence for interpretation and processing of complex geomatic data*. EGU General Assembly.

Pinte, D., Jones, E., Kern, R., & Berkes, P. (2012). Python for geophysical data processing and visualization. *74th EAGE Conference and Exhibition Incorporating SPE EUROPEC 2012*. IEEE. 10.3997/2214-4609.20149884

Read, J., Reutemann, P., Pfahringer, B., & Holmes, G. (2016). MEKA: A multi-label/multi-target extension to WEKA. *Journal of Machine Learning Research*, 17.

Reichstein, M., Camps-Valls, G., Stevens, B., Jung, M., Denzler, J., Carvalhais, N., & Prabhat. (2019). Deep learning and process understanding for data-driven Earth system science. *Nature*, *566*(7743), 195–204. doi:10.1038/s41586-019-0912-1 PMID:30760912

Rey, S. J., Anselin, L., Li, X., Pahle, R., Laura, J., Li, W., & Koschinsky, J. (2015). Open geospatial analytics with PySAL. *ISPRS International Journal of Geo-Information*, *4*(2), 815–836. doi:10.3390/ijgi4020815

Roscher, R., Bohn, B., Duarte, M. F., & Garcke, J. (2020). Explainable Machine Learning for Scientific Insights and Discoveries. *IEEE Access : Practical Innovations, Open Solutions*, *8*, 42200–42216. doi:10.1109/ACCESS.2020.2976199

Rozemberczki, B., Scherer, P., He, Y., Panagopoulos, G., Riedel, A., Astefanoaei, M., Kiss, O., Beres, F., López, G., Collignon, N., & Sarkar, R. (2021). PyTorch Geometric Temporal: Spatiotemporal Signal Processing with Neural Machine Learning Models. *International Conference on Information and Knowledge Management, Proceedings*. ACM. 10.1145/3459637.3482014

Saibi, H., Bersi, M., Mia, M. B., Saadi, N. M., Al Bloushi, K. M. S., & Avakian, R. W. (2018). Applications of Remote Sensing in Geoscience. In Recent Advances and Applications in Remote Sensing. IEEE. doi:10.5772/intechopen.75995

Santosuosso, U., Cini, A., & Papini, A. (2020). Tracing outliers in the dataset of Drosophila suzukii records with the Isolation Forest method. *Journal of Big Data, 7*(1), 14. doi:10.1186/s40537-020-00288-8

Soliman, A., & Terstriep, J. (2019). Keras Spatial Extending deep learning frameworks for preprocessing and on-the-fly augmentation of geospatial data. *Proceedings of the 3rd ACM SIGSPATIAL International Workshop on AI for Geographic Knowledge Discovery, GeoAI 2019*. ACM. 10.1145/3356471.3365240

Stewart, A. J., Robinson, C., Corley, I. A., Ortiz, A., Ferres, J. M. L., & Banerjee, A. (2022). TorchGeo: deep learning with geospatial data. *GIS: Proceedings of the ACM International Symposium on Advances in Geographic Information Systems*. ACM. 10.1145/3557915.3560953

Tamiminia, H., Salehi, B., Mahdianpari, M., Quackenbush, L., Adeli, S., & Brisco, B. (2020). Google Earth Engine for geo-big data applications: A meta-analysis and systematic review. In *ISPRS Journal of Photogrammetry and Remote Sensing* (Vol. 164, pp. 152–170). Elsevier B.V., doi:10.1016/j.isprsjprs.2020.04.001

Terstriep, J. (2023). *Keras Spatial*. https://pypi.org/project/keras-spatial/#description

Ueckermann, M. P., Bieszczad, J., Entekhabi, D., Shapiro, M. L., Callendar, D. R., Sullivan, D., & Milloy, J. (2020). PODPAC. Open-source Python software for enabling harmonized, plug-and-play processing of disparate earth observation data sets and seamless transition onto the serverless cloud by earth scientists. *Earth Science Informatics, 13*(4), 1507–1521. doi:10.1007/s12145-020-00506-0

Valentini, G., & Masulli, F. (2002). NEURObjects: An object-oriented library for neural network development. In *Neurocomputing, 48,* 1–4. doi:10.1016/S0925-2312(01)00632-4

Vatsavai, R. R., & Bhaduri, B. (2011). A hybrid classification scheme for mining multisource geospatial data. *GeoInformatica, 15*(1), 29–47. doi:10.1007/s10707-010-0113-4

Verma, S., Gupta, S., Shin, H., Panigrahi, A., Goswami, S., Pardeshi, S., Exe, N., Dutta, U., Joshi, T. R., & Bhojwani, N. (2022). GeoEngine: A Platform for Production-Ready Geospatial Research. *Proceedings of the IEEE Computer Society Conference on Computer Vision and Pattern Recognition, 2022-June*. IEEE. 10.1109/CVPR52688.2022.02073

Vopham, T., Hart, J. E., Laden, F., & Chiang, Y. Y. (2018). Emerging trends in geospatial artificial intelligence (geoAI): Potential applications for environmental epidemiology. In Environmental Health: A Global Access Science Source, 17(1). doi:10.1186/s12940-018-0386-x

Whigham, P. A., McKay, R. I., & Davis, J. R. (1992). Machine induction of geospatial knowledge. Lecture Notes in Computer Science (Including Subseries Lecture Notes in Artificial Intelligence and Lecture Notes in Bioinformatics), 639 LNCS. doi:10.1007/3-540-55966-3_24

Wu, Q. (2020). geemap: A Python package for interactive mapping with Google Earth Engine. *Journal of Open Source Software*, *5*(51), 2305. doi:10.21105/joss.02305

Wu, Q. (2021a). Interactive mapping and geospatial analysis with Leafmap and Jupyter. *Proceedings of the 3rd ACM SIGSPATIAL International Workshop on APIs and Libraries for Geospatial Data Science, SpatialAPI 2021 - In Conjunction with the 29th International Conference on Advances in Geographic Information Systems, ACM SIGSPATIAL 2021*. ACM. 10.1145/3486189.3490015

Wu, Q. (2021b). Leafmap: A Python package for interactive mapping and geospatial analysis with minimal coding in a Jupyter environment. *Journal of Open Source Software*, *6*(63), 3414. doi:10.21105/joss.03414

Xie, Y., Eftelioglu, E., Ali, R. Y., Tang, X., Li, Y., Doshi, R., & Shekhar, S. (2017). Transdisciplinary foundations of geospatial data science. *ISPRS International Journal of Geo-Information*, *6*(12), 395. doi:10.3390/ijgi6120395

Yadav, A., Saraswat, S., & Faujdar, N. (2022). Geological Information Extraction from Satellite Imagery Using Machine Learning. *2022 10th International Conference on Reliability, Infocom Technologies and Optimization (Trends and Future Directions), ICRITO 2022*. IEEE. 10.1109/ICRITO56286.2022.9964623

Yan, X., Liu, S., Liu, W., & Dai, Q. (2022). An improved coverage-oriented retrieval algorithm for large-area remote sensing data. *International Journal of Digital Earth*, *15*(1), 606–625. doi:10.1080/17538947.2022.2030816

Zhang, Z., & Sejdić, E. (2019). Radiological images and machine learning: Trends, perspectives, and prospects. In *Computers in Biology and Medicine* (Vol. 108, pp. 354–370). Elsevier Ltd. doi:10.1016/j.compbiomed.2019.02.017

Chapter 11
Big Data Analytics for Geospatial Application Using Python

Assefa Senbato Genale
https://orcid.org/0000-0002-7411-9456
Bule Hora University, Ethiopia

ABSTRACT

Numerous organizations regularly produce enormous volumes of geospatial data due to the widespread use of sensors and location-based services. However, traditionally collecting, storing, managing, exploring, analyzing, and visualization of geospatial data has been a complex and time-consuming task. This study proposed a big data analytics approach to collect, store, manage, explore, process, and analyze massive amounts of geospatial data. A comprehensive literature review, various Python libraries for geospatial big data, challenges in geospatial big data analytics, and big data analytics techniques such as spatial clustering, spatial regression analysis, and spatial-temporal analysis, were presented. In addition, geospatial big data analytics algorithms like K-means clustering, ordinary least squares (OLS), geographically weighted regression (GWR), Spatio-temporal clustering algorithms, Spatio-temporal regression models, and others were discussed. Finally, case studies on performing geospatial big data analytics using Pyspark were addressed.

INTRODUCTION

The amount of data being produced at this time is increasing at an unprecedented rate because of the digital revolution. With the process of gathering real-time data using

DOI: 10.4018/979-8-3693-6381-2.ch011

sensor devices, geoinformation systems produce vast and complicated geospatial data(Li, Tang, Huang, Shook, & Guan, 2020). Big data is a term used to describe datasets that are too huge, complex, or exceed standard datasets in terms of their ability to be collected, stored, managed, and analyzed. Geospatial applications are one domain in which big data analytics has been demonstrated to be very beneficial. (Pappula & Cheetoori, 2016). Big data analytics can transform geospatial science by harnessing the power of big data analytics, allowing businesses to gain valuable insights and make informed decisions based on spatial information.

Urban planning, environmental management, transportation and logistics, agriculture, natural resource management, disaster management, and public health are just a few industries that frequently use geospatial applications(Lanka, 2023). Combining several types of geospatial data, such as satellite imagery, aerial pictures, GPS data, and digital maps, allows users to comprehend spatial relationships, patterns, and trends. However, analyzing geospatial data has traditionally been a complex and time-consuming task. Typically, it relies on manual interpretation or rule-based methods, which are time-consuming and incapable of handling enormous datasets(Roy, Fouché, Morales, & Möhler, 2019). Big data analytics and potent programming languages like Python have made this procedure more accessible and efficient. Due to technological improvements, geospatial data is becoming more and more prevalent, so it is essential to create effective techniques for processing and deriving insights from this data. Big data analytics approaches can find patterns, trends, and relationships in geographical datasets that may not be visible through conventional analysis techniques(Tamiminia, Salehi, Mahdianpari, Quackenbush, & Adeli, 2020).

To effectively provide relevant real-time analysis on the enormous amount and heterogeneity of spatial data, big data analytics (BDA) leveraging open-source frameworks like SpatialHadoop and GeoSpark are crucial(Lenka et al., 2017). Big data provides many techniques and algorithms, but choosing and using the appropriate ones to address issues in many real-world industries, including geospatial applications, can be difficult. Apache Spark and Hadoop are the two most widely used big data processing and computing solutions. Hadoop MapReduce has several flaws; one is that it only supports batch processing and is unsuitable for iterative, interactive, or real-time stream processing(Senbato, 2019). Apache Spark is a general-purpose data processing engine that can be placed on top of Hadoop to improve in-memory computation for batch and stream data processing, which is difficult for MapReduce. Because of its flexibility, simplicity, and robust libraries, Python is a well-liked programming language for geospatial data processing.

This chapter mainly aims to provide an in-depth understanding of big data analytics methods utilizing Python for geospatial applications, providing professionals working with geospatial data with insightful information and useful suggestions. Additionally, the chapter seeks to delve into big data analytics and how it applies to geospatial analysis. It also aims to examine various big data analytics methods and algorithms for geospatial applications and comprehend the opportunities and difficulties of handling and analyzing massive geographical datasets. To offer a comprehensive approach to utilizing Python for big data analytics in geospatial applications. Finally, this chapter aims to contribute to the understanding and advancement of big data analytics techniques specifically tailored for geospatial analysis using Python as a programming language and to provide insights and recommendations for future research and development in the field of big data analytics for geospatial applications.

LITERATURE REVIEW

Big data analytics plays a great role in collecting, storing, managing, processing, and analyzing geospatial big data in various organizations to generate meaningful insights from huge data. This section has reviewed the most recent and relevant literature for applying big data analytics in geospatial analysis.

A geospatial data generating and preprocessing system was proposed(Alexey Golubev, Ilya Chechetkin, Danila Parygin, Alexander Sokolov, 2016) to generate and preprocess urban computing system development. The authors performed a comprehensive review to develop tools for generating and preprocessing geospatial data for urban computing systems. They used tools like scatter, clustering, 0-Network, spatial indexing, and routing to generate geospatial data and perform preprocessing. The authors analyzed existing tools' limitations in generating and preprocessing geospatial data with practical experiments. The study's findings demonstrate the effectiveness of the proposed tools in generating and preprocessing geospatial data for urban computing system development. Finally, the authors recommended technologies like artificial Intelligence, machine learning, the Internet of Things, big data, and others to generate geospatial data and perform preprocessing tasks in advance.

Authors(Lenka et al., 2017) presented the two most popular big data analytics platforms (SpatialHadoop and GeoSpark) and their strengths and weaknesses for geospatial big data analytics. The authors performed a comparative analysis with examples for both platforms to assess their capabilities to analyze geospatial big data. According to the author's suggestion, both platforms have high capabilities for performing analytics for geospatial big data. GeoSpark can perform both batch processing and real-time processing with high speed. SpatialHadoop is limited to batch processing, but it has advantages over GeoSpark. The authors evaluated the performance of both platforms in processing large volumes of geospatial data. In the end, the study showed that both SpatialHadoop and GeoSpark can efficiently process geospatial big data, but each has its strengths and weaknesses depending on the specific analytical tasks.

Authors in(Li, n.d.) proposed a high-performance computing approach to overcome challenges in geospatial big data. The authors conducted a comprehensive literature review to identify current and future directions for collecting, managing, processing, analyzing, and visualizing large-scale geospatial data using high-performance computing techniques. The authors discussed various algorithms used in existing approaches, including parallel processing algorithms for large-scale spatial analysis and distributed storage algorithms for managing massive geospatial data. In addition, the author examined the opportunities and challenges of high-performance computing techniques by reviewing existing high-performance computing platforms and tools for geospatial big data processing. Current technologies like geospatial big data, artificial intelligence, cloud computing, and fog computing are reshaping geospatial data processing. Ultimately, the author recommended current and future directions in leveraging high-performance computing techniques for handling geospatial big data.

In(Ali & Gupta 2017) authors discussed geospatial big data visualization tools to encounter challenges in extracting insight from geospatial big data. This study aimed to investigate and review visualization tools for geospatial big data. The authors conducted experiments using different visualization techniques on various large-scale geospatial data to evaluate the effectiveness of existing visualization tools for geospatial big data. They used the following visualization tools, GeoMesa, GRASS GIS, HadoopViz, ArcMap, and Whitebox GAT, to visualize geospatial data. The authors concluded that while existing visualization tools can handle large-scale geospatial datasets effectively, there is still room for improvement in performance optimization and user interaction. Finally, they recommended various parameters like desktop processing, code availability, online processing, online course availability, mobile client processing, and various API compatibilities according to the requirements.

A big data framework for geographical data analysis in the environment was proposed(Pappula & Cheetoori, 2016) to develop a framework that can handle geographical information using Big Data platforms like Hadoop or Spark. As the amount of geospatial data generated from various sources increases, proposing suitable or appropriate big data platforms for processing geospatial data is the most crucial task. In this study, the authors performed comprehensive reviews of the available big data analytics platforms to collect, manage, process, and analyze geospatial big data easily to extract hidden from the data. The platform used by the authors included distributed computing techniques such as MapReduce algorithms, Hadoop distributed file system, real-time processing, and parallel processing methods.

A long short-term memory modeling (LSTM) deep learning algorithm was proposed(Zhichao Li, Helen Gurgel, Lei Xu, 2022) to improve dengue forecasts by utilizing geospatial big data and historical dengue information. The authors focused on the historical information on dengue epidemics in the federal district of Brazil from 2007 to 2019 for implementation work. The algorithms used for geospatial big data analysis in Google Earth Engine were multi-step-ahead Long Short-Term Memory modeling to predict historical dengue information. In addition, the authors identified challenges faced during predicting dengue forecasts, such as time-effectiveness due to the time-consuming nature of downloading satellite data and processing, performance issues without historical dengue information, and weak spatial representation due to data independence. The study showed improved accuracy in dengue forecasts, demonstrating the effectiveness of utilizing geospatial big data and historical information in LSTM modeling to predict historical dengue information.

In(Kovacs-györi et al., 2020), the authors assessed the prospects and challenges of geospatial analysis for urban livability using machine learning and big data approach. The authors explored the prospects and challenges of using big data and machine learning for geospatial analysis to promote urban livability. They comprehensively reviewed existing literature on geospatial analysis, urban livability, big data, and machine learning. In addition, the authors discussed the limitations of the conventional approach to collect, store, manage, process, and analyze urban information for geospatial analysis. Big data and the machine learning era brought great opportunities for urban livability in geospatial analysis with their powerful tools and algorithms to collect, manage, process, and analyze huge amounts of geospatial big data to extract insight for effective decisions. Finally, the authors emphasized the potential benefits of utilizing geospatial analysis to promote urban livability using big data and machine learning while addressing challenges such as data privacy concerns and algorithm biases.

In(Tamiminia et al., 2020), authors presented a systematic review and a meta-analysis of the Google Earth engine for geospatial big data applications. The authors conducted a systematic review and a meta-analysis of research papers 349 published in the different journals related to the Google Earth engine for geospatial big data application. They discussed the performance of Google Earth Engine in analyzing geospatial big data with various algorithms like machine learning over cloud computing with incredible storage. Random forest and linear regression algorithms were used mostly to process satellite images. Finally, they explored the versatility of Google Earth Engine in handling diverse geospatial big data applications while also identifying areas for future research improvements, such as scalability issues when dealing with extremely large datasets based on data processing.

Distributed Computing Environments for processing geospatial big data was proposed by(A. Olasz a, B. Nguyen Thai b, 2016) authors as a new initiative for collecting, managing, and processing geospatial big data. The authors presented a new approach to managing geospatial big data over distributed platforms or environments with fast data processing speed. They performed practical works to process geospatial big data over distributed computing with demos. In addition, the authors proposed an opportunity available regarding distributed computing tools for processing geospatial data. Big data platforms or frameworks can process geospatial big data over distributed environments with high-speed processing data. Finally, the authors emphasized a new initiative for tiling, stitching, and processing geospatial big data that could significantly improve geospatial big data processing efficiency and scalability.

The potential of geospatial big data analytics for promoting sustainable smart cities was proposed by(Mete, 2023). The authors reviewed current technology, such as Geographic Information Systems, the Internet of Things, cloud computing, and geospatial big data, to collect, manage, process, analyze, and predict enormous amounts of data generated from smart cities. Big data plays a great role in the storage, processing, analytics, and analysis of green energy, green building, and other services in smart cities. The authors designed models for the components of Smart Environment and Smart Governance. In addition, the authors performed comparison analyses for two platforms, big data (Dask-GeoPandas and Apache Sedona), based on their performance of parallel processing systems. The proposed Apache Sedona is a sustainable smart city based on its performance. Ultimately, the authors addressed opportunities for leveraging geospatial analysis and big data analytics to promote sustainable urban development.

Authors (Priyashani, Kankanamge, and Yigitcanlar, 2023) proposed a Multisource Open Geospatial Big Data Fusion approach to develop a method for fusing multisource open geospatial big data to demarcate urban agglomeration footprints. The authors used the Southern Coastal Belt of Sri Lanka as the testbed to validate the abilities of the proposed approach. The algorithm used by the authors involved a fusion technique that combined different types of geospatial data to demarcate urban agglomeration footprints accurately. The study showed that the method effectively created detailed and accurate urban agglomeration footprints by fusing multisource open geospatial big data. Finally, the authors applied a multisource geospatial big data fusion approach to demarcate urban agglomeration footprints and inform urban authorities of emerging appropriate procedures for managing urban growth.

Authors(Koh, Hyder, Karale, & Boulos, 2022) conducted a systematic narrative review for the various published research in different databases like Scopus, Web Science, and others to leverage the use of Spatial Data Infrastructures for Big Geospatial Sensing Data in Public Health. In addition to the authors reviewing existing literature, authors discussed some case studies on the use of spatial data infrastructures for managing big geospatial sensing data in public health applications. The study focused on analyzing how spatial data infrastructures can effectively manage and analyze geospatial big-sensing data for public health applications. They discussed geospatial sensing data related to public health, such as disease outbreak information, environmental monitoring data, and healthcare facility locations. According to their study, spatial data infrastructures are crucial in managing and analyzing big geospatial sensing data for public health purposes.

In(Loukili, Lakhrissi, Elhaj, and Ali, 2022), the authors comprehensively reviewed the geospatial big data platform. According to the authors' study, collecting, storing, managing, processing, analyzing, and extracting insight from big geospatial data

Table 1. Literature review summary

Title	Author/s	Method/s	Result/Gaps
Geospatial Data Generation and Preprocessing Tools for Urban Computing System Development	Alexey Golubev, Ilya Chechetkin, Danila Parygin, Alexander Sokolov, and M. S.	Scatter, clustering, 0-Network, spatial indexing, and routing methods were used.	The study demonstrates the effectiveness of the proposed tools in generating and preprocessing geospatial data for urban computing system development. Current technologies like artificial intelligence, machine learning, the Internet of Things, big data, and others that generate geospatial data and perform preprocessing tasks in advance are recommended.
Geospatial Big Data Analytics Applications Trends, Challenges & Opportunities(	Ali & Gupta	They used the following visualization tools: GeoMesa, GRASS GIS, HadoopViz, ArcMap, and Whitebox GAT.	They used the following visualization tools, GeoMesa, GRASS GIS, HadoopViz, ArcMap, and Whitebox GAT, to visualize geospatial data. The authors concluded that while existing visualization tools can handle large-scale geospatial datasets effectively, there is still room for improvement in performance optimization and user interaction.
Big Data Environment for Geospatial Data	Pappula & Cheetoori	MapReduce algorithms, Hadoop distributed file system, real-time processing, and parallel processing methods.	In this study, the authors performed comprehensive reviews of the available big data analytics platforms to collect, manage, process, and analyze geospatial big data easily to extract hidden from the data. It is better to consider advanced big data tools for geospatial data.
Improving Dengue Forecasts by Using Geospatial Big Data Analysis in Google Earth Engine and the Historical Dengue	Zhichao Li and Helen Gurgel, Lei Xu	Multi-step-ahead Long Short-Term Memory Modeling.	The study showed improved accuracy in dengue forecasts, demonstrating the effectiveness of utilizing geospatial big data and historical information in LSTM modeling to predict historical dengue information. The authors considered only one algorithm.
Opportunities and Challenges of Geospatial Analysis for Promoting Urban Livability in the Era of Big Data and Machine Learning	Kovacs-györi et al.	They comprehensively reviewed existing literature on geospatial analysis, urban livability, big data, and machine learning.	The authors emphasized the potential benefits of utilizing geospatial analysis to promote urban livability using big data and machine learning while addressing challenges such as data privacy concerns and algorithm biases. The simulation or implemented part was missed for further understanding.
Big Data Analytics For Sustainable Smart Cities	Mete, 2023	The proposed Apache Sedona for sustainable smart cities	The authors designed models for the components of Smart Environment and Smart Governance. In addition, the authors performed comparison analysis for two big data platforms (Dask-GeoPandas and Apache Sedona) based on their performance of parallel processing systems. Considering more big data frameworks GeoSpark is recommended for real-time and memory-based processing.

using traditional methods is difficult. Big data provides the most popular and capable platforms to handle challenges related to volume, variety, and speed of geospatial data to extract the right insight for the right decision. The authors evaluated and compared different geospatial big data platforms based on their functionality, scalability, and usability for handling large volumes of geospatial data. They explored several robust geospatial big data platforms with diverse features and capabilities suitable for application domains such as urban planning, environmental monitoring, disaster management, and transportation analysis. Table 1 provides a summary of the literature review.

BIG DATA ANALYTICS FOR GEOSPATIAL DATA

The process of understanding patterns and relationships within a geographic area through analyzing spatial data such as maps, satellite imagery, and GPS coordinates is known as geospatial analysis(Li et al., 2020). Due to the widespread use of sensors and location-based services, organizations often produce enormous volumes of geospatial data(Lawhead, 2015). This data includes information on transportation routes, weather patterns, land use, population density, etc. Analyzing this data at scale requires advanced Big data analytics techniques to derive meaningful insights. Big data analytics(Alfadul & Ali, 2022) refers to examining large and complex data sets to uncover hidden patterns, unknown correlations, market trends, customer preferences, and other useful information that can help organizations make more informed decisions. This type of analysis is particularly valuable in geospatial data, which involves information about the Earth's surface and objects, such as buildings, roads, and natural features. Geospatial big data analytics involves processing and analyzing large volumes of geospatial data to extract meaningful insights that can be used in various applications.

Big data analytics for geospatial analysis offers significant potential for organizations seeking valuable insights from their large-scale geospatial datasets(Lanka, 2023). Companies can use advanced analytic methods like artificial intelligence and machine learning to reveal previously undetected patterns, trends, and connections within their geographic information. The benefits of using big data analytics for geospatial analysis are numerous. By applying advanced analytic tools, like machine learning algorithms and spatial clustering approaches, to extensive geospatial information, organizations can better understand intricate spatial linkages that were previously challenging to discern. This can lead to more informed decision-making processes across various industries, including urban planning, environmental management, disaster response, and management. The Geospatial Big Data Analytics concept is illustrated in Figure 1, sourced from the authors' elaboration.

Figure 1. Geospatial big data analytics concepts
(Source: Authors' Elaboration)

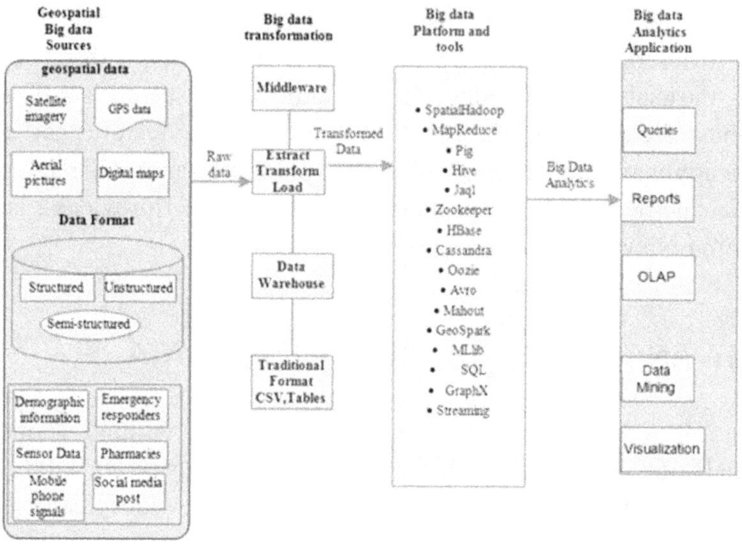

Application of Big Data Analytics in Geospatial Analysis

Urban planning and development(Alexey Golubev, Ilya Chechetkin, Danila Parygin, Alexander Sokolov, 2016) is one of the main fields in which big data analytics is used for geospatial analysis. City planners can better understand how people travel across metropolitan regions, where infrastructure upgrades are needed, and how to maximize public services by evaluating vast amounts of geospatial data(Mete, 2023). For instance, city planners might identify locations with significant congestion and prioritize funding for road building or public transportation by examining traffic patterns and population density statistics.

Environmental monitoring and conservation(Pappula & Cheetoori, 2016) is another area where big data analytics for geospatial analysis is used. Environmental scientists can monitor changes in land cover, deforestation rates, wildlife habitats, and other related issues by examining satellite photos and sensor data. Making educated decisions regarding conservation initiatives and sustainable land use practices requires knowledge of this information. Big data analytics for geospatial analysis also has uses in catastrophe management and response(Alfadul & Ali, 2022). Organizations can more efficiently allocate resources and promptly identify damaged areas during natural disasters or humanitarian crises by evaluating geospatial data in real time from sources like social media feeds or remote sensing platforms.

In addition, big data analytics in geospatial data has broader implications for various industries. For instance, big data analytics can be applied to agriculture to optimize farming operations, boost food production, and lessen environmental impact by analyzing geographical data on crop yields, weather patterns, and soil quality(Alsaedi, Sadeq, & Jalal, 2023; Loukili et al., 2022).

In transportation(Alfadul & Ali, 2022; Saputra & Radam, 2023; Zhao, Wang, & Pham, 2023), big data analytics can be used to examine spatial data on traffic patterns, the use of public transportation, and the state of the roads to increase safety, lessen traffic, and improve urban mobility. To optimize store locations, target marketing campaigns, and raise consumer happiness, big data analytics can be applied to the retail industry to evaluate geographic data on customer demographics, shopping behavior, and rival locations. Overall, big data analytics has the potential to revolutionize our understanding of the world around us by providing valuable insights from large volumes of geospatial data.

Geospatial Big Data Analytics Activities

Some activities of big data analytics in geospatial data are presented in the following diagram. Figure 2 depicts the activities of Geospatial Big Data Analytics, which is sourced from the authors' elaboration.

Figure 2. Geospatial big data analytics activities
(Source: Authors' Elaboration)

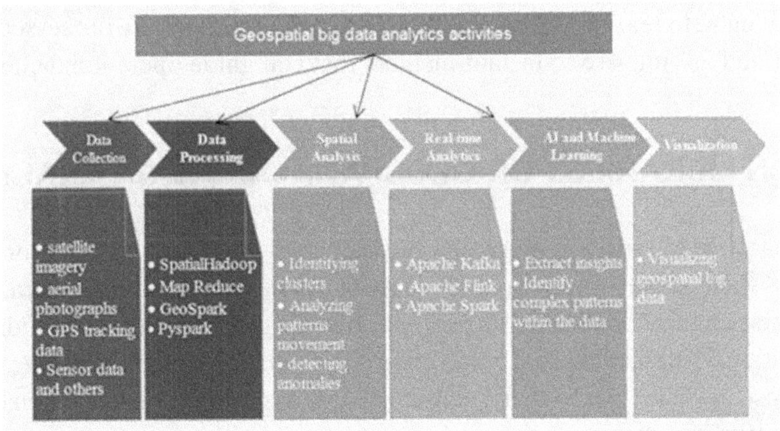

Data Collection: collecting pertinent geospatial data from several sources is the initial stage of geospatial big data analytics. This can include sensor readings from Internet of Things devices, satellite photography, aerial imagery, GPS tracking information, and location-tagged social media posts. Robust systems are needed for collecting and storing geospatial data due to its vast volume and diversity.

Data Processing: The geographical data must be analyzed to extract valuable insights after gathering it. This entails removing any flaws or inconsistencies from the data and formatting it so that it may be readily evaluated. Integrating many datasets to produce a holistic view of a certain region or phenomenon is another aspect of data processing.

Spatial Analysis: Spatial analysis techniques play a crucial role in geospatial big data analytics as they help uncover patterns and linkages within the data. This may entail locating activity hubs in a particular region, tracking movement patterns over time, or seeing anomalies pointing to possible problems or possibilities.

Visualization: Visualizing geospatial big data is crucial for making sense of complex patterns and trends. Maps are often used as a primary visualization tool for geospatial data analysis because they provide an intuitive way to understand spatial relationships. Interactive maps can be used to overlay different layers of information and explore how they interact.

Machine Learning and AI: With the increasing volume of geospatial big data being generated, machine learning and artificial intelligence (AI) techniques are becoming essential for extracting insights from this vast amount of information. These technologies can help identify complex patterns within the data that may not be immediately apparent through traditional analysis methods.

Real-time analytics: Making prompt decisions in several applications, like disaster relief or transportation management, depends on real-time analysis of large geospatial data. In real-time analytics, geospatial data streams are processed as they are generated, giving rise to instant insights that can guide operational choices.

PYTHON LIBRARIES FOR GEOSPATIAL BIG DATA ANALYTICS

Python has a wide range of geospatial big data analytics libraries used for various purposes such as mapping, spatial analysis, and visualization(Westra, 2010). These libraries provide different functionalities for geospatial data, including data manipulation, visualization, and performing analysis using scalable machine learning algorithms(Lawhead, 2015). This section will discuss some popular Python libraries for geospatial big data analytics and examples of their usage. Some of Python's most popular libraries for geospatial data analysis are GeoPandas, Pyspark, H3-Py, Disk-Geospatial, Rasterio, and GeoAlchemy.

GeoPandas: an open-source library that provides a high-level, easy-to-use interface for working with geospatial data in Python. It extends the popular Pandas

Figure 3. Reading Shapefile using Geopandas
(Source: Authors' Demo)

```
#read shape file using geopanda
geo_df = gpd.read_file('/content/ne_110m_admin_0_countries.shp')
```

library to include geospatial data types and operations, making it easy to manipulate and analyze large geospatial datasets. GeoPandas also integrates with other Python libraries for advanced geospatial analytics. Figure 3 demonstrates the process of reading Shapefile using Geopandas.

PySpark: a powerful library for big data processing and analytics, including geospatial data. It provides a distributed computing framework that allows users to process large-scale geospatial datasets in parallel, making it suitable for handling big data analytics tasks. Figure 4 demonstrates the application of PySpark for handling geospatial data.

H3-Py: a Python binding for the H3 library, which provides a hierarchical hexagonal grid indexing system for geospatial data. This library is particularly useful for big data analytics tasks that require spatial indexing and aggregation of large-scale geospatial datasets. Figure 5 illustrates an example of using H3-Py with PySpark.

Dask-GeoSpatial: an extension of Dask, a parallel computing library in Python that supports the distributed processing of large-scale geospatial datasets. It allows users to perform complex geospatial analytics tasks on big data using Dask's parallel computing capabilities. Figure 6 showcases an example utilizing the Dask-geospatial Python library for handling geospatial big data.

Rasterio: a Python library for reading and writing raster datasets, making it suitable for processing and analyzing large-scale satellite imagery and other raster-based geospatial data. It provides efficient tools for working with big raster datasets

Figure 4. An Example of Pyspark for Geospatial Data
(Source: Authors' Demo)

```
[11] from pyspark.sql import SparkSession
     from pyspark.sql.functions import col
     import geopandas as gpd

     # Create a Spark session
     spark = SparkSession.builder.appName("geospatial-analytics").getOrCreate()

     # Read a shapefile into a Spark DataFrame
     df = spark.read.format("shapefile").load('/content/ne_110m_admin_0_countries.shp')

     # Perform spatial join with another dataset
     joined_df = df.join(geo_df, col("geometry").contains(geo_df.geometry))
```

Figure 5. An Example of H3-Py with Pyspark
(Source: Authors' Demo)

```
[46] from pyspark.sql import SparkSession, functions as F
     import h3_pyspark

     spark = SparkSession.builder.getOrCreate()
     df = spark.createDataFrame([{"lat": 37.769377, "lng": -122.388903, 'resolution': 9}])

[48] df = df.withColumn('h3_9', h3_pyspark.geo_to_h3('lat', 'lng', 'resolution'))
     df.show()

     +---------+-----------+----------+---------------+
     |      lat|        lng|resolution|           h3_9|
     +---------+-----------+----------+---------------+
     |37.769377|-122.388903|         9|89283082e73ffff|
     +---------+-----------+----------+---------------+
```

using its integration with NumPy and other scientific computing libraries. Figure 7 demonstrates an example of using Rasterio.

GeoAlchemy: an extension of SQLAlchemy, a popular SQL toolkit and Object-Relational Mapping (ORM) library in Python, that adds support for working with geospatial databases such as PostGIS or SpatiaLite. It enables users to perform complex spatial queries and analytics on big geospatial datasets stored in relational databases.

Figure 8 depicts an example of employing H3-Py in conjunction with PySpark.

BIG DATA ANALYTICS TECHNIQUES FOR GEOSPATIAL APPLICATIONS

Big data analytics techniques for geospatial data using Python involve various libraries and tools to process, store, analyze, and visualize large volumes of geospatial data. Python is a popular programming language for big data analytics due to its

Figure 6. An example of Dask-geospatial Python Library for Geospatial big data
(Source: Authors' Demo)

```
import dask.dataframe as dd
import dask_geopandas as dask_gpd

# Read a shapefile into a Dask DataFrame
ddf = dask_gpd.read_file('/content/ne_110m_admin_0_countries.shp')

# Perform spatial join with another dataset in parallel
joined_ddf = ddf.spatial.join(geo_df)
```

Figure 7. An example of Rasterio
(Source: Authors' Demo)

```
import rasterio
# Open a raster dataset
with rasterio.open('path_to_raster.tif') as src:
    # Read the raster data into a NumPy array
    array = src.read()

    # Perform analysis on the raster data using NumPy operations
    result = array.mean()
```

simplicity, flexibility, and extensive library support(Lawhead, 2015)(Rey et al., 2015). Several Python libraries, including GeoPandas, Pyspark, Shapely, Fiona, and Folium, are commonly used for geospatial big data analytics. Various big data analytics techniques are used for geospatial applications. This section discusses the most common big data analytics techniques, such as spatial indexing, spatial clustering, spatial regression analysis, and spatial-temporal analysis.

Spatial indexing Techniques: a fundamental method in big data analytics for geospatial data(Alsaedi et al., 2023). It is a method for organizing geospatial data to optimize queries based on location. This technique is essential for efficiently processing large volumes of geospatial data. Spatial Widely used in various domains, including Geographic Information Systems (GIS), location-based services, spatial databases, and spatial data mining, Python provides several libraries that support spatial indexing such as R-tree, K-d tree, Geohash, Spatial Hashing, and others(Zeng, Zhang, & Luo, 2022).

R-tree: is a tree data structure optimized for indexing multi-dimensional spatial data. It organizes the spatial objects into a hierarchy of bounding rectangles, with

Figure 8. An example of H3-Py with Pyspark
(Source: Authors' Demo)

```
from sqlalchemy import create_engine
from sqlalchemy.orm import sessionmaker
from geoalchemy2 import Geometry

# Create an engine and session
engine = create_engine('postgresql://user:password@localhost/mydatabase')
Session = sessionmaker(bind=engine)
session = Session()

# Define a spatial table
class PointTable(Base):
    __tablename__ = 'point_table'
    id = Column(Integer, primary_key=True)
    geom = Column(Geometry('POINT'))

# Create the table in the database
Base.metadata.create_all(engine)
```

each node encompassing a set of children. R-trees are suitable for representing spatial data with varying shapes and sizes.

Spatial Hashing: This technique involves dividing the space into a grid of cells and mapping the spatial objects to their respective cells based on their coordinates. Spatial hashing can be efficient for proximity searches and spatial aggregations.

K-d tree: is a binary tree structure that organizes spatial data in k-dimensional space. It partitions the space into regions and assigns data points to different nodes based on their spatial coordinates. K-d trees are often used for nearest-neighbor searches and range queries.

Spatial Clustering Techniques: Another important technique in big data analytics for geospatial data is spatial clustering. Spatial clustering(Priambodo et al., 2018) involves grouping similar geographic features based on their proximity or other spatial characteristics. This technique can identify patterns or hotspots within the geospatial data. Python provides several libraries that support spatial clustering algorithms, such as K-means Clustering and DBSCAN Clustering.

Spatial clustering with K-Means clustering is a popular unsupervised machine-learning algorithm for clustering data points into a specified number of groups or clusters(Priambodo et al., 2018). It can also organize spatial data into clusters in spatial indexing. The algorithm works by iteratively assigning data points to the nearest cluster centroid and then recalculating the centroids based on the assigned data points. The process continues until the centroids no longer change significantly.

How Does a K-Means Clustering Algorithm Work?

Algorithm: K-means clustering algorithm

Input: Sample of spatial data

Step 1: Initialize k(randomly select) cluster centroids (C1, C2, ..., Ck) in the space.

Step 2: Assign each data point (Xi) to the nearest cluster centroid based on a distance metric (e.g., Euclidean distance).

Step 3: Recalculate the cluster centroids based on the assigned data points:

$$Cj= (1/Nj)*\sum Xi \quad (1)$$

Where Nj is the number of data points assigned to the jth cluster.

Step 4: Repeat steps 2 and 3 until convergence (i.e., when the cluster centroids no longer change significantly).

Table 2. Comparison between two spatial clustering

	K-means Clustering	DBSCAN clustering
Advantages	Simple and easy to implement Efficient for large datasets Works well with spherical clusters	Can identify arbitrarily shaped clusters Robust to noise and outliers Does not require specifying the number of clusters beforehand
Drawback	Sensitive to initial centroid selection Assumes equal variance among clusters May not perform well with non-linear or irregularly shaped clusters	Sensitivity to parameter selection (epsilon and MinPts) Computationally intensive for large datasets May struggle with varying density clusters

Spatial clustering with DBSCAN(Density-Based Spatial Clustering of Applications with Noise) clustering: It is another popular method for spatial clustering that groups points that are closely packed together while marking outliers as noise(Huang, Ma, Liu, & Liu, 2022). Unlike K-means, DBSCAN does not require the number of clusters to be predefined, and it is based on the density of data points within the space. The algorithm defines two parameters, epsilon (ε) and minimum points (MinPts), and then identifies core points, border points, and noise points based on their density within the defined neighborhood.

The DBSCAN algorithm can be summarized as follows:

Step 1: Find the core points and their corresponding clusters based on the MinPts and Epsilon parameters.

Step 2: Expand the clusters by assigning border points to the cluster of their respective core points.

Step 3: Label the noise points as outliers.

Table 2 outlines a comparison between two spatial clustering methods.

Spatial Regression Analysis Techniques: Another big data analytics technique used to model the relationship between a dependent variable and one or more independent variables that are geographically referenced(Xu & Zhang, 2023). This method accounts for the spatial autocorrelation in the data, meaning that nearby locations may have values similar to those of the studied variables. These methods allow for incorporating spatial weight matrices that capture the degree of spatial dependence among observations. Spatial regression analysis has applications in geography, economics, and environmental science. For instance, it can be used to model the relationship between housing prices and neighborhood characteristics or to analyze patterns of land use change in response to environmental factors. There are several methods for conducting spatial regression analysis: Ordinary Least Squares (OLS) and Geographically Weighted Regression (GWR).

Ordinary Least Squares (OLS) is a statistical method used to estimate the relationship between a dependent variable and one or more independent variables(Isazade, Baser, Pinliang, Gordana, & Esmail, 2023). It is a widely used technique in econometrics, finance, and other fields to analyze and predict the behavior of a dependent variable based on the values of independent variables.

The OLS method aims to minimize the sum of the squared differences between the observed values of the dependent variable and the values predicted by the regression equation.

The regression equation:

$$Y = \beta 0 + \beta 1X1 + \beta 2X2 + \ldots + \beta nXn + \varepsilon \quad (2)$$

Where: Y = Dependent variable, X1, X2, ..., Xn = Independent variables, β0, β1, ..., βn = Coefficients, ε = Error term. The coefficients (β0, β1, ..., βn) are estimated using OLS to find the best-fitting line that minimizes the sum of squared errors. The OLS method provides estimates for these coefficients that best fit the data and can be used to make predictions about the dependent variable.

Geographically Weighted Regression (GWR): It is a more advanced method for conducting spatial regression analysis that allows for varying relationships between variables across space(Anismuslim, Pramoedyo, & Andarini, 2023). It recognizes that relationships between variables may differ from one location to another due to local factors such as geography, climate, or socio-economic conditions. GWR estimates separate sets of coefficients for different locations within a dataset, considering local variations in the relationships between variables(Isazade et al., 2023). The equation of GWR:

$$Yi = \beta 0(i) + \beta 1(i)X1i + \beta 2(i)X2i + \ldots + \beta n(i)Xni + \varepsilon i \quad (3)$$

Where: Yi is the dependent variable at the location I, Xi represents independent variables at location I, β0(i), β1(i), ..., βn(i) are coefficients specific to location I, and εi is an error term specific to location i. In GWR(Anismuslim et al., 2023), separate local coefficients are estimated for each location within a study area instead of estimating global coefficients like in OLS. These local coefficients capture spatial relationship variations between variables by considering neighboring observations with higher weights than distant observations.

Spatial-temporal analysis is a method used to analyze data that varies over space and time(Deb & Karmakar, 2023). It involves examining how phenomena change over geographical locations and periods, allowing for a deeper understanding of patterns and trends in data. Several techniques are used in spatial-temporal analysis, including spatial autocorrelation analysis, spatiotemporal clustering analysis, space-time modeling, and others. These techniques help researchers identify spatial patterns in data and how these patterns change over time.

Spatio-temporal clustering algorithms: It used to identify clusters or groupings of spatial and temporal data points that share similar characteristics or behaviors(Zhang, Zhang, Wang, & Zhang, 2023). These algorithms consider the data's temporal and spatial dimensions to find patterns and relationships. ST-DBSCAN algorithm is frequently used for spatiotemporal clustering (Spatio-Temporal Density-Based Spatial Clustering of Applications with Noise). Adding time as a new dimension expands upon the classic DBSCAN algorithm. Every data point has a neighborhood defined by the ST-DBSCAN algorithm, depending on how close it is to other points both in space and time. After that, it connects nearby spots that meet specific density requirements to identify clusters. The equation for ST-DBSCAN:

$$ST - DBSCAN(D,\varepsilon,MinPts,T)= \{C1,C2,\ldots,Ck\} \tag{4}$$

Where: D is the spatiotemporal dataset, ε is the maximum distance threshold for defining a neighborhood, MinPts is the minimum number of points required to form a cluster, T is the maximum time threshold for defining a temporal neighborhood and C1, C2, ..., Ck are the resulting clusters.

Spatio-temporal regression models: They are statistical models used to analyze and predict how spatial and temporal factors influence a dependent variable(Hodges & Schnell, 2023). These models consider both the spatial location and period associated with each observation. One commonly used spatiotemporal regression model is the space-time autoregressive (STAR) model(Gao & Ding, 2022). The STAR model extends traditional regression models by incorporating spatial and temporal autocorrelation terms. It assumes that observations nearby in space and time are more likely to be similar than those further apart.

The equation for a basic STAR model:

$$Y= X\beta + W\lambda + \varepsilon \tag{5}$$

Where: Y is the dependent variable, X represents the independent variables, β represents their corresponding coefficients, W represents a matrix of weights capturing spatial and temporal relationships between observations, λ represents their corresponding coefficients, and ε represents the error term.

CASE STUDIES: PERFORMING GEOSPATIAL BIG DATA ANALYTICS USING SPARK

In order to create coastal marine predictions and analyses, it is necessary to define the coastal marine zones. Four geospatial datasets are available freely on the national

Figure 9. Loading geospatial dataset using Geopandas Dataframe.
Source:(Abanihi, n.d.)

```
# Load the Coastal Marine Zones data
mz_df = GeoDataFrame.from_file('/Users/abanihi/Documents/shapefiles/marine-zones/mz11au16/')

world = gpd.read_file(gpd.datasets.get_path('naturalearth_lowres'))
world.head(2)

hz_df = GeoDataFrame.from_file('/Users/abanihi/Documents/shapefiles/marine-zones/hz04jn14/')

oz_df = GeoDataFrame.from_file('/Users/abanihi/Documents/shapefiles/marine-zones/oz01apl4b/')
```

weather service https://www.weather.gov/, and https://www.naturalearthdata.com/ and are used to perform geospatial big data analytics using Spark. Sample codes are discussed here:

Figure 9 demonstrates the process of loading a geospatial dataset using a Geopandas DataFrame. Source: Abanihi (n.d.).

Preprocessing data for Spark: Spark expects a geospatial column to be a WKT (Well-Known Text)string. Geometries uses This internally via Java Topology Suite(JTS) to construct an Open Geospatial Consortium (OGC). Therefore, we will add the WKT column to our data so that we may use Spatial Spark. Figure 10 illustrates the preparation of data for Spark. Source: Abanihi (n.d.).

Figure 11 showcases the creation of a Spark DataFrame using a Geopandas DataFrame. Figure 12 demonstrates the process of obtaining all countries from the 'world_spark_df' as a Spark DataFrame. Source: Abanihi (n.d.). Figure 13 presents a visualization of regions from the Natural Earth dataset. Source: Abanihi (n.d.).

CONCLUSION AND FUTURE WORK

The amount of data being produced at this time is increasing at an unprecedented rate because of the digital revolution. With the proliferation of sensors and location-based services, organizations generate massive amounts of geospatial data daily. However, collecting, storing, managing, exploring, and analyzing geospatial data has traditionally been a complex and time-consuming task. Typically, it relies on manual interpretation or rule-based methods, which are time-consuming and incapable of handling enormous datasets. Thus, analyzing geospatial big data at scale requires advanced big data analytics techniques to derive meaningful insights. In this chapter, big data analytics was proposed to collect, store, manage, explore, process, and analyze massive amounts of geospatial data. To provide relevant real-time analysis

Figure 10. Preparing data for Spark,
Source:(Abanihi, n.d.)

```
In [27]: world['wkt'] = pd.Series(
                     map(lambda geom: str(geom.to_wkt()), world['geometry']),
                     index=world.index, dtype='string')

In [28]: oz_df['wkt'] = pd.Series(
                     map(lambda geom: str(geom.to_wkt()), oz_df['geometry']),
                     index=oz_df.index, dtype='string')
         hz_df['wkt'] = pd.Series(
                     map(lambda geom: str(geom.to_wkt()), hz_df['geometry']),
                     index=hz_df.index, dtype='string')
         mz_df['wkt'] = pd.Series(
                     map(lambda geom: str(geom.to_wkt()), mz_df['geometry']),
                     index=mz_df.index, dtype='string')

In [29]: # drop the geometry column because Spark can't infer
         # a schema for it when it's a nested geometry shape
         def drop_geometry_column(dataframe, col_drop="geometry"):
             return dataframe.drop(col_drop, axis=1)
```

of the enormous amount and heterogeneity of spatial data, big data analytics used the most popular platforms like spatial Hadoop and GeoSpark. One point addressed in this chapter is that we performed a comprehensive literature review of recent and relevant research papers on big data analytics for geospatial applications. We discussed the most common geospatial big data analytics Python libraries like GeoPandas, Pyspark, H3-Py, Disk-Geospatial, Rasterio, and GeoAlchemy with their example. Geospatial big data analytics activities like data collection, data

Figure 11. By using geopandas dataframe creating a Spark Dataframe,
Source: (Abanihi, n.d.)

```
In [30]: world_spark_df = spark.createDataFrame(drop_geometry_column(world)).cache()

In [31]: oz_spark_df = spark.createDataFrame(drop_geometry_column(oz_df)).cache()

In [32]: hz_spark_df = spark.createDataFrame(drop_geometry_column(hz_df)).cache()

In [33]: mz_spark_df = spark.createDataFrame(mz_df).cache()

In [34]: mz_spark_df.printSchema()

         root
          |-- GL_WFO: string (nullable = true)
          |-- ID: string (nullable = true)
          |-- LAT: double (nullable = true)
          |-- LON: double (nullable = true)
          |-- NAME: string (nullable = true)
          |-- WFO: string (nullable = true)
          |-- geometry: struct (nullable = true)
          |    |-- __geom__: long (nullable = true)
          |    |-- _is_empty: boolean (nullable = true)
          |    |-- _ndim: long (nullable = true)
          |-- wkt: string (nullable = true)
```

Figure 12. Getting all countries from world_spark_df as Spark dataframe,
Source: (Abanihi, n.d.)

```
In [38]: # Select all countries
         countries = spark.sql(
                     """
                     SELECT name, wkt as geometry, pop_est, continent
                     FROM world
                     ORDER BY continent, name
                     """)

In [39]: countries.show()

         +--------------------+--------------------+------------+---------+
         |                name|            geometry|     pop_est|continent|
         +--------------------+--------------------+------------+---------+
         |             Algeria|POLYGON ((11.9995...|3.4178188E7|   Africa|
         |              Angola|MULTIPOLYGON (((1...|1.2799293E7|   Africa|
         |               Benin|POLYGON ((2.69170...|  8791832.0|   Africa|
         |            Botswana|POLYGON ((25.6491...|  1990876.0|   Africa|
         |        Burkina Faso|POLYGON ((-2.8274...|1.5746232E7|   Africa|
         |             Burundi|POLYGON ((29.3399...|  8988091.0|   Africa|
         |            Cameroon|POLYGON ((13.0758...|1.8879301E7|   Africa|
         |Central African Rep.|POLYGON ((15.2794...|  4511488.0|   Africa|
         |                Chad|POLYGON ((14.4957...|1.0329208E7|   Africa|
         |               Congo|POLYGON ((12.9955...|  4012809.0|   Africa|
         |       Côte d'Ivoire|POLYGON ((-2.8561...|2.0617068E7|   Africa|
         |     Dem. Rep. Congo|POLYGON ((30.8338...|6.8692542E7|   Africa|
         |            Djibouti|POLYGON ((43.0812...|   516055.0|   Africa|
         |               Egypt|POLYGON ((34.9226...|8.3082869E7|   Africa|
         |          Eq. Guinea|POLYGON ((9.49288...|   650702.0|   Africa|
         |             Eritrea|POLYGON ((42.3515...|  5647168.0|   Africa|
         |            Ethiopia|POLYGON ((37.9060...|8.5237338E7|   Africa|
         |               Gabon|POLYGON ((11.0937...|  1514993.0|   Africa|
         |              Gambia|POLYGON ((-16.841...|  1782893.0|   Africa|
         |               Ghana|POLYGON ((1.06012...|2.3832495E7|   Africa|
         +--------------------+--------------------+------------+---------+
         only showing top 20 rows
```

Figure 13. Visualization of Region from Natural Earth dataset,
Source: (Abanihi, n.d.)

```
In [57]: # Set the region of interest color to red
         world.plot(ax=gdf.plot(cmap='Set2', alpha=1, color='red'), alpha=1)
         plt.show()
```

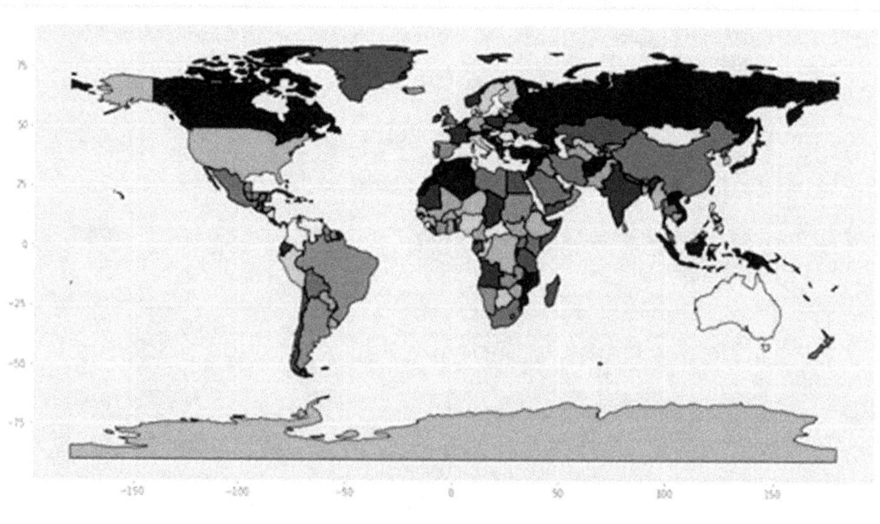

processing, spatial analysis, visualization, machine learning and AI, and real-time analytics were addressed.

In addition, Challenges in Geospatial Big Data Analytics: Volume, Velocity, and Variety were addressed; most common big data analytics techniques such as Spatial Indexing, Spatial Clustering, Spatial Regression Analysis, and Spatial-Temporal Analysis were discussed, and some algorithms used in the geospatial big data analytics like K-means clustering, DBSCAN clustering, Ordinary Least Squares(OLS), Geographically Weighted Regression(GWR), Spatio-temporal clustering algorithms and Spatio-temporal regression models were presented. Finally, case studies on performing geospatial big data analytics using Pyspark and visualization of geospatial data were performed for geospatial datasets taken from an open-source database. Therefore, big data analytics has revolutionized geospatial analytics using powerful platforms like Spark and Python libraries to process, store, analyze, and visualize large volumes of geospatial data that can process large volumes of geospatial data to extract meaningful insights. Overall, big data analytics has the potential to revolutionize our understanding of the world around us by providing valuable insights from large volumes of geospatial data.

Future work in big data analytics for geospatial Applications will focus on the following areas:

- Examining the application of big data analytics to real-time geospatial analysis, such as the surveillance of natural disasters or the observation of traffic patterns.
- Designing scalable machine learning models that can handle big geospatial data more efficiently and accurately
- Performing real-time big data analytics on cloud computing for geospatial big data, and investigating challenges relates to the interpretability of information from huge and complex geospatial data.

REFERENCES

Abanihi. (n.d.). *Tutorials PySparkGeoAnalysis Coastal Marine Zones*. NCAR. https://ncar.github.io/PySpark4Climate/tutorials/pyspark-geo-analysis/coastal-marine-zones/

Alfadul, T., & Ali, T. (2022). Geospatial Big Data Analytics Applications Trends. *Challenges & Opportunities*, *4*(1), 140–144.

Ali, S. M., & Gupta, N. (2017). Investigation into the efficacy of geospatial big data visualization tools, (May). doi:10.1109/CCAA.2017.8229777

Alsaedi, N., Sadeq, A., & Jalal, A. (2023). Big Spatial Data Systems -. *RE:view*, (August). doi:10.1109/IICETA54559.2022.9888481

Anismuslim, M., Pramoedyo, H., & Andarini, S. (2023). Modeling of Risk Factors of Childhood Stunting Cases in Malang Regency using Geographically Weighted Regression (GWR) Abstract :, 1–11. https://doi.org/ doi:10.2174/18749445-v17-e230420-2022-165

Deb, S., & Karmakar, S. (2023). A novel spatio-temporal clustering algorithm with applications on COVID-19 data from the United States. *Computational Statistics & Data Analysis*, *188*, 107810. doi:10.1016/j.csda.2023.107810

Gao, Z., & Ding, M. (2022). Application of convolutional neural network fused with machine learning modeling framework for geospatial comparative analysis of landslide susceptibility. *Natural Hazards*, *113*(2), 833–858. doi:10.1007/s11069-022-05326-7

Golubev, A., Chechetkin, I., Parygin, D., Sokolov, A., & Shcherbakov, M. (2016). Geospatial Data Generation and Preprocessing Tools for Urban Computing System Development *. *Procedia -. Procedia Computer Science*, *101*, 217–226. doi:10.1016/j.procs.2016.11.026

Hodges, J. S., & Schnell, P. M. (2023). Alleviating confounding in spatio-temporal areal models with an application on crimes against women in India, *23*(x), 9–30. doi:10.1177/1471082X211015452

Huang, X., Ma, T., Liu, C., & Liu, S. (2022). GriT-DBSCAN : A Spatial Clustering Algorithm for Very Large Databases, 1–18.

Isazade, V., Baser, A., Pinliang, Q., Gordana, D., & Esmail, K. (2023). Integration of Moran' s I, geographically weighted regression (GWR), and ordinary least square (OLS) models in spatiotemporal modeling of COVID - 19 outbreak in Qom and Mazandaran Provinces, Iran. *Modeling Earth Systems and Environment*, *9*(4), 3923–3937. doi:10.1007/s40808-023-01729-y PMID:36820101

Koh, K., Hyder, A., Karale, Y., & Boulos, M. N. K. (2022). Big Geospatial Data or Geospatial Big Data ? *A Systematic Narrative Review on the Use of Spatial Data Infrastructures for Big Geospatial Sensing Data in Public Health.*

Kovacs-györi, A., Ristea, A., Havas, C., Meha, M., Hochmair, H. H., Resch, B., & Blaschke, T. (2020). Opportunities and Challenges of Geospatial Analysis for Promoting Urban Livability in the Era of Big Data and Machine Learning.

Lanka, S. (2023). *Agriculture for Developing Nations*. Springer. doi:10.1007/978-981-19-8738-0

Lawhead, J. (2015). *Learning Geospatial Analysis with Python* (2nd ed.).

Lenka, R. K., Barik, R. K., Gupta, N., Ali, S. M., Rath, A., & Dubey, H. (2017). Data Analytics.

Li, Z. (2022). *Helen Gurgel, Lei Xu, L. Y. and J. D.* Improving Dengue Forecasts by Using Geospatial Big Data Analysis in Google Earth Engine and the Historical Dengue.

Li, Z. (n.d.). *Geospatial Big Data Handling with High Performance Computing : Current Approaches and Future Directions*.

Li, Z., Tang, W., Huang, Q., Shook, E., & Guan, Q. (2020). *Introduction to Big Data Computing for Geospatial Applications*.

Loukili, Y., Lakhrissi, Y., Elhaj, S., & Ali, B. (2022). Geospatial Big Data Platforms : A Comprehensive Review. *KN - Journal of Cartography and Geographic Information*, *72*(4), 293–308. doi:10.1007/s42489-022-00121-7

Mete, M. O. (2023). *GEOSPATIAL BIG DATA ANALYTICS FOR SUSTAINABLE SMART CITIES*, *XLVIII*(July), 141–146.

Olasz, A., Nguyen Thai, B., & Kristóf, D.A. Olasz a. (2016). A NEW INITIATIVE FOR TILING, STITCHING AND PROCESSING GEOSPATIAL BIG DATA IN DISTRIBUTED COMPUTING ENVIRONMENTS A. *ISPRS Annals of the Photogrammetry, Remote Sensing and Spatial Information Sciences*, *III*(July), 12–19. doi:10.5194/isprsannals-III-4-111-2016

Priyashani, N., Kankanamge, N., & Yigitcanlar, T. (2023). *Multisource Open Geospatial Big Data Fusion : Application of the Method to Demarcate Urban Agglomeration Footprints*.

Rey, S. J., Anselin, L., Li, X., Pahle, R., Laura, J., & Li, W. (2015). *Open Geospatial Analytics with PySAL*, 815–836. doi:10.3390/ijgi4020815

Roy, A., Fouché, E., Morales, R. R., & Möhler, G. (2019). *In-Database Geospatial Analytics using Python*.

Saputra, H. Y., & Radam, I. F. (2023). International Journal of Transportation Accessibility model of BRT stop locations using Geographically Weighted regression (GWR): A case study in Banjarmasin, Indonesia. *International Journal of Transportation Science and Technology, 12*(3), 779–792. doi:10.1016/j.ijtst.2022.07.002

Srivastava, N., & Saxena, N. (2023). Applications of Artificial Intelligence and Machine Learning in Geospatial Data. *Advances in Geospatial Technologies*, (May), 196–219. doi:10.4018/978-1-6684-7319-1.ch010

Tamiminia, H., Salehi, B., Mahdianpari, M., Quackenbush, L., Adeli, S., & Brisco, B. (2020). Google Earth Engine for geo-big data applications : A meta-analysis and systematic review ISPRS Journal of Photogrammetry and Remote Sensing Google Earth Engine for geo-big data applications : A meta-analysis and systematic review. *ISPRS Journal of Photogrammetry and Remote Sensing, 164*(May), 152–170. doi:10.1016/j.isprsjprs.2020.04.001

Westra, E. (2010). *Python Geospatial Development.*

Xu, H., & Zhang, C. (2023). Development and applications of GIS-based spatial analysis in environmental geochemistry in the big data era. *Environmental Geochemistry and Health, 45*(4), 1079–1090. doi:10.1007/s10653-021-01183-8 PMID:35066745

Zeng, C., Zhang, F., & Luo, M. (2022). A deep neural network-based decision support system for intelligent geospatial data analysis in intelligent agriculture system. *Soft Computing, 26*(20), 10813–10826. doi:10.1007/s00500-022-07018-7

Zhang, R., Zhang, J., Wang, Q., & Zhang, H. (2023). *DOIDS : An Intrusion Detection Scheme Based on DBSCAN for.*

Zhao, W., Wang, M., & Pham, V. T. (2023). *Unmanned Aerial Vehicle and Geospatial Analysis in Smart Irrigation and Crop Monitoring on IoT Platform*. Research Gate.

Chapter 12
Topological Insights Into Weather Dynamics in the Indian Context:
An Application of Clustering and Mapper Algorithm

Azarudheen S.
Christ University, India

Daphne Julia Menezes
https://orcid.org/0009-0009-9897-8966
Christ University, India

Vivan Clements
Northeastern University, Canada

ABSTRACT

Analysis of day-to-day weather patterns is critical and essential in daily life. Although traditional methods exist, in modern times, we have developed realistic and reliable methods to provide better insights and understanding of complex weather patterns for various surges, especially in these times of global warming. Implementing clustering and topological data analysis in this analysis has looked into a vast understanding of how regions with similar characteristics behave when weather changes occur due to heat, pressure, or wind-related phenomena. The classification model developed using Mapper analysis has produced 95.8% accuracy, concluding that weather follows a transient weather pattern due to various resources and how stagnant conditions affect transient weather patterns, causing rise in sub-clusters. Thus, fitting and interpreting newer models helps us understand weather analysis and classification.

DOI: 10.4018/979-8-3693-6381-2.ch012

INTRODUCTION

Motivation and Background

Topological data analysis (TDA) has emerged as a powerful tool for studying complex systems, including weather dynamics. TDA techniques like the Mapper algorithm can provide valuable insights into meteorological data's intricate patterns and structures. In the context of Indian weather dynamics, these methods have been applied to unravel the underlying topological features and gain a deeper understanding of the complex phenomena at play. Applying clustering algorithms to meteorological data allows for identifying distinct weather regimes or patterns based on similarities in various atmospheric variables, such as temperature, humidity, wind speed, and pressure. By grouping similar weather conditions, these techniques can reveal the dominant modes of variability and the transitions between different weather states.

The Mapper algorithm, a topological technique, has also been utilized in the Indian context to study weather dynamics. This algorithm constructs a simplicial complex (a high-dimensional generalization of a network) from the data, capturing the underlying shape and connectivity of the system *(Singh et al., 2007)*. Researchers can identify intrinsic patterns, persistent features, and potential transitions between weather states by visualizing and analyzing the resulting topological structure. In recent research, the Mapper algorithm was applied to rainfall data over the Indian subcontinent. The resulting topological representation revealed distinct regions exhibiting similar rainfall patterns and the connections and transitions between these regions. This approach allowed for a comprehensive understanding of rainfall's spatial and temporal variability, which is crucial for water resource management, agricultural planning, and disaster preparedness *(Nanda et al., 2018)*. Furthermore, clustering and Mapper techniques have been employed to investigate the complex interactions between atmospheric variables and their influence on weather patterns *(Chintakunta et al., 2019)*.

By integrating multiple variables into a high-dimensional data space, these methods can uncover intricate relationships and unravel the underlying dynamics driving weather phenomena. Thus, applying clustering techniques and topological data analysis (TDA) methods presents a unique opportunity to conduct a collaborative study that harnesses the power of these cutting-edge approaches within the realm of data analysis. By integrating the strengths of clustering algorithms and the topological insights offered by the Mapper algorithm, we can explore weather dynamics comprehensively in the Indian context. This interdisciplinary endeavor has the potential to unravel intricate patterns, uncover intrinsic relationships, and elucidate the underlying complexities that govern atmospheric processes. By leveraging the synergies between data-driven clustering and the topological lens provided by TDA, we can unlock novel perspectives and gain a deeper understanding of the interplay between various meteorological variables. Ultimately, this collaborative study has the potential to push the boundaries of our knowledge, fostering sustainable innovations in data analysis techniques and their applications in the field of weather and climate research.

Objective of the Study

The objective of this study (summarized in Table 1) is to harness the combined power of clustering techniques and topological data analysis (TDA) methods, particularly the Mapper algorithm, to gain a comprehensive understanding of the intricate weather dynamics in the Indian context. Specifically, we aim to identify distinct weather regimes and patterns over the Indian subcontinent by applying advanced clustering algorithms to meteorological data spanning multiple years and regions *(Muñoz-Díaz & Rodrigo, 2004)*. This will involve incorporating a wide range of atmospheric variables, such as temperature, precipitation, humidity, wind speed, and pressure, to capture the multidimensional nature of weather patterns *(Nanda et al., 2018)*. Additionally, we will employ the Mapper algorithm to construct a topological representation of the weather data, capturing the underlying shape and connectivity of the system *(Chintakunta et al., 2019)*.

This topological lens will facilitate the visualization and analysis of persistent features, holes, and potential transitions between different weather states, unveiling insights that may be obscured by traditional analytical methods *(Carlsson, 2009)*. Furthermore, we will investigate the complex interactions and interdependencies between various atmospheric variables by integrating them into high-dimensional data space, leveraging the combined strengths of clustering and Mapper techniques to unravel the intricate relationships and key drivers influencing weather phenomena *(Nanda et al., 2018; Chintakunta et al., 2019)*. To ensure the practical utility of our findings, we will assess the methods' robustness and applicability across different spatial and temporal scales, ranging from local to regional levels and from short-term to long-term weather patterns *(Nowotarski & Jensen, 2019)*. This will enable the development of a framework for integrating topological insights and clustering-based weather regimes into existing meteorological models and decision-support systems, ultimately contributing to improved climate resilience and sustainable development strategies in the Indian context *(Zscheischler et al., 2018)*.

Hence, the objective of integrating the best clustering algorithms, topological data analysis techniques, and machine learning classification models presents an ambitious and comprehensive approach to unraveling the complexities of weather dynamics in the Indian context. By synergizing these powerful analytical tools, it aims to achieve a holistic understanding that transcends the limitations of any general traditional models. The synergy of these three approaches has the potential to yield a comprehensive and multi-faceted understanding of Indian weather dynamics. The clustering algorithms will identify the distinct patterns, the topological data analysis will reveal the intricate topological structure, and the machine learning models will enable accurate classification and prediction. This integration will not only contribute to advancing theoretical knowledge but also pave the way for practical applications in weather forecasting, risk assessment, resource management, and climate change adaptation strategies, ultimately enhancing our resilience and preparedness in the face of environmental challenges.

Objectives of the study are:

- Identify distinct weather regimes and patterns
- Employ the Mapper algorithm to construct a topological representation
- Investigate the complex interactions and interdependencies
- Assess the robustness and applicability
- Develop a framework for integrating topological insights and clustering

Scope and Limitations

The scope of this study is defined by its ambitious aim to unravel the intricate patterns and dynamics of weather systems across the diverse geographic expanse of the Indian subcontinent. With a focus on harnessing the power of advanced analytical techniques, the study provides a comprehensive understanding of the complex interplay between various atmospheric variables and their manifestations in the form of distinct weather regimes.

- The study will focus on the Indian subcontinent, which exhibits diverse weather patterns influenced by monsoons, topography, and ocean proximity. This broad geographical scope will allow for a comprehensive understanding of the various weather regimes across different regions of India.
- The study will analyze meteorological data spanning multiple years, enabling the identification of both short-term and long-term weather patterns. This will provide insights into the temporal variability of weather dynamics, including seasonal and inter-annual variations.
- By combining clustering techniques, Topological Data Analysis (TDA), and machine learning classification models, the study will offer a holistic perspective on weather dynamics. This integrative approach will leverage the strengths of each method, allowing for a more robust and comprehensive analysis.

- The study's findings can have far-reaching implications for various sectors, including weather forecasting, disaster management, agriculture, water resource management, and climate change adaptation strategies. The developed framework and decision-support tools can contribute to enhanced preparedness and resilience in the face of environmental challenges.

While the proposed study aims to provide valuable insights and practical applications, it is essential to acknowledge potential limitations:

- The study's effectiveness heavily relies on the availability of high-quality and comprehensive meteorological data encompassing a wide range of atmospheric variables. Incomplete or inconsistent data could affect the findings' accuracy and reliability.
- The integration of multiple advanced analytical techniques, such as clustering, TDA, and machine learning models, may introduce computational challenges, mainly when dealing with large-scale datasets. Efficient algorithms and robust computational resources may be required to handle the complexity of the analyses.
- While the study focuses on the Indian subcontinent, the applicability of the findings to other geographic regions may be limited due to unique local weather patterns and environmental factors. Caution should be exercised when extrapolating the results to regions with significantly different climatic conditions.
- Certain aspects of the study, particularly those involving topological data analysis and machine learning models, may present challenges regarding interpretability and quantification of uncertainties. Effective communication and visualization techniques will be crucial to convey the findings accurately to stakeholders and decision-makers.
- The study's success relies on effective collaboration among experts from various disciplines, including atmospheric sciences, data science, topology, and domain-specific knowledge. Bridging the gaps between these fields and ensuring seamless communication may pose challenges.

Despite these limitations, the proposed study can contribute significantly to our understanding of weather dynamics in the Indian context, provided that appropriate measures are taken to address potential challenges and limitations.

LITERATURE REVIEW

Weather Dynamics in India

Weather dynamics is the study of the physical processes and interactions that govern the behaviour and evolution of atmospheric phenomena. It encompasses the complex interplay of various atmospheric variables, including temperature, pressure, humidity, wind patterns, precipitation, and their interactions with the Earth's surface and other environmental factors. At the core of weather dynamics lies fluid mechanics, thermodynamics, and atmospheric physics principles. The atmosphere is a complex fluid system driven by the uneven distribution of solar radiation and the Earth's rotation. These factors cause variations in temperature, pressure, and density, leading to weather patterns and systems forming. Weather dynamics study involves analyzing large-scale atmospheric circulation patterns, such as mid-latitude cyclones, tropical cyclones, and monsoon systems. It also delves into smaller-scale phenomena like thunderstorms, tornados, and local wind patterns, which can significantly impact local environments.

The Indian monsoon is a critical weather system that governs most of the country's annual rainfall. Researchers have dedicated substantial efforts to understanding monsoon rainfall's spatial and temporal patterns, its variability, and the underlying mechanisms driving the monsoon *circulation (Rajeevan et al., 2010; Krishnan et al., 2019)*. These studies aim to improve forecasting capabilities and provide insights into the potential impacts of monsoon variability on sectors like agriculture and water resources. Additionally, the influence of large-scale atmospheric phenomena, such as the El Niño-Southern Oscillation (ENSO) and the Indian Ocean Dipole (IOD), on monsoon dynamics has been extensively explored *(Ashok et al., 2004; Ratna et al., 2016)*, highlighting the complex interactions between global climate patterns and regional weather systems. India is prone to various extreme weather events, including cyclones, droughts, and heatwaves, which can devastate communities and ecosystems.

Studies have focused on understanding the formation, intensification, and tracking of cyclones in the Indian Ocean region *(Mohapatra et al., 2017; Sharma et al., 2019)*, providing valuable insights for disaster preparedness and risk mitigation strategies. Additionally, researchers have investigated the impact of climate change on the frequency and severity of extreme weather events *(Goswami et al., 2006; Mishra et al., 2014)*, as these phenomena are expected to intensify in the future, posing significant challenges for adaptation and resilience-building efforts. India's diverse topography, including the Himalayas, Western Ghats, and other mountain ranges, is crucial in shaping weather patterns. Researchers have explored the influence of topography on precipitation distribution, orographic effects, and the formation of local weather systems *(Das et al., 2002; Dimri et al., 2017)*. These studies are particularly relevant for regions with complex terrain, where topographic features can significantly impact local microclimates and weather patterns, affecting activities like agriculture, water resource management, and disaster risk assessment. With rapid urbanization, the urban heat island effect has become a significant concern in Indian cities. Studies have investigated the extent and impact of the urban heat island effect on local weather patterns, energy consumption, and human health *(Mallick et al., 2013; Dhiman et al., 2019)*. Understanding this phenomenon is crucial for urban planning, energy management, and public health strategies as urban areas continue to expand and face the challenges of climate change and increasing temperatures.

Recent studies have explored the application of topological data analysis (TDA) and machine learning techniques to unravel the complex patterns and dynamics of weather systems in India. These advanced analytical approaches have shown promise in identifying weather regimes, capturing nonlinear relationships, and improving forecasting capabilities *(Nanda et al., 2018; Chintakunta et al., 2019)*. By leveraging the power of these techniques, researchers aim to gain deeper insights into the intricate structure and behavior of weather phenomena, potentially leading to more robust and accurate predictive models. It's important to note that the literature review presented here is not exhaustive, and numerous other studies and research efforts focused on specific regions, weather phenomena, or analytical techniques in the context of Indian weather dynamics. Additionally, interdisciplinary collaborations and the integration of diverse data sources continue to drive advancements in this field, contributing to our understanding of the complex interplay between atmospheric processes, climate change, and human activities.

Topological Data Analysis (TDA) for Weather Analysis

Topological data analysis (TDA), a relatively new field born in the early 2000s, draws upon applied topology, algebraic topology, and computational geometry *(Edelsbrunner et al., 2002; Zomorodian & Carlsson, 2005)*. While geometric approaches to data analysis have a long history, TDA's emergence as a distinct field can be credited to pioneering work on persistent homology *(Edelsbrunner et al., 2002; Zomorodian & Carlsson, 2005)*, further popularized by Carlsson's influential article in 2009 *(Carlsson, 2009)*. The core idea behind TDA is that topology and geometry offer a powerful framework to extract robust qualitative and potentially quantitative information about the underlying structure of data (often represented as point clouds in Euclidean or more general metric spaces) *(Chazal, 2017)*. TDA aims to develop well-founded mathematical, statistical, and algorithmic methods to uncover, analyze, and leverage the complex topological and geometric structures hidden within data. In recent years, significant advancements have been made in creating robust and efficient data structures and algorithms specifically for TDA. These tools are now implemented and readily available in popular libraries like GUDHI (C++ and Python) *(Maria et al., 2014)* with its R interface *(Fasy et al., 2014)*, alongside other libraries such as Dionysus, PHAT, DIPHA, and Giotto. While TDA is still under rapid development, it offers a mature and efficient toolkit that can be used in conjunction with or to complement existing data science techniques.

TDA is a young and flourishing field that harnesses the power of topology and geometry to extract valuable insights and reveal the underlying structure of complex data, particularly point cloud data. It provides a robust mathematical and computational framework for data analysis, acting as a valuable complement to existing data science approaches. While the field of Topological Data Analysis (TDA) boasts a wide variety of methods, this section focuses on a standard pipeline that serves as its foundation. Here's a breakdown of each step:

Data Input: We begin with a set of data points, each having a defined distance or similarity measure. This distance can be based on the surrounding space (like Euclidean distance in Rn) or come from a separate "intrinsic" distance matrix. Choosing the right distance metric is crucial for revealing important topological features in the data.

Shape Building: To highlight the underlying structure, a "continuous" shape, often called a simplicial complex, is constructed from the data points. Think of it as a higher-dimensional version of the neighbor graphs used in many data analysis algorithms. This simplicial complex captures the data structure at different scales through a nested family called a filtration. The key here is to define these structures effectively and efficiently while ensuring they accurately reflect the data's structure.

Information Extraction: Once the shape is built, we can extract topological or geometric information. This might involve reconstructing the underlying shape (like a triangulation) for easy feature extraction or using methods like persistent homology to extract summaries or approximations from the data. A crucial aspect is ensuring the extracted information is relevant and stable, meaning it does not drastically change with slight variations or noise in the original data. Statistical analysis of these features is also essential.

Data Enrichment: The extracted topological and geometric information becomes a new set of features for the data. These features can be used for visualization and better data understanding or combined with other features for further analysis or machine learning tasks. They can even be used to design more suitable data analysis and machine learning models. This step's key aspect is demonstrating the value and complementarity (compared to other features) of information obtained through TDA tools.

Since continuous spaces are typically associated with topological and geometric aspects, data expressed as finite sets of observations do not inherently disclose topological information. "Connecting" nearby data points to show a continuous global shape beneath the data is a natural technique that draws attention to some topological pattern. TDA is typically quantified using a distance (or a dissimilarity measure). General notions for geometric and topological inference are introduced; it is frequently convenient to treat data sets as discrete metric spaces or as samples of metric spaces (Boissonant et al., *2018)*.

Metric Spaces

A set M with the function ρ: $M*M \rightarrow R+$ is a metric space (M, ρ) referred to as a distance, such that the following is true for all x, y, and z ∈ M:

i) $\rho(x, y) \geq 0$ and $\rho(x, y) = 0$ if and only if $x = y$,
ii) $\rho(x, y) = \rho(y, x)$, and
iii) $\rho(x, z) \leq \rho(x, y) + \rho(y, z)$

The set K(M) in a metric space (M, ρ). A Hausdorff distance can be applied to one of its compact subsets; given two compact subsets $A, B \subseteq M$, the Hausdorff distance $\delta H(A,B)$ between A and B is defined as the smallest nonnegative number δ such that, for any a ∈ A, there exists b ∈ B such that ρ(a, b) ≤ δ, and for any b ∈ B, there exists an a ∈ A such that $\rho(a, b) \leq \delta$. The Hausdorff distance is, in fact, a distance on the set of compact subsets of a metric space, which is a fundamental and classical conclusion. It offers a practical means of measuring the distance between various data sets released from the same ambient metric space from a TDA standpoint.

Geometric and Abstract Simplicial Complexes

One way to think of simplicial complexes is as a higher-dimensional graph generalization. They are combinatorial and topological mathematical objects, which is a feature that makes them especially helpful for TDA.

The convex hull of a set $X = \{x0,\ldots,xk\} \subset Rd$ of k + 1 affinely independent points is represented by the k-dimensional simplex $\sigma = [x0,\ldots, xk]$ spanned by X. The simplices that the subsets of X span are known as the faces of σ, and the points of X are known as the vertices of σ. A group of simplices such that the following hold is called a geometric simplicial complex *K in Rd*.

i) Every face of a K simplex is a K simplex,
ii) Any two simplices of K that cross over are either empty or have a common face.

The underlying space of K is a subset of Rd that derives from the topology of Rd and is the union of the simplices of K. Thus, by its underlying space, K can also be viewed as a topological space. Observe that K can be fully described by the combinatorial description of a set of simplices that meet certain incidence criteria once its vertices are known.

Given a set V, a set $\tilde{K}$ of finite subsets of V such that the elements of V belong to an abstract simplicial complex with the vertex set V. The underlying space of K is a subset of Rd that derives from the topology of Rd and is the union of the simplices of K. Thus, by its underlying space, K can also be viewed as a topological space. Observe that K can be fully described by the combinatorial description of a set of simplices that meet specific incidence criteria once its vertices are known.

An abstract simplicial complex with a vertex set V, given a set V, is a set $\tilde{K}$ of finite subsets of V such that all subsets of σ that belong to $\tilde{K}$ and all elements of V belong to $\tilde{K}$ for each σ that is a member of $\tilde{K}$. The faces or simplices of $\tilde{K}$ are the elements that make up $\tilde{K}$. The maximum dimension of an abstract simplice is called $\tilde{K}$, whereas the dimension of an abstract simplex is just its cardinality minus 1. Observe that dimension one simplicial complexes are graphs. It is evident that an abstract simplicial complex $\tilde{K}^{-1}$ arises from the combinatorial description of any geometric simplicial K. Conversely, it is always possible to correlate a topological space |K-|| with an abstract simplicial complex $\tilde{K}^{-1}$ such that, in the case that K is a geometric complex with the same combinatorial description as $\tilde{K}^{-1}$, then K's underlying space is homeomorphic to |K~|. We refer to such a K as a geometric realisation of K^. Consequently, geometric complexes can be viewed as geometric spaces and abstract simplicial complexes as topological spaces.

Building Symbolical Complexes From Data

Simplicial complexes are constructed using various methods, such as data collection or, more generally, a topological or metric space. An instantaneous expansion of the concept of the α-neighboring graph serves as the first example. Let us assume we have an accurate number α > 0 and a set of points X in a metric space (M, ρ). The Rips\(X) Vietoris–Rips complex is, as shown in Figure 1, the set of simplices *[x0,..., xk]* such that *dX(xi,xj)≤ϱ* for every (i, j). The definition immediately makes it clear that this is an abstract simplicial complex. However, generally speaking, Ripsα(X) does not accept a geometric realization in Rd, even if X is a finite subset of Rd;; specifically, it may have a dimension greater than d.

The Čech complex Cechα(X) is closely associated with the Vietoris–Rips complex. It can be defined as the set of simplices *[x0,..., xk]* such that the k + 1 closed balls *B(xi, α)* have a non-empty intersection.

Figure 1. Simplicial complex formations

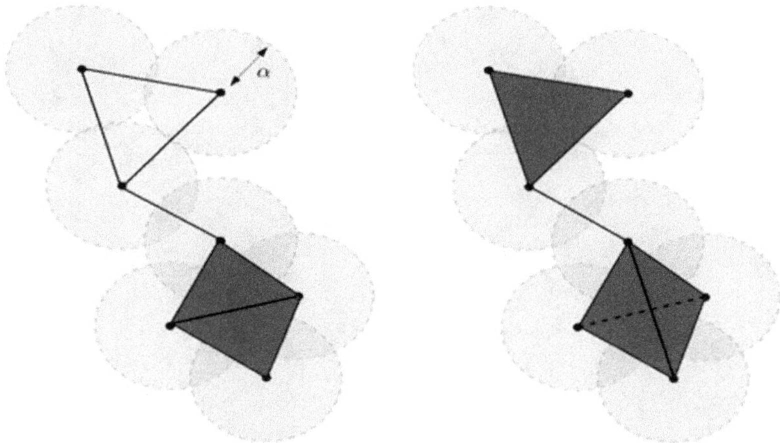

Note the relationship between these two complexes:

Cechα(X) and Rips2α(X) share the same one-dimensional skeleton, or collection of vertices and edges, if X⊂Rd. This means that

Ripsα(X)⊆Cechα(X)⊆Rips2α(X)

Mapper Algorithm

The Mapper algorithm's concept uses the nerve of the refined pull-back of a cover U of f(X) to summarise X given a data set X and a well-chosen real-valued function *f:X→Rd*. For well selected covers *U* (see below), this nerve is a graph that offers a simple and practical means of visualising the data summary. Figure 2(A) depicts the height function's refined pull-back cover on a surface in R^3 and its nerve. Figure 2(B) Mapper algorithm using the height function and a point cloud sampled around a circle. Initially, the height function defined on the point cloud's pull-back cover is calculated and fine-tuned (left). Second, a graph (right) represents the nerve of the refined pull-back.

Terminologies

Filter Function (f): This function highlights specific data features. Here are some common choices:

Figure 2. Mapper algorithm representation

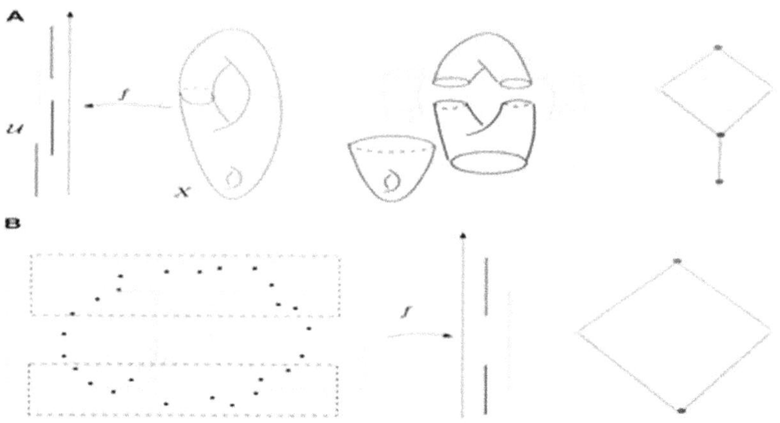

- Useful for understanding high-density regions (clusters) in the data. Common density estimation techniques include kernel density estimation or nearest neighbor methods.
- Can help reveal hidden structures when dealing with non-linear data. Specific formulas for these techniques depend on the chosen NLDR method.
- This measures the sum of distances between a point x and all other points in the data set X.
- This measures the maximum distance between a point x and any other point in the data set X. Here, d(x, y) represents the distance metric chosen for your data.

Data Cover (U): This cover partitions the data based on the filter function's output. A common approach uses equally spaced intervals with adjustable resolution and overlap:

- $U = \{ [a^i, a^{(i+1)}) \mid i = 1, ..., k \}$: This represents a cover where each interval is defined by a lower bound a^i and an upper bound.
- This parameter controls the number of intervals (k) in the cover. A higher resolution (smaller k) leads to more intervals with finer granularity.

Clustering: The algorithm requires grouping data points within each interval of the cover. There are two main strategies:

- **Individual clustering per interval:** Apply a chosen clustering algorithm (e.g., k-means clustering) to each interval's data points $(f^{-1})(U))$ separately.

- **Global clustering:** Build a neighbor graph based on the entire data set X (e.g., k-nearest neighbors graph or ε-graph). For each interval U, identify the connected components of the subgraph induced by the points in $f^{-1}(U)$.

It's important to note that the Mapper output is sensitive to these choices, and small changes can lead to significantly different results. Finding the most informative settings often involves exploring a range of parameter values.

Clustering Algorithms for TDA

K-Means: Arguably the most popular and extensively applied clustering technique. The approach minimizes the distance between each point and its cluster across a total of k clusters (where the user specifies k) until a stopping point (a specific value, number of iterations, etc.) is attained *(Lloyd, 1957) (Ghosh, 2009).*

Spectral Clustering: According to *Singh (2010)*, spectral Clustering performs better on connected data rather than compact data. Spectral clustering clusters the points following the eigenvectors of matrices by employing an alternative clustering technique such as k-means or k-nearest neighbor (KNN) *(VonLuxburg, 2007).* The number of clusters in a spectrum cluster is user-specified, and it functions best with data that includes few clusters *(Scikit, 2022).*

DBSCAN (Density-Based Spatial Clustering of Applications with Noise): As its name suggests, DBSCAN was first published in 1996 and employs a novel density method *(Ester, 1996).* Two parameters are needed for DBSCAN: epsilon, often referred to as "eps" or ε, which gives the minimum distance the points must lie in order to be included in the & "sample" and a minimum number of points per "core" (called "minPts"). "Core" refers to a location or sample in a densely populated area that is anticipated to create a cluster *(Scikit, 2022).*

BIRCH (Balanced Iterative Reducing and Clustering using Hierarchies): Rather than employing distance to cluster data, BIRCH uses tree topologies (Zhang, 1996). A Clustering Feature Tree (CFT) is created, used for global Clustering, and then compressed into a smaller CFT if desired. The Clustering Feature nodes created from the divided data comprise subclusters of the Clustering Feature (Scikit, 2022). The branching factor ("limits the number of sub-clusters within a node") and the threshold ("limits the distance between the entering sample and the existing sub-clusters") control how Scikit-learn is implemented.

The choice of clustering algorithm depends on the characteristics of the data, the desired cluster properties, and the application's specific requirements. In many cases, exploring and comparing the results of multiple clustering techniques may be beneficial to gain a more comprehensive understanding of the data's structure and leverage the strengths of different approaches. Additionally, recent advances in clustering techniques, such as ensemble methods, hybrid approaches, and the integration of domain knowledge, offer promising avenues for achieving more robust and interpretable clustering results, particularly in complex and high-dimensional data scenarios.

METHODOLOGY

Data Collection and Pre-Processing

This analysis delves into a comprehensive weather dataset from Climate Research and Services (a subsidiary of the Indian Meteorological Department's Pune site). This valuable resource offers real-time weather information for major Indian cities, providing a snapshot of current conditions rather than future forecasts. Unlike forecasts, this dataset boasts over 40 unique features, encompassing essential weather parameters like temperature, wind speed and direction, air pressure, precipitation, visibility, and even air quality measurements. This wealth of data presents a fantastic opportunity to explore India's weather trends and uncover the intricate relationships between weather elements. The data acquisition process began on August 29th, 2023, and continues to be updated daily. This ensures we are working with the most recent weather picture for India. However, during data pre-processing, some interesting insights emerged. The process likely involved checking for missing data points in 40+ features. Missing values can skew analysis and must be addressed through imputation (filling in missing values) or removing data points with excessive missingness. But out dataset has no missing values.

Analyzing temperature, humidity, and wind speed data through boxplots revealed a deviation from "expected" weather conditions in most states. This suggests potential outliers or unique weather patterns specific to certain regions. With over 40 features encompassing various units (e.g., temperature in degrees Celsius, wind speed in kilometers per hour), the dataset might benefit from feature scaling. This process transforms features to a standard scale, ensuring all features contribute equally during analysis. We can gain valuable insights into India's current weather landscape by delving deeper into this dataset. We can identify patterns, correlations, and potential anomalies that might not be readily apparent through individual weather reports. This comprehensive analysis has the potential to improve our understanding of India's current weather dynamics and provide a foundation for further exploration of weather phenomena across the country.

Evaluation Metrics

Comprehensive data cleaning procedures were executed to ensure data quality, targeting missing values and inconsistencies within the dataset. The dataset consisted of no missing values but included anomalies leading to the requirement of further analysis using clustering. After that, a range of clustering algorithms, including K-Means, Hierarchical Clustering, Spectral Clustering, and DBSCAN, were employed to compare variables. The performance of each clustering algorithm was rigorously evaluated based on silhouette scores and computational time, enabling the identification of the most effective algorithm for variable comparison. The silhouette score offers a clear and intuitive measure of clustering quality. Ranging from -1 to 1, it tells how well data points are matched to their assigned clusters. Values closer to 1 indicate good separation, while values near -1 suggest points might be in the wrong cluster.

This ease of interpretation makes it a valuable tool for understanding the effectiveness of your clustering. Additionally, the silhouette score is flexible and can be applied to various clustering algorithms without requiring specific assumptions about the data distribution. On the other hand, computational time is a crucial factor in real-world applications, especially when dealing with large datasets like yours. By evaluating the running time of different algorithms, you can ensure you choose one that's efficient and doesn't take unreasonable time to complete the clustering process. This is particularly important for analyzing weather data, where timely insights are often essential. While other metrics like cluster validity indices might also be considered, silhouette score and computational time provide a strong foundation for initial evaluation. They offer a balance between understanding the quality of your clusters (silhouette score) and ensuring the chosen algorithm is practical for your data size (computational time). This combination allows to make informed decisions about which clustering algorithm is best suited for your analysis of India's current weather patterns.

Subsequently, a comprehensive model assessment was conducted by employing five diverse classification algorithms, including Decision Trees, Random Forest, and Support Vector Machines (SVM), on the clustered dataset. This multi-algorithm approach aimed to leverage the strengths of different classification techniques and provide a robust and comprehensive analysis. Evaluating Machine Learning (ML) models comprehensively is critical in building trustworthy and dependable tools. Metrics go beyond a simple "accuracy" score. They provide insights into how well a model performs on unseen data, which is crucial for real-world applications. Imagine training a model to predict weather patterns. A high accuracy of training data does not guarantee success with new weather scenarios. Evaluation metrics quantitatively measure this generalizability, ensuring the model's effectiveness in real-world situations. Evaluation is not just about achieving a high score but about identifying weaknesses. These metrics can pinpoint areas where the model might struggle. For instance, a weather prediction model might excel on sunny days but falter with thunderstorms. We can identify these shortcomings by analyzing evaluation results and taking corrective actions, such as gathering more data on specific weather events or refining the model's architecture.

Furthermore, evaluation allows for objective comparisons between different models. Developing ML models often involves exploring various algorithms and approaches. Evaluation metrics provide a standardized way to compare these models and choose the best performance for your task. Imagine building two weather prediction models - a simple decision tree and a complex neural network. Evaluation metrics will tell you which delivers more accurate and reliable forecasts for your weather data. Finally, comprehensive evaluation helps prevent overfitting. This occurs when a model memorizes the training data too well and fails to adapt to unseen data. Evaluation techniques help us detect signs of overfitting and take steps to prevent it, like applying regularization (adding constraints to the model). Comprehensive evaluation using ML techniques is about building robust, generalizable, and trustworthy models. It is a systematic process that goes beyond a single score, ensuring the models we create are reliable and deliver accurate results.

Topological Data Analysis (TDA) methodologies were incorporated to deepen insights and comprehension, unveiling intricate patterns and behaviors within the dataset that traditional analytical methods might have overlooked. The integration of TDA techniques facilitated a more profound exploration of the Indian Weather dataset by revealing the formation of interconnected clusters, capturing the underlying topological structure and connectivity within the data. By amalgamating clustering, classification, and TDA, this multi-dimensional approach enriched the dataset's comprehensibility and bolstered its usability for a myriad of analytical purposes. From the given evaluation metrics, the synergistic combination of these techniques allowed for a holistic and insightful exploration of the dataset's analysis and interpretation.

Evaluation Metrics

Step 1: Data Cleaning

a. Checking for missing values and conducting a through pre-processing step.

Step 2: Clustering

a. Apply various algorithms (K-Means, Hierarchical, Spectral, and DBSCAN)
b. Evaluate performance with silhouette score and computational time

Step 3: Model Assessment

a. Employ multiple classification algorithms (Decision Trees, Random Forest, SVM).

b. Evaluate model performance using various metrics (generalizability, weaknesses).
c. Compare models and select the best one for the task.

Step 4: Insights

a. Apply Topological Data Analysis (TDA) techniques
b. Uncover hidden patterns and connections within the data

RESULTS AND DISCUSSION

Identified Weather Patterns and Clusters

The analysis leveraged a specific technique within the Mapper algorithm, a method rooted in topological data analysis (TDA). This technique acts like a density-based clustering tool, identifying localized regions across India that share similar weather patterns. Imagine a map of India where areas with comparable weather conditions are grouped. Figure 3 offers a visual representation of this analysis using the Mapper technique. It reveals a fascinating picture – 8,580 mini-clusters, each with an average size of 10 data points, are scattered throughout the country.

The color assigned to each mini-cluster provides a quick visual cue about the average weather conditions within that region. Green indicates areas experiencing irregular weather, while blue represents regions with expected weather fluctuations (typical for the season). This visualization highlights a key takeaway – India's weather patterns exhibit significant spatial variation. The mini-clusters showcase a diverse range of weather regimes across the country. Interestingly, one particular cluster stands out as an anomaly, showcasing distinct and unusual weather patterns compared to the others. Figure 3 delves deeper into this outlier cluster. It encompasses states like Samastipur (Bihar), Kangpokpi (Manipur), Kapurthala (Punjab), Mathura (Uttar Pradesh), Saharanpur (Uttar Pradesh), and Wanaparthy (Andhra Pradesh). The weather patterns in these regions deviate significantly from the norm observed in other clusters. This anomaly could be attributed to two main possibilities:

- **Data Anomalies:** The data might have potential errors or inconsistencies for these specific locations. Further investigation into the data for these regions is crucial to confirm this possibility.
- **Rare Weather Events:** This cluster could represent uncommon and unusual weather phenomena in these areas. These events could be extreme weather occurrences or localized weather patterns not typically observed elsewhere in India.

Figure 3. Mapper representation of clusters with similar condition-like weather patterns

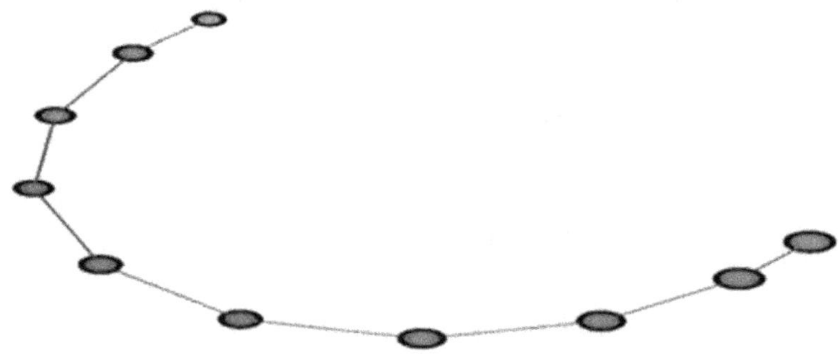

By leveraging the Mapper technique and TDA, this analysis sheds light on the rich tapestry of weather patterns across India. The spatial variations and potential anomalies highlight the need for further exploration to understand these weather variations' underlying causes and potential implications.

Interpretation of Clustering Results

From Table 1 it is noted that K-Means exhibited moderate silhouette scores (0.124) and relatively moderate computational usage (0.351), while BIRCH displayed a higher silhouette score (0.180) with notably low computational demands (0.138). Spectral Clustering outperformed others with a superior silhouette score (0.253) but required significantly higher computational resources (4.303). Conversely, DBSCAN showed a negative silhouette score (-0.564) with minimal computational requirements (0.056), indicating poor clustering results. This analysis highlights that Spectral Clustering offers superior cluster definition but consumes more resources. BIRCH strikes a better balance, showcasing efficient clustering with moderate computational costs. Under Spectral Clustering for pre-determined cluster size values (n=2,4,6,8), a decreasing silhouette score with increasing clusters suggests that fewer clusters may better represent the underlying structure of weather data, indicating stronger patterns and relationships in fewer cluster groupings. This insight aids in identifying the most effective clustering approach for grouping weather features efficiently.

The Spectral Clustering with a high silhouette score and calculation score is shown in Figure 4. The traits of each cluster that was found were then examined. This required creating a profile of the meteorological elements or characteristics that support the development of each cluster. We extract or build new features that capture

Table 1. Clustering algorithms with their silhouette score and computational scores

Clustering Algorithm	Silhouette Score	Computational Scores
K-Means	0.1242	0.3510
BIRCH	0.1805	0.1384
Spectral Cluster	0.25303	4.3027
DBSCAN	-0.5639	0.0556

the unique qualities of each cluster based on the cluster profiles. These characteristics can be used as model inputs for classification. The temperature, pressure, humidity, visibility, UV index, precipitation, and wind are the seven primary characteristics of this model. In order to understand cluster formations, a classification of cloud circumstances was added to analyze cluster profiling.

Gradient Boosting was shown to be the most successful machine learning model after many models—Random Forest, Decision Tree, Gradient Boosting, and Logistic Regression—were evaluated and used for classification tasks. Compared to other models, it performed better at classifying different weather situations, with an accuracy of 95.80%. When interpreting the data, it was found that this model performed exceptionally well in forecasting a variety of weather classes, including clear, fog, mist, partly cloudy, sunny, and thundery outbreaks possible. Despite the difficulties seen in specific weather classes—such as light rain, moderate rain, patchy light drizzle, and others—where memory and precision were low, the issue here is one of chance or uncertainty, which can impede the normal course of events. Overall performance showed how reliable the model is at differentiating between weather patterns based on the influence of current circumstances, indicating that weather patterns in India are cyclical and have unknown sources of uncertainty.

APPLICATIONS AND IMPLICATIONS

Potential Applications of the Findings

Unveiling India's intricate weather tapestry through BIRCH clustering and Topological Data Analysis (TDA) paves the way for many applications that can bolster India's resilience. Disaster preparedness takes center stage. Identifying outlier regions like Samastipur and Mathura, susceptible to extremes in temperature and erratic rainfall, allows for targeted early warnings and resource allocation. This proactive approach can significantly minimize the impact of floods, droughts, and heatwaves in these vulnerable areas. Sustainable agricultural practices can flourish with a

Figure 4. Cluster division based on spectral clustering analysis (n=4) with efficient silhouette score and computational score

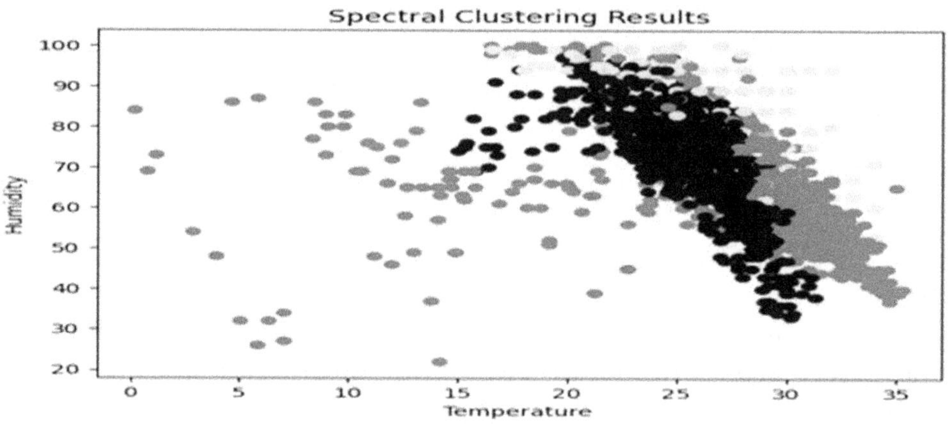

deeper understanding of regional weather variations. Cluster analysis empowers farmers to adapt their strategies. Knowing Kapurthala's potentially hotter and drier conditions can guide farmers towards drought-resistant crops and water conservation techniques, ensuring long-term food security and minimizing crop failures due to unexpected weather shifts.

Infrastructure development can also benefit from these insights. Cities like Mathura and Saharanpur, experiencing more extreme temperatures, can leverage this knowledge to incorporate heat-resistant building materials and improved drainage systems to manage potential flash floods. By understanding regional weather patterns, we can build infrastructure resilient to future weather events. Water resource management strategies can also be optimized. Understanding unpredictable rainfall patterns, particularly in outliers like Wanaparthy, allows for informed decision-making. Building additional reservoirs or implementing rainwater harvesting systems in these areas can help mitigate the impact of droughts and ensure water security for local communities. Public health protection efforts can be further strengthened. Identifying regions prone to extreme weather events empowers public health officials to issue targeted advisories and implement preventive measures. Warnings about heatwaves in Mathura can prompt heatstroke prevention campaigns, while heavy rainfall alerts in Kangpokpi can lead to vector-borne disease control initiatives.

By translating these findings into actionable strategies, we can build a more resilient India that is prepared to handle the challenges of diverse and evolving weather patterns. This empowers stakeholders – from farmers and disaster management teams to infrastructure planners and public health officials – to make informed decisions safeguarding lives, livelihoods, and resources across the country.

Implications for Weather Forecasting and Prediction

The identification of distinct weather clusters highlights the limitations of one-size-fits-all forecasts. Regions like Samastipur and Mathura, flagged as outliers with extreme temperatures and unpredictable rainfall, expose the need for more spatially nuanced predictions. Weather models can be refined to incorporate these regional variations, leading to more accurate forecasts for specific locations. TDA techniques like ToMATo Model and Persistence can delve deeper into the data, uncovering hidden relationships between weather patterns. This knowledge can improve the accuracy of weather models by incorporating these previously unseen connections. For instance, the model might identify a link between high-pressure systems in the Himalayas and increased rainfall patterns in Kangpokpi, allowing for more precise forecasts in that region. The outliers identified in the analysis, like Mathura and Wanaparthy, warrant special attention. These regions, prone to extreme temperatures, unpredictable rainfall, or altered monsoon patterns, are particularly vulnerable to weather-related disasters. By focusing on these outliers, we can develop targeted early warning systems that provide timely and location-specific alerts, allowing communities to prepare for potential hazards.

Combining the insights from TDA clustering with time series modeling holds immense potential. Time series models excel at capturing temporal trends. Integrating these models with the cluster analysis allows for the development of more dynamic forecasts. For instance, a model might predict a period of increased heatwaves in Mathura based on historical data and current cluster trends, providing valuable lead time for heatstroke prevention efforts. Understanding regional weather variations provides a more comprehensive picture of India's climate. This knowledge can be fed into long-term climate change prediction models, leading to more accurate forecasts about future weather patterns. This information is crucial for policymakers to develop effective nationwide climate change adaptation strategies for different regions. By leveraging the findings from this analysis and exploring advanced TDA techniques, we can move beyond generalized forecasts and towards a future of more precise, spatially nuanced, and temporally dynamic weather predictions for India. This will empower communities to better prepare for weather events, leading to a more resilient nation.

CONCLUSION

Summary of Key Findings

The analysis of weather patterns using the Mapper technique revealed six intriguing outliers – Samastipur, Kangpokpi, Kapurthala, Mathura, Saharanpur, and Wanaparthy. These locations stand out within their respective clusters due to distinct and potentially concerning weather peculiarities. These outliers might experience extreme or atypical temperatures, drastic temperature swings, and unpredictable rainfall patterns with seasonal anomalies. For instance, Samastipur could face intense heat, while Kangpokpi might see higher than usual rainfall. Kapurthala could be hotter and drier, and both Mathura and Saharanpur might experience more extreme temperatures with less predictable rainfall. Wanaparthy, on the other hand, could be grappling with altered monsoon patterns.

The key takeaway here is the significant difference in consistency and predictability compared to other regions. These outliers highlight the crucial importance of understanding regional weather variations across India. This knowledge is essential for sustainable development, disaster preparedness, and the overall well-being of local communities. Ignoring or misinterpreting the unique weather patterns of these outliers could have a cascading effect of potential hazards. Agricultural practices that do not account for atypical rainfall and temperature shifts can lead to disastrous crop failures and food insecurity. Misinterpreting water availability based on unpredictable rainfall patterns can cause droughts or floods, jeopardizing water resources and infrastructure. Unprepared communities face increased health risks from extreme weather events like heat waves and heavy rainfall. Unusual weather patterns can further disrupt livelihoods, infrastructure, and transportation, triggering economic losses and social unrest. These outliers serve as a stark reminder of the critical need to understand and adapt to regional weather variations across India. By incorporating this knowledge into planning and development strategies, we can build resilience, ensure food security, and create a more sustainable future for all regions of the country.

Limitations and Future Work

However, this analysis scratches the surface of what is possible. Topological Data Analysis (TDA) offers a treasure trove of techniques for further exploration. Techniques like the ToMATo Model and Persistence can be applied to delve deeper into the data's underlying structures and identify hidden relationships between weather patterns. These advanced models might reveal previously unseen connections within the data, leading to a more comprehensive understanding of India's weather dynamics. Furthermore, a comparative study combining TDA clustering with other established methods can prove highly valuable. By integrating spatial data analysis, which considers the geographical location of weather data points, alongside time series modeling that captures the temporal aspect of weather changes, we can potentially uncover even richer insights. This combined approach could provide a more holistic view of India's weather patterns, encompassing both spatial variations and temporal trends.

REFERENCES

Brown, M. (2020). Categorification of the probabilistic convergence of Mapper. *Journal of Topology and Analysis*, *12*(3), 891–912.

Carlsson, G. (2009). Topology and data. *Bulletin of the American Mathematical Society*, *46*(2), 255–308. doi:10.1090/S0273-0979-09-01249-X

Carrière, M. (2018). Towards a statistically well-founded version of Mapper. *Journal of Machine Learning Research*, *19*(1), 2321–2351.

Carrière, M., & Michel, B. (2019). Stochastic and multivariate filters in Mapper. *Proceedings of the AAAI Conference on Artificial Intelligence*, *33*(01), 1882–1889.

Carrière, M., & Oudot, S. (2017). Stability of Mapper in the small sample regime. *Advances in Neural Information Processing Systems*, *30*, 590–600.

Chazal, F. (2017). Robust topological inference: Distance to a measure and kernel distance. *Journal of Machine Learning Research*, *18*(1), 1–53.

Das, S., Ashrit, R., & Moncrieff, M. W. (2002). Simulation of a Himalayan cloudbursting event. *Journal of the Meteorological Society of Japan*, *80*(3), 505–528.

Dey, T. K. (2016). Instabilities of the Mapper algorithm. *Journal of Machine Learning Research*, *17*(5), 1–32.

Dey, T. K. (2017). Dealing with instabilities in the Mapper algorithm. *Advances in Neural Information Processing Systems*, *30*, 4045–4055.

Dhiman, G., Sayeed, U., Yunus, J., Singh, N., Ray, R. S., Gandhi, R. M., & Pathan, S. K. (2019). Human health risk assessment via heterogeneous microbial aerosols: A case study of Indian cities. *Urban Climate*, *27*, 214–226.

Dimri, A. P., Niyogi, D., Sarkar, C., & Jain, V. (2017). An Observational Study of the Impact of Topography in influencing the Weather Systems over Western Himalayas and the Associated Indo-Gangetic Plains. *Meteorology and Atmospheric Physics*, *129*(2), 199–217.

Edelsbrunner, H., Letscher, D., & Zomorodian, A. (2002). Topological persistence and simplification. *Discrete & Computational Geometry*, *28*(4), 511–533. doi:10.1007/s00454-002-2885-2

Goswami, B. N., Deshpande, M., Mukhopadhyay, P., Saha, S. K., Rao, S. A., Murthugudde, R., & Goswami, B. V. (2010). Simulation of monsoon intraseasonal variability in NCEP CFSv2 and its role on influencing seasonal mean. *Climate Dynamics*, *35*(7), 1089–1113.

Goswami, B. N., Venugopal, V., Sengupta, D., Madhusoodanan, M. S., & Xavier, P. K. (2006). Increasing trend of extreme rain events over India in a warming environment. *Science*, *314*(5804), 1442–1445. doi:10.1126/science.1132027 PMID:17138899

Gupta, A., & Arulkumar, N. (2023). An Exploratory Study of Python's Role in the Advancement of Cryptocurrency and Blockchain Ecosystems. In *Advanced Applications of Python Data Structures and Algorithms* (pp. 236–252). IGI Global. doi:10.4018/978-1-6684-7100-5.ch012

Krishnan, R., Shrestha, A. B., Ren, G., Rajbhandari, R., Saeed, S., Sanjay, J., & Refsgaard, J. C. (2019). Unravelling climate change in the Hindu Kush Himalaya: rapid warming in the mountains and increasing extremes. In *The Hindu Kush Himalaya Assessment* (pp. 57-97). Springer, Cham.

Kumar, T. S., Jothilakshmi, S., James, B. C., Prakash, M., Arulkumar, N., & Rekha, C. (2023). HHO-based vector quantization technique for biomedical image compression in cloud computing. *International Journal of Image and Graphics*, *23*(03), 2240008. doi:10.1142/S0219467822400083

Mallick, J., Kant, Y., & Bharath, B. D. (2013). Estimation of land surface temperature over Delhi using Landsat-7 ETM+. *Photonirvachak (Dehra Dun)*, *36*(2), 131–140.

Manimaran, A., Goundar, S., Chandramohan, D., & Arulkumar, N. (2023). Application Areas, Benefits, and Research Challenges of Converging Blockchain and Machine Learning Techniques. *Integrating Blockchain and Artificial Intelligence for Industry 4.0 Innovations*, 1-16.

Maria, C. (2014). GUDHI: Computational homology. *Computational Geometry*, *47*(3), 373–378.

Mishra, V., Mukherjee, S., Kumar, R., & Stone, D. A. (2014). *Heat wave exposure in India in current climate and their expected amplification under climate change.*

Mohan, P., Neelakandan, S., Mardani, A., Maurya, S., Arulkumar, N., & Thangaraj, K. (2023). Eagle strategy arithmetic optimisation algorithm with optimal deep convolutional forest based fintech application for hyper-automation. *Enterprise Information Systems, 17*(10), 2188123. doi:10.1080/17517575.2023.2188123

Mohapatra, M., Srivastava, A. K., Balachandran, S., & Geetha, B. (2017). Inter-annual variation and tendencies of tropical cyclones over the North Indian Ocean. *Mausam (New Delhi), 68*(2), 263–272.

Muñoz-Díaz, D., & Rodrigo, F. S. (2004). Spatio-temporal patterns of seasonal rainfall in Spain (1912-2000) using cluster and principal component analysis: Relationship to the North Atlantic oscillation. *Annales Geophysicae, 22*(5), 1471–1489. doi:10.5194/angeo-22-1435-2004

Nanda, A., Kumari, A., Gupta, S., & Bhattacharya, S. (2018). Topological analysis of Indian monsoon rainfall data using Mapper. In *International Conference on Pattern Recognition and Machine Intelligence* (pp. 268-275). Springer, Cham.

Nowotarski, C. J., & Jensen, C. (2019). Topology and topological data analysis for Big Data: Observations and a conceptual treatise. *Big Data and Cognitive Computing, 3*(3), 33.

Rajeevan, M., Bhate, J., Kale, J. D., & Lal, B. (2010). High resolution daily gridded rainfall data for the Indian region: Analysis of break and active monsoon spells. *Current Science*, 296–306.

Ratna, S. B., Behera, S., Ratnam, J. V., Takahashi, K., & Yamagata, T. (2016). An index for tropical temperate troughs over southern Asia. *Climate Dynamics, 46*(9-10), 3173–3192.

Sharma, K., Chakraborty, A., & Nair, A. (2019). Cyclone risk analysis and cyclone adaptation measures for South Asia. In *Regional Climate Hazards and Adaptations in South Asia* (pp. 19–62). Elsevier.

Singh, G., Mémoli, F., & Carlsson, G. E. (2007). Topological methods for the analysis of high dimensional data sets and 3d object recognition. *PBG@ Eurographics*, 2, 091-100.

Compilation of References

Aaron Jabbour, R. B. (2023). Geospatial Data, Information and Intelligence. In R. B. Aaron Jabbour, Geospatial Data, Information and Intelligence (pp. 20-50). USA.

Abanihi. (n.d.). *Tutorials PySparkGeoAnalysis Coastal Marine Zones.* NCAR. https://ncar.github.io/PySpark4Climate/tutorials/pyspark-geo-analysis/coastal-marine-zones/

Abujayyab, S. K. M. (2019). Geospatial machine learning datasets structuring and classification tool. *Case Study For Mapping Lulc From Rasat Satellite Images*, *42*(October), 1–3.

Acharya, A., Mondal, B. K., Bhadra, T., Abdelrahman, K., & Mishra, P. K. (2022). *Geospatial Analysis of Geo-Ecotourism Site Suitability Using AHP and GIS for Sustainable and Resilient Tourism Planning in West Bengal.* doi:10.3390/su14042422

Aggarwal, K., Mijwil, M. M., Alomari, S., Gök, M., Alaabdin, A. M. Z., & Abdulrhman, S. H. (2022). Has the Future Started? The Current Growth of Artificial Intelligence, Machine Learning, and Deep Learning. *Iraqi Journal for Computer Science and Mathematics*, 115–123. doi:10.52866/ijcsm.2022.01.01.013

Agrawal, R. (2020). Fundamentals of Machine Learning. In Machine Learning for Healthcare (pp. 1–16). doi:10.1201/9780429330131-1

Aksha, S. K., Resler, L. M., Juran, L., & Carstensen, L. W. Jr. (2020). A geospatial analysis of multi-hazard risk in Dharan, Nepal. *Geomatics, Natural Hazards & Risk*, *11*(1), 88–111. doi:10.1080/19475705.2019.1710580

Al Jawarneh, I. M., Bellavista, P., Corradi, A., Foschini, L., & Montanari, R. (2020). Big spatial data management for the Internet of Things: A survey. *Journal of Network and Systems Management*, *28*(4), 990–1035. doi:10.1007/s10922-020-09549-6

Al-Azizi, J. I., Shafri, H. Z. M., Hashim, S. J., & Mansor, S. B. (2022). DeepAutoMapping: Low-cost and real-time geospatial map generation method using deep learning and video streams. *Earth Science Informatics*, *15*(3), 1481–1494. doi:10.1007/s12145-020-00529-7

Albrecht, C. M., Elmegreen, B., Gunawan, O., Hamann, H. F., Klein, L. J., Lu, S., Mariano, F., Siebenschuh, C., & Schmude, J. (2020). Next-generation geospatial-temporal information technologies for disaster management. *IBM Journal of Research and Development*, *64*(1/2), 5–1. doi:10.1147/JRD.2020.2970903

Alfadul, T., & Ali, T. (2022). Geospatial Big Data Analytics Applications Trends. *Challenges & Opportunities*, *4*(1), 140–144.

Al-Hasani, B. (2023), *Integrated Geospatial approach for adaptive rainwater harvesting site selection under the impact of climate change.*

Ali, S. M., & Gupta, N. (2017). Investigation into the efficacy of geospatial big data visualization tools, (May). doi:10.1109/CCAA.2017.8229777

Alkhatib, R., Sahwan, W., Alkhatieb, A., & Schütt, B. (2023). A Brief Review of Machine Learning Algorithms in Forest Fires Science. *Applied Sciences (Basel, Switzerland)*, *13*(14), 8275. doi:10.3390/app13148275

Alsaedi, N., Sadeq, A., & Jalal, A. (2023). Big Spatial Data Systems -. *RE:view*, (August). doi:10.1109/IICETA54559.2022.9888481

Andrienko, G. L., Andrienko, N., Keim, D. A., MacEachren, A. M., & Wrobel, S. (2011). *Challenging problems of geospatial visual analytics.*

Andronie, M., & Iatagan, M. (2022). *Deep Learning-Assisted Smart Process Planning*. Robotic Wireless Sensor Networks, and Geospatial Big Data Management Algorithms in the Internet of Manufacturing Things.

Anismuslim, M., Pramoedyo, H., & Andarini, S. (2023). Modeling of Risk Factors of Childhood Stunting Cases In Malang Regency using Geographically Weighted Regression (GWR) Abstract :, 1–11. https://doi.org/ doi:10.2174/18749445-v17-e230420-2022-165

Anitha Elavarasi, S., & Jayanthi, J. (2022). Programming language support for implementing machine learning algorithms. In *Research Anthology on Machine Learning Techniques*. Methods, and Applications. doi:10.4018/978-1-6684-6291-1.ch028

Antenucci, J. C., Brown, K., Croswell, P. L., Kevany, M. J., & Archer, H. (1991). *Geographic Information Systems: a guide to the technology* (Vol. 115). Van Nostrand Reinhold.

Araujo, V., Pacifico, M. J., & Pinheiro, M. (2013). Adapted random perturbations for non-uniformly expanding maps. *Stochastics and Dynamics*, *14*(4), 1450007. doi:10.1142/S0219493714500075

Arundel, S. T., Li, W., & Wang, S. (2020). GeoNat v1.0: A dataset for natural feature mapping with artificial intelligence and supervised learning. *Transactions in GIS*, *24*(3), 556–572. doi:10.1111/tgis.12633

Austin, L. M. (2006). *Reviewing PIPEDA: Control.* Privacy and the Limits of Fair Information Practices.

Barber, D. (2012). Bayesian Reasoning and Machine Learning. In Bayesian Reasoning and Machine Learning. doi:10.1017/CBO9780511804779

Bassi, S. (2007). A primer on Python for life science researchers. In PLoS Computational Biology, 3(11). doi:10.1371/journal.pcbi.0030199

Bédard, Y., Wright, E., & Rivest, S. (2015), *A Guide to Geospatial Data Quality.*

Bendechache, M., Le-Khac, N., & Kechadi, M. T. (2016). Hierarchical Aggregation Approach for Distributed Clustering of Spatial Datasets. *2016 IEEE 16th International Conference on Data Mining Workshops (ICDMW),* (pp. 1098-1103). IEEE.

Bera, A., Taloor, A. K., Meraj, G., Kanga, S., Singh, S. K., Đurin, B., & Anand, S. (2021). Climate vulnerability and economic determinants: Linkages and risk reduction in Sagar Island, India; A geospatial approach. *Quaternary science advances, 4*, 100038.

Bertino, E., Thuraisingham, B., Gertz, M., & Damiani, M. L. (2008, November). Security and privacy for geospatial data: concepts and research directions. In *Proceedings of the SIGSPATIAL ACM GIS 2008 International Workshop on Security and Privacy in GIS and LBS* (pp. 6-19). ACM. 10.1145/1503402.1503406

Beuthner, C., Breuer, J., & Jünger, S. (2021). *Data linking—linking survey data with geospatial, social media, and sensor data.* GESIS Survey Guidelines.

Bickler, S. H. (2021). Machine Learning Arrives in Archaeology. In Advances in Archaeological Practice, 9(2). doi:10.1017/aap.2021.6

Bilina, R., & Lawford, S. (2012). Python for Unified Research in Econometrics and Statistics. *Econometric Reviews*, *31*(5), 558–591. doi:10.1080/07474938.2011.553573

Blatt, A. J. (2012). Ethics and Privacy Issues in the Use of GIS. *Journal of Map & Geography Libraries*, *8*(1), 80–84. doi:10.1080/15420353.2011.627109

Bonald, T., De Lara, N., Lutz, Q., & Charpentier, B. (2020). Scikit-network: Graph analysis in python. *Journal of Machine Learning Research*, 21.

Bonney, R., Ballard, H., Jordan, R., McCallie, E., Phillips, T., Shirk, J., & Wilderman, C. C. (2009). *Public Participation in Scientific Research: Defining the Field and Assessing Its Potential for Informal Science Education. A CAISE Inquiry Group Report.*

Borcherds, P. H. (2007). Python: a language for computational physics. *Computer Physics Communications, 177*(1-2 SPEC. ISS.). doi:10.1016/j.cpc.2007.02.019

Breunig, M., Bradley, P. E., Jahn, M., Kuper, P., Mazroob, N., Rösch, N., & Jadidi, M. (n.d.). Geospatial Data Management Research : Progress and Future Directions.

Breunig, M., Bradley, P., Jahn, M., & Kuper, P. (2020). Geospatial Data Management Research: Progress and Future Directions. *ISPRS International Journal of Geo-Information.*

Breunig, M., Bradley, P. E., Jahn, M. W., Kuper, P. V., Mazroob, N., Rösch, N., Al-Doori, M., Stefanakis, E., & Jadidi, M. (2020). Geospatial Data Management Research: Progress and Future Directions. *ISPRS International Journal of Geo-Information*, *9*(2), 95. doi:10.3390/ijgi9020095

Brown, M. (2020). Categorification of the probabilistic convergence of Mapper. *Journal of Topology and Analysis*, *12*(3), 891–912.

Brundage, M., Avin, S., Clark, J., Toner, H., Eckersley, P., Garfinkel, B., & Amodei, D. (2018). The malicious use of artificial intelligence: Forecasting, prevention, and mitigation. *arXiv preprint arXiv:1802.07228*.

Bystriakova, N., Peregrym, M., Erkens, R., Bezsmertna, O., & Schneider, H. (2012). Sampling bias in geographic and environmental space and its effect on the predictive power of species distribution models. *Systematics and Biodiversity*, *10*(3), 305–315. doi:10.1080/14772000.2012.705357

Cao, Q. (2022). Reshaping Geostatistical Modeling and Prediction for Extreme-Scale Environmental Applications. *SC22: International Conference for High Performance Computing, Networking, Storage and Analysis*, Dallas, TX, USA. 10.1109/SC41404.2022.00007

Cao, R., Tu, W., Cai, J., Zhao, T., Xiao, J., Cao, J., Gao, Q., & Su, H. (2022). Machine learning-based economic development mapping from multi-source open geospatial data. *ISPRS Annals of the Photogrammetry, Remote Sensing and Spatial Information Sciences*, *5*(4), 259–266. doi:10.5194/isprs-annals-V-4-2022-259-2022

Cao, Y., Xu, C., Aziz, N. M., & Kamaruzzaman, S. N. (2023). BIM–GIS integrated utilization in urban disaster management: The contributions, challenges, and future directions. *Remote Sensing (Basel)*, *15*(5), 1331. doi:10.3390/rs15051331

Carleo, G., Cirac, I., Cranmer, K., Daudet, L., Schuld, M., Tishby, N., Vogt-Maranto, L., & Zdeborová, L. (2019). Machine learning and the physical sciences. *Reviews of Modern Physics*, *91*(4), 045002. doi:10.1103/RevModPhys.91.045002

Carlsson, G. (2009). Topology and data. *Bulletin of the American Mathematical Society*, *46*(2), 255–308. doi:10.1090/S0273-0979-09-01249-X

Carrière, M. (2018). Towards a statistically well-founded version of Mapper. *Journal of Machine Learning Research*, *19*(1), 2321–2351.

Carrière, M., & Michel, B. (2019). Stochastic and multivariate filters in Mapper. *Proceedings of the AAAI Conference on Artificial Intelligence*, *33*(01), 1882–1889.

Carrière, M., & Oudot, S. (2017). Stability of Mapper in the small sample regime. *Advances in Neural Information Processing Systems*, *30*, 590–600.

Cartography: Visualization of Geospatial Data. In (2013). *M.-J. K. ORMELING, Cartography: Visualization of Geospatial Data* (pp. 34–70). Pearson Education Limited.

Chai, X., Yang, T., Gu, H., Tang, G., Cao, W., & Wang, Y. (2023). Geophysics-steered self-supervised learning for deconvolution. *Geophysical Journal International, 234*(1), 40–55. doi:10.1093/gji/ggad015

Chang, K. T. (2008). *Introduction to geographic information systems* (Vol. 4). McGraw-Hill.

Chang, S., Wang, Z., Mao, D., Guan, K., Jia, M., & Chen, C. (2020). Mapping the essential urban land use in changchun by applying random forest and multi-source geospatial data. *Remote Sensing (Basel), 12*(15), 2488. doi:10.3390/rs12152488

Chang, V., Mou, Y., & Xu, Q. A. (2021). The ethical issues of location-based services on big data and IoT. In *Modern Industrial IoT, Big Data and Supply Chain* [Springer Singapore.]. *Proceedings of the IIoTBDSC, 2020*, 195–205.

Chazal, F. (2017). Robust topological inference: Distance to a measure and kernel distance. *Journal of Machine Learning Research, 18*(1), 1–53.

Cheng, Q., Zhu, Y., Zeng, H., Song, J., Wang, S., Zhang, J., Qian, L., & Qi, Y. (2021). A method for identifying geospatial data sharing websites by combining multi-source semantic information and machine learning. *Applied Sciences (Basel, Switzerland), 11*(18), 8705. doi:10.3390/app11188705

Chrisman, N. (2002). *Exploring geographic information systems* (2nd ed.). John Wiley & Sons.

Cinnamon, J. (2020). *Geographic Information Systems; Ethics*. International Encyclopedia of Human Geography. doi:10.1016/B978-0-08-102295-5.10554-2

Cobb, S. (2016). *Data privacy and data protection: US law and legislation.*

Cobb, C. D. (2003). GEOGRAPHIC METHODS & POLICY: Using Geographic Information Systems to Inform Education Policy. *Educational Research Quarterly, 27*, 28.

Cobb, C. D. (2020). Geospatial Analysis: A New Window Into Educational Equity, Access, and Opportunity. *Review of Research in Education, 44*(1), 129–197. doi:10.3102/0091732X20907362

Coetzee, S., Ivánová, I., Mitasova, H., & Brovelli, M. A. (2020). Open geospatial software and data: A review of the current state and a perspective into the future. *ISPRS International Journal of Geo-Information, 9*(2), 90. doi:10.3390/ijgi9020090

Coleman, D., Georgiadou, Y., & Labonte, J. (2009). Volunteered geographic information: The nature and motivation of produsers. *International Journal of Spatial Data Infrastructures Research, 4*(4), 332–358.

Colonna, L. (2020). Privacy, Risk, Anonymization and Data Sharing in the Internet of Health Things. *Pittsburgh Journal of Technology Law & Policy, 20*(1), 148–175. doi:10.5195/tlp.2020.235

Crampton, J. (1995). The ethics of GIS. *Cartography and Geographic Information Systems, 22*(1), 84–89. doi:10.1559/152304095782540546

Crickard, P. III. (2014). *Leaflet. js essentials*. Packt Publishing Ltd.

Crompvoets, J., Rajabifard, A., Van Loenen, B., & Fernández, T. D. (2008). A multi-view framework to assess SDIs. *Wageningen University, RGI.*

Curry, M. R. (1997). Digital people, digital places: Rethinking privacy in a world of geographic information. *Ethics & Behavior*, *7*(3), 253–263. doi:10.1207/s15327019eb0703_7

Dai, L., Wu, L., Li, H., Cai, C., Wu, Q., Kong, H., Liu, R., Wang, X., Hou, X., Liu, Y., Long, X., Wen, Y., Lu, L., Shen, Y., Chen, Y., Shen, D., Yang, X., Zou, H., Sheng, B., & Jia, W. (2021). retinopathy across the disease spectrum. *Nature Communications*, *12*(1), 3242. doi:10.1038/s41467-021-23458-5 PMID:34050158

Dalla Mura, M., Prasad, S., Pacifici, F., Gamba, P., Chanussot, J., & Benediktsson, J. A. (2015). Challenges and Opportunities of Multimodality and Data Fusion in Remote Sensing. *Proceedings of the IEEE*, *103*(9), 1585–1601. doi:10.1109/JPROC.2015.2462751

Dangermond, J., & Goodchild, M. F. (2020). Building geospatial infrastructure. *Geo-Spatial Information Science*, *23*(1), 1–9. doi:10.1080/10095020.2019.1698274

Dano, U. L., Balogun, A. L., Abubakar, I. R., & Aina, Y. A. (2020). Transformative urban governance: Confronting urbanization challenges with geospatial technologies in Lagos, Nigeria. *GeoJournal*, *85*(4), 1039–1056. doi:10.1007/s10708-019-10009-1

Das, S. K. (n.d.). *Geospatial Data Analytics- A Deep Learning Perspective.*

Das, S., Ashrit, R., & Moncrieff, M. W. (2002). Simulation of a Himalayan cloudbursting event. *Journal of the Meteorological Society of Japan*, *80*(3), 505–528.

David. In (2023). *Designing Geodatabases: Case Studies in GIS Data Modeling* (pp. 300–350). ESRI.

Dc Smith, M. J., Goodchild, M. F., & Longley, P. (2007). *Geospatial analysis: a comprehensive guide to principles, techniques and software tools.* Troubador publishing ltd.

Deb, S., & Karmakar, S. (2023). A novel spatio-temporal clustering algorithm with applications on COVID-19 data from the United States. *Computational Statistics & Data Analysis*, *188*, 107810. doi:10.1016/j.csda.2023.107810

Dey, T. K. (2016). Instabilities of the Mapper algorithm. *Journal of Machine Learning Research*, *17*(5), 1–32.

Dey, T. K. (2017). Dealing with instabilities in the Mapper algorithm. *Advances in Neural Information Processing Systems*, *30*, 4045–4055.

Dhand, S., Singh, A., Litoria, P., & Pateriya, B. (2021). *Geo-smart City Planning Using Geospatial Techniques.*

Dhiman, G., Sayeed, U., Yunus, J., Singh, N., Ray, R. S., Gandhi, R. M., & Pathan, S. K. (2019). Human health risk assessment via heterogeneous microbial aerosols: A case study of Indian cities. *Urban Climate*, *27*, 214–226.

Dimri, A. P., Niyogi, D., Sarkar, C., & Jain, V. (2017). An Observational Study of the Impact of Topography in influencing the Weather Systems over Western Himalayas and the Associated Indo-Gangetic Plains. *Meteorology and Atmospheric Physics*, *129*(2), 199–217.

Dodd, L., Muir, R., Noll, B., Cubero, A. R., Solodkov, N., & Koraktis, K. (2023). State of the Developer Nation Q3 2023. In SlashData (Issue August).

Donovan, J., Caplan, R., Matthews, J., & Hanson, L. (2018). Algorithmic accountability. *PRiMER: Peer-Reviewed Reports in Medical Education Research.*

Du, P., Bai, X., Tan, K., Xue, Z., Samat, A., Xia, J., Li, E., Su, H., & Liu, W. (2020). Advances of Four Machine Learning Methods for Spatial Data Handling: a Review. In Journal of Geovisualization and Spatial Analysis, 4(1). doi:10.1007/s41651-020-00048-5

Dubois, P. F. (1997). Is Java for Scientific Programming? *Computers in Physics*, *11*(6), 611–617. doi:10.1063/1.4822611

Dueker, K. J., & Butler, J. A. (2000). A geographic information system framework for transportation data sharing. *Transportation Research Part C, Emerging Technologies*, *8*(1-6), 13–36. doi:10.1016/S0968-090X(00)00006-1

Eckman, S., & Himelein, K. (2019). *Methods of Geospatial Sampling*. Data Collection in Fragile States.

Edelsbrunner, H., Letscher, D., & Zomorodian, A. (2002). Topological persistence and simplification. *Discrete & Computational Geometry*, *28*(4), 511–533. doi:10.1007/s00454-002-2885-2

Ekeu-Wei, I. T., & Blackburn, G. A. (2020). Catchment-scale flood modelling in data-sparse regions using open-access geospatial technology. *ISPRS International Journal of Geo-Information*, *9*(9), 512. doi:10.3390/ijgi9090512

El Hamdani, S., Loudari, S., Novotny, S., Bouchner, P., & Benamar, N. (2021). *A Markov Decision Process Model for a Reinforcement Learning-based Autonomous Pedestrian Crossing Protocol.* 2021 3rd IEEE Middle East and North Africa COMMunications Conference (MENACOMM), Agadir, Morocco. 10.1109/MENACOMM50742.2021.9678310

Elwood, S., & Cope, M. (2009). *Introduction: Qualitative Gis: Forging Mixed Methods Through Representations*. Analytical Innovations, and Conceptual Engagements.

Elwood, S., & Leszczynski, A. (2011). Privacy, reconsidered: New representations, data practices, and the geoweb. *Geoforum*, *42*(1), 6–15. doi:10.1016/j.geoforum.2010.08.003

Emam, W. W., & Soliman, K. M. (2022). Geospatial analysis, source identification, contamination status, ecological and health risk assessment of heavy metals in agricultural soils from Qallin city, Egypt. *Stochastic Environmental Research and Risk Assessment*, *36*(9), 2437–2459. doi:10.1007/s00477-021-02097-8

Fabrègue, B. F., & Bogoni, A. (2023). *Privacy and Security Concerns in the Smart City*. Smart Cities. doi:10.3390/smartcities6010027

FoliniA.LenziE.BiraghiC. A. (2022). CLUSTER ANALYSIS: A COMPREHENSIVE AND VERSATILE QGIS PLUGIN FOR PATTERN RECOGNITION IN GEOSPATIAL DATA. *International Archives of the Photogrammetry, Remote Sensing and Spatial Information Sciences - ISPRS Archives, 48*(4/W1-2022). doi:10.5194/isprs-archives-XLVIII-4-W1-2022-151-2022

Foody, G. M. (2006). Privacy in the public eye: Exploring the phenomenon of geospatial stigma. *Area, 38*(3), 311–321.

Forghani, A., Kazemi, S., & Bruce, D. (2021). A Machine-Learning Approach to Generalisation of GIS Data. *International Journal of Geoinformatics, 17*(2), 41–59. doi:10.52939/ijg.v17i2.1757

Franch-Pardo, I., Napoletano, B. M., Rosete-Verges, F., & Billa, L. (2020). Spatial analysis and GIS in the study of COVID-19. A review. *The Science of the Total Environment, 739*, 140033. doi:10.1016/j.scitotenv.2020.140033 PMID:32534320

Freire, S., Macmanus, K., Pesaresi, M., Doxsey-Whitfield, E., & Mills, B. (2016). *Development of new open and free multi-temporal global population grids at 250 m resolution.*

Gao, K., Mei, G., Piccialli, F., Cuomo, S., Tu, J., & Huo, Z. (2020). Julia language in machine learning: Algorithms, applications, and open issues. In Computer Science Review (Vol. 37). doi:10.1016/j.cosrev.2020.100254

Gao, Z., & Ding, M. (2022). Application of convolutional neural network fused with machine learning modeling framework for geospatial comparative analysis of landslide susceptibility. *Natural Hazards, 113*(2), 833–858. doi:10.1007/s11069-022-05326-7

Garfinkel, S.L. (2015). De-Identification of Personal Information.

Garrard, C. (2016). *Geoprocessing with python*. Simon and Schuster.

Gerdan, S. (2018). GIS-based decision-support system applications in disaster management. *Yönetim ve Ekonomi Dergisi, 25*(3), 961–979. doi:10.18657/yonveek.306383

Ghinita, G. (2013). *Privacy for location-based services* (Vol. 4). Morgan & Claypool Publishers. doi:10.1007/978-3-031-02336-1

Giannopoulos, A., Spantideas, S., Kapsalis, N., Gkonis, P., Sarakis, L., Capsalis, C., Vecchio, M., & Trakadas, P. (2022). Supporting Intelligence in Disaggregated Open Radio Access Networks: Architectural Principles, AI/ML Workflow, and Use Cases. *IEEE Access : Practical Innovations, Open Solutions, 10*, 39580–39595. doi:10.1109/ACCESS.2022.3166160

Gillies, S. (2023a). *The fiona user manual*. Read the Docs. https://fiona.readthedocs.io/en/stable/manual.html

Golubev, A., Chechetkin, I., Parygin, D., Sokolov, A., & Shcherbakov, M. (2016). Geospatial Data Generation and Preprocessing Tools for Urban Computing System Development. *Procedia Computer Science*, *101*, 217–226. doi:10.1016/j.procs.2016.11.026

Gomes, V. C., Queiroz, G. R., & Ferreira, K. R. (2020). An overview of platforms for big earth observation data management and analysis. *Remote Sensing (Basel)*, *12*(8), 1253. doi:10.3390/rs12081253

Gómez, P. M., Santiago, A. R., Seco, G. G., Casanova, R., MacKenzie, D., & Tucker, C. (2022). Ethics in the use of geospatial information in the Americas. Technology in Society, 69, 101964.

Goodchild, M. F. (2009). Geographic information systems and science: Today and tomorrow. *Annals of GIS*, *15*(1), 3–9. doi:10.1080/19475680903250715

Goodchild, M. F., Steyaert, L. T., Parks, B. O., Johnston, C., Maidment, D., Crane, M., & Glendinning, S. (Eds.). (1996). *GIS and environmental modeling: progress and research issues.*

Gordon, E., Elwood, S., & Mitchell, K. (2016). Critical spatial learning: Participatory mapping, spatial histories, and youth civic engagement. *Children's Geographies*, *14*(5), 558–572. doi:10.1080/14733285.2015.1136736

Goswami, B. N., Deshpande, M., Mukhopadhyay, P., Saha, S. K., Rao, S. A., Murthugudde, R., & Goswami, B. V. (2010). Simulation of monsoon intraseasonal variability in NCEP CFSv2 and its role on influencing seasonal mean. *Climate Dynamics*, *35*(7), 1089–1113.

Goswami, B. N., Venugopal, V., Sengupta, D., Madhusoodanan, M. S., & Xavier, P. K. (2006). Increasing trend of extreme rain events over India in a warming environment. *Science*, *314*(5804), 1442–1445. doi:10.1126/science.1132027 PMID:17138899

Grekousis, G. (2020). *Spatial analysis methods and practice: describe–explore–explain through GIS*. Cambridge University Press. doi:10.1017/9781108614528

Guillame-Bert, M., Bruch, S., Stotz, R., & Pfeifer, J. (2023). Yggdrasil Decision Forests: A Fast and Extensible Decision Forests Library. *Proceedings of the ACM SIGKDD International Conference on Knowledge Discovery and Data Mining*. ACM. 10.1145/3580305.3599933

Gunawardena, N. K. (2014). *Introduction to geographic information system, (August).*

Güney, S., & Erdaş, Ç. B. (2019). *A Deep LSTM Approach for Activity Recognition*. 2019 42nd International Conference on Telecommunications and Signal Processing (TSP), Budapest, Hungary. 10.1109/TSP.2019.8768815

Gupta, V., Mishra, V. K., Singhal, P., & Kumar, A. (2022). *An Overview of Supervised Machine Learning Algorithm.* 2022 11th International Conference on System Modeling & Advancement in Research Trends (SMART), Moradabad, India. 10.1109/SMART55829.2022.10047618

Gupta, A., & Arulkumar, N. (2023). An Exploratory Study of Python's Role in the Advancement of Cryptocurrency and Blockchain Ecosystems. In *Advanced Applications of Python Data Structures and Algorithms* (pp. 236–252). IGI Global. doi:10.4018/978-1-6684-7100-5.ch012

Haining, R. (2022). *Spatial Data Analysis Theory and Practice*. University of Cambridge.

Hall, O., Ohlsson, M., & Rögnvaldsson, T. (2022). A review of explainable AI in the satellite data, deep machine learning, and human poverty domain. In Patterns, 3(10). Cell Press. doi:10.1016/j.patter.2022.100600

Hao, J., & Ho, T. K. (2019). Machine Learning Made Easy: A Review of Scikit-learn Package in Python Programming Language. In Journal of Educational and Behavioral Statistics, 44(3). doi:10.3102/1076998619832248

Hassan, A., & Vijayaraghavan, J. (2019). *Geospatial Data Science Quick Start Guide: Effective techniques for performing smarter geospatial analysis using location intelligence*. Packt Publishing Ltd.

Heinold, B. (2018). *A Practical Introduction to Python Programming*.

Helm, J. M., Swiergosz, A. M., Haeberle, H. S., Karnuta, J. M., Schaffer, J. L., Krebs, V. E., Spitzer, A. I., & Ramkumar, P. N. (2020). Machine Learning and Artificial Intelligence: Definitions, Applications, and Future Directions. *Current Reviews in Musculoskeletal Medicine*, *13*(1), 69–76. doi:10.1007/s12178-020-09600-8 PMID:31983042

Hodges, J. S., & Schnell, P. M. (2023). Alleviating confounding in spatio-temporal areal models with an application on crimes against women in India, *23*(x), 9–30. doi:10.1177/1471082X211015452

Hoffimann, J., Zortea, M., de Carvalho, B., & Zadrozny, B. (2021). Geostatistical Learning: Challenges and Opportunities. *Frontiers in Applied Mathematics and Statistics*, *7*, 689393. doi:10.3389/fams.2021.689393

Holt, D., Steel, D. G., Tranmer, M., & Wrigley, N. (2010). Aggregation and Ecological Effects in Geographically Based Data. *Geographical Analysis*, *28*(3), 244–261. doi:10.1111/j.1538-4632.1996.tb00933.x

Hoofnagle, C. J., van der Sloot, B., & Borgesius, F. J. (2019). The European Union general data protection regulation: What it is and what it means*. *Information & Communications Technology Law*, *28*(1), 65–98. doi:10.1080/13600834.2019.1573501

Huang, X., Ma, T., Liu, C., & Liu, S. (2022). GriT-DBSCAN : A Spatial Clustering Algorithm for Very Large Databases, 1–18.

Huang, H., Yao, X. A., Krisp, J. M., & Jiang, B. (2021). Analytics of location-based big data for smart cities: Opportunities, challenges, and future directions. Computers, Environment and Urban Systems, 90, 101712. Chang, V., Mou, Y., & Xu, Q. A. (2021). The ethical issues of location-based services on big data and IoT. In Modern Industrial IoT, Big Data and Supply Chain [Springer Singapore.]. *Proceedings of the IIoTBDSC*, *2020*, 195–205.

Hu, N., Meng, X., Liu, H., & Li, Y. (2022, June). Exploring Geospatial Data Visualization Based on Python. In *Proceedings of the 2022 2nd International Conference on Control and Intelligent Robotics* (pp. 858-862). IEEE. 10.1145/3548608.3559322

Hu, Y. (2017). *Spatial Data Infrastructures*. Geographic Information Science & Technology Body of Knowledge. doi:10.22224/gistbok/2017.2.1

Iftikhar, N., & Pedersen, T. (2010). Using a Time Granularity Table for Gradual Granular Data Aggregation. *Fundamenta Informaticae*.

Ighile, E. H., Shirakawa, H., & Tanikawa, H. (2022). *Application of GIS and Machine Learning to Predict Flood Areas in Nigeria*.

Imambi, S., Prakash, K. B., & Kanagachidambaresan, G. R. (2021). PyTorch. *Programming with TensorFlow: Solution for Edge Computing Applications*, 87–104.

Isazade, V., Baser, A., Pinliang, Q., Gordana, D., & Esmail, K. (2023). Integration of Moran' s I, geographically weighted regression (GWR), and ordinary least square (OLS) models in spatiotemporal modeling of COVID - 19 outbreak in Qom and Mazandaran Provinces, Iran. *Modeling Earth Systems and Environment*, *9*(4), 3923–3937. doi:10.1007/s40808-023-01729-y PMID:36820101

Jafar, A., Dollah, R., Sakke, N., Mapa, M. T., Hua, A. K., Eboy, O. V., Joko, E. P., Hassan, D., & Hung, C. V. (2022). Assessing the challenges of e-learning in Malaysia during the pandemic of Covid-19 using the geo-spatial approach. *Scientific Reports*, *12*(1), 17316. doi:10.1038/s41598-022-22360-4 PMID:36243784

Jalil, N. A., Hwang, H. J., & Dawi, N. M. (2019). Machines learning trends, perspectives and prospects in education sector. *ACM International Conference Proceeding Series*, (pp. 201–205). ACM. 10.1145/3345120.3345147

Jena, R., Pradhan, B., Prasanajit, S., & Alamri, A. M. (2021). Geoscience Frontiers Earthquake risk assessment in NE India using deep learning and geospatial analysis. *Geoscience Frontiers*, *12*(3), 101110. doi:10.1016/j.gsf.2020.11.007

Jiang, B. (2015). Geospatial analysis requires a different way of thinking: The problem of spatial heterogeneity. *GeoJournal*, *80*(1), 1–13. doi:10.1007/s10708-014-9537-y

Jiang, S., Bevacqua, E., & Zscheischler, J. (2022). River flooding mechanisms and their changes in Europe revealed by explainable machine learning. *Hydrology and Earth System Sciences*, *26*(24), 6339–6359. doi:10.5194/hess-26-6339-2022

Jiang, Y., Li, Y., Yang, C., Hu, F., Armstrong, E. M., Huang, T., Moroni, D., McGibbney, L. J., & Finch, C. J. (2018). Towards intelligent geospatial data discovery: A machine learning framework for search ranking. *International Journal of Digital Earth*, *11*(9), 956–971. doi:10.1080/17538947.2017.1371255

Jing, F., Li, Z., Qiao, S., Zhang, J., Olatosi, B., & Li, X. (2023). Using Geospatial social media data for infectious disease studies: A systematic review. *International Journal of Digital Earth*, *16*(1), 130–157. doi:10.1080/17538947.2022.2161652 PMID:37997607

Jobin, A., Ienca, M., & Vayena, E. (2019). The global landscape of AI ethics guidelines. *Nature Machine Intelligence*, *1*(9), 389–399. doi:10.1038/s42256-019-0088-2

Jordan, M. I., & Mitchell, T. M. (2015). Machine learning: Trends, perspectives, and prospects. In Science, 349(6245). doi:10.1126/science.aaa8415

Jordan, D. S. (2023). *Applied Geospatial Data Science with Python: Leverage geospatial data analysis and modeling to find unique solutions to environmental problems.* Packt Publishing Ltd.

Joseph, S. (2023). AI in Wildlife Conservation: Poaching Prevention and Ethics. *Medium.* https://medium.com/@staneyjoseph.in/ai-in-wildlife-conservation-poaching-prevention-and-ethics-03e2076f1a3a

Kande, Y., Diogoul, N., Brehmer, P., Dabo-Niang, S., Ngom, P., & Perrot, Y. (2024). Demonstrating the relevance of spatial-functional statistical analysis in marine ecological studies: The case of environmental variations in micronektonic layers. *Ecological Informatics, 81*, 102547. doi:10.1016/j.ecoinf.2024.102547

Kanevski, A. P. (2024). *Machine learning algorithms for geo spatial data. Applications and software*. Research Gate.

Kanevski, M., Pozdnukhov, A., & Timonin, V. (2008). *Machine Learning Algorithms for GeoSpatial Data*. Applications and Software Tools Machine Learning Algorithms for GeoSpatial.

Kang, S., Kim, D., & Cho, S. (2019). Approximate training of one-class support vector machines using expected margin. *Computers & Industrial Engineering, 130.* doi:10.1016/j.cie.2019.03.029

Kantabutra, V. (2021). GeoDataLinks: A Suggestion for a Replacement for the ESRI Shapefile. In Advances in Data Science and Information Engineering: Proceedings from ICDATA 2020 and IKE 2020 (pp. 685-694). Springer International Publishing. doi:10.1007/978-3-030-71704-9_50

Kanungo, T., Mount, D. M., Netanyahu, N. S., Piatko, C. D., Silverman, R., & Wu, A. Y. (2002, July). An efficient k-means clustering algorithm: Analysis and implementation. *IEEE Transactions on Pattern Analysis and Machine Intelligence, 24*(7), 881–892. doi:10.1109/TPAMI.2002.1017616

Karamelaki. (2023). *Guide on Geospatial Data Integration.* ESRI. https://support.esri.com/en/otherresources/gis-dictionary/term/7cbd3f7c-e17f-4bb0-a51a-318ccf5b68f1#:~:text=and%20

Karpatne, A., Ebert-Uphoff, I., Ravela, S., Babaie, H. A., & Kumar, V. (2019). Machine Learning for the Geosciences: Challenges and Opportunities. *IEEE Transactions on Knowledge and Data Engineering, 31*(8), 1544–1554. doi:10.1109/TKDE.2018.2861006

Kedron, P., & Trgovac, A. B. (2021). Assessing connections and tradeoffs between geospatial data ethics, privacy, and the effectiveness of digital contact tracing technologies. *Mapping COVID-19 in Space and Time: Understanding the Spatial and Temporal Dynamics of a Global Pandemic,* 115-136.

Kedron, P. J., Li, W., Fotheringham, S. A., & Goodchild, M. F. (2020). Reproducibility and replicability: Opportunities and challenges for Geospatial research. *International Journal of Geographical Information Science, 35*(3), 427–445. doi:10.1080/13658816.2020.1802032

KELLEHER, J. D. (2019). *Deepl learning.*

Kempeneers, P. (2023). A Python Package for the Analysis of Geospatial Data. *International Journal of Geo-Information.*

Kerski, J. J. (2003). The Implementation and Effectiveness of Geographic Information Systems Technology and Methods in Secondary Education. *The Journal of Geography*, *102*(3), 128–137. doi:10.1080/00221340308978534

Kitchin, R., & Dodge, M. (2014). *Code/space: Software and everyday life*. Mit Press.

Kiwelekar, A. W., Mahamunkar, G. S., Netak, L. D., & Nikam, V. B. (2020). *Deep Learning Techniques for Geospatial Data Analysis*. doi:10.1007/978-3-030-49724-8_3

Kiwelekar, A. W., Mahamunkar, G. S., Netak, L. D., & Nikam, V. B. (2020). Deep learning techniques for geospatial data analysis. *Machine Learning Paradigms: Advances in Deep Learning-based Technological Applications,* 63-81.

Kliegr, T., Bahník, Š., & Fürnkranz, J. (2020). Advances in Machine Learning for the Behavioral Sciences. *The American Behavioral Scientist*, *64*(2), 145–175. doi:10.1177/0002764219859639

Koh, K., Hyder, A., Karale, Y., & Boulos, M. N. K. (2022). Big Geospatial Data or Geospatial Big Data ? *A Systematic Narrative Review on the Use of Spatial Data Infrastructures for Big Geospatial Sensing Data in Public Health.*

Kolesnikov, A., Kikin, P., & Panidi, E. (2021). Modelling of traffic flows and supply chains based on geospatial knowledge. *Journal of Physics: Conference Series*, *2068*(1), 012042. doi:10.1088/1742-6596/2068/1/012042

Kölle, M., Walter, V., Schmohl, S., & Soergel, U. (2020). Hybrid Acquisition of High Quality Training Data for Semantic Segmentation of 3D Point Clouds Using Crowd-Based Active Learning. *ISPRS Annals of the Photogrammetry, Remote Sensing and Spatial Information Sciences*, *5*(2), 501–508. doi:10.5194/isprs-annals-V-2-2020-501-2020

Korosov, A. A., Hansen, M. W., Dagestad, K.-F., Yamakawa, A., Vines, A., & Riechert, M. (2016). Nansat: A Scientist-Orientated Python Package for Geospatial Data Processing. *Journal of Open Research Software*, *4*(1), 39. doi:10.5334/jors.120

Korstanje, J. (2022). Creating Maps. In *Machine Learning on Geographical Data Using Python: Introduction into Geodata with Applications and Use Cases* (pp. 79–101). Apress. doi:10.1007/978-1-4842-8287-8_4

Kovacs-györi, A., Ristea, A., Havas, C., Meha, M., Hochmair, H. H., Resch, B., & Blaschke, T. (2020). Opportunities and Challenges of Geospatial Analysis for Promoting Urban Livability in the Era of Big Data and Machine Learning.

Kress, A. C., Asberry, A., Taillepierre, J. D., Johns, M. M., Tucker, P., & Penman-Aguilar, A. (2021). Collection of Data on Sex, Sexual Orientation, and Gender Identity by US Public Health Data and Monitoring Systems, 2015–2018. *International Journal of Environmental Research and Public Health*, 18. PMID:34831945

Krishnan, R., Shrestha, A. B., Ren, G., Rajbhandari, R., Saeed, S., Sanjay, J., & Refsgaard, J. C. (2019). Unravelling climate change in the Hindu Kush Himalaya: rapid warming in the mountains and increasing extremes. In *The Hindu Kush Himalaya Assessment* (pp. 57-97). Springer, Cham.

Kross, A., Kaur, G., & Jaeger, J. A. (2022). A geospatial framework for the assessment and monitoring of environmental impacts of agriculture. *Environmental Impact Assessment Review*, *97*, 106851. doi:10.1016/j.eiar.2022.106851

Krygier, J., & Wood, D. (2016). *Making maps: a visual guide to map design for GIS*. Guilford Publications.

Kucklick, J., Müller, J., Beverungen, D., & Mueller, O. (2021). *AIS Electronic Library (AISeL)*. Quantifying The Impact Of Location Data For Real Estate Appraisal – A Gis-Based Deep Learning Approach Quantifying The Impact Of Location Data For Real Estate Appraisal – A Gis-Based Deep.

Kumar, M., Ali Khan, S., Bhatia, A., Sharma, V., & Jain, P. (2023). *A Conceptual introduction of Machine Learning Algorithms*. 2023 1st International Conference on Intelligent Computing and Research Trends (ICRT), Roorkee, India. 10.1109/ICRT57042.2023.10146676

Kumar, T. S., Jothilakshmi, S., James, B. C., Prakash, M., Arulkumar, N., & Rekha, C. (2023). HHO-based vector quantization technique for biomedical image compression in cloud computing. *International Journal of Image and Graphics*, *23*(03), 2240008. doi:10.1142/S0219467822400083

Kuner, C., Bygrave, L., Docksey, C., Tosoni, L., & Drechsler, L. (2021, May 10). *The EU General Data Protection Regulation: A Commentary/Update of Selected Articles*. Oxford University Press.

Lampa, S., Dahlö, M., Alvarsson, J., & Spjuth, O. (2019). Scipipe: A workflow library for agile development of complex and dynamic bioinformatics pipelines. *GigaScience*, *8*(5). doi:10.1093/gigascience/giz044 PMID:31029061

Lanka, S. (2023). *Agriculture for Developing Nations*. Springer. doi:10.1007/978-981-19-8738-0

Lary, D. J., Alavi, A. H., Gandomi, A. H., & Walker, A. L. (2016). Machine learning in geosciences and remote sensing. *Geoscience Frontiers*, *7*(1), 3–10. doi:10.1016/j.gsf.2015.07.003

Laube, S. T. (2023). Advances in Geographic Information Science. In *Advances in Spatial Data Handling* (pp. 120–160). Springer.

Lawhead, J. (2015). *Learning geospatial analysis with Python*. Packt Publishing Ltd.

Lawhead, J. (2023). *Learning Geospatial Analysis with Python*.

Lee, T., Singh, V. P., & Cho, K. H. (2021). *Tensorflow and Keras Programming for Deep Learning*. doi:10.1007/978-3-030-64777-3_11

Lee, W.W., Zankl, W.D., & Chang, H.K. (2016). *An Ethical Approach to Data Privacy Protection*.

Lee, J. G., & Kang, M. (2015). Geospatial big data: Challenges and opportunities. *Big Data Research*, *2*(2), 74–81. doi:10.1016/j.bdr.2015.01.003

Lee, P. D. (2023). *Machine Learning Techniques Applied to Geospatial Big Data*. MDPI.

Lemenkova, P., & Debeir, O. (2023). GDAL and PROJ libraries integrated with grass GIS for terrain modelling of the georeferenced raster image. *Technologies, 11*(2), 46. doi:10.3390/technologies11020046

Lenka, R. K., Barik, R. K., Gupta, N., Ali, S. M., Rath, A., & Dubey, H. (2017). Data Analytics.

Li, J., Sun, F., & Chen, Y. (2022). *Research on spatial data management and application*. 2022 2nd International Conference on Electronic Information Engineering and Computer Technology (EIECT), Yan'an, China. 10.1109/EIECT58010.2022.00018

Li, Z. (2020). Geospatial big data handling with high performance computing: Current approaches and future directions. *High Performance Computing for Geospatial Applications*, 53-76.

Li, Z. (n.d.). *Geospatial Big Data Handling with High Performance Computing : Current Approaches and Future Directions.*

Li, Z., Tang, W., Huang, Q., Shook, E., & Guan, Q. (2020). *Introduction to Big Data Computing for Geospatial Applications.*

Li, D., & Lu, M. (2018). Integrating geometric models, site images and GIS based on Google Earth and Keyhole Markup Language. *Automation in Construction, 89*, 317–331. doi:10.1016/j.autcon.2018.02.002

Li, M., Perrier, E., & Xu, C. (2019). Deep hierarchical graph convolution for election prediction from geospatial census data. *33rd AAAI Conference on Artificial Intelligence, AAAI 2019, 31st Innovative Applications of Artificial Intelligence Conference, IAAI 2019 and the 9th AAAI Symposium on Educational Advances in Artificial Intelligence, EAAI 2019.* IEEE. 10.1609/aaai.v33i01.3301647

Liu, X., Chen, M., Claramunt, C., Batty, M., Kwan, M.-P., Senousi, A. M., Cheng, T., Strobl, J., Cöltekin, A., Wilson, J., Bandrova, T., Konecny, M., Torrens, P. M., Zhang, F., He, L., Wang, J., Ratti, C., Kolditz, O., Klippel, A., & Lü, G. (2022). Geographic information science in the era of geospatial big data: A cyberspace perspective. *Innovation (Cambridge (Mass.)), 3*(5), 100279. doi:10.1016/j.xinn.2022.100279 PMID:35967836

Li, W. (2020). GeoAI: Where machine learning and big data converge in GIScience. *Journal of Spatial Information Science, 20*(20). doi:10.5311/JOSIS.2020.20.658

Li, Y., Jiang, Y., Yang, C., Yu, M., Kamal, L., Armstrong, E. M., Huang, T., Moroni, D., & McGibbney, L. J. (2020). Improving search ranking of geospatial data based on deep learning using user behavior data. *Computers & Geosciences, 142*, 104520. doi:10.1016/j.cageo.2020.104520

Li, Z. (2022). *Helen Gurgel, Lei Xu, L. Y. and J. D.* Improving Dengue Forecasts by Using Geospatial Big Data Analysis in Google Earth Engine and the Historical Dengue.

Loukili, Y., Lakhrissi, Y., Elhaj, S., & Ali, B. (2022). Geospatial Big Data Platforms : A Comprehensive Review. *KN - Journal of Cartography and Geographic Information*, *72*(4), 293–308. doi:10.1007/s42489-022-00121-7

Lu, J., Qin, M., & Tong, Y. (2022). Research on Reinforcement Learning algorithms in Computer Vision. *2022 International Symposium on Advances in Informatics, Electronics and Education (ISAIEE)*, Frankfurt, Germany. 10.1109/ISAIEE57420.2022.00057

Lu, Y., Ye, T., & Zheng, J. (2022). Decision Tree Algorithm in Machine Learning. *2022 IEEE International Conference on Advances in Electrical Engineering and Computer Applications (AEECA)*. IEEE. 10.1109/AEECA55500.2022.9918857

Madhugiri, D. (2023). *Python Packages For Geospatial Data Analysis*. ResearchGate.

Majeed, A., & Lee, S. (2020). Anonymization techniques for privacy preserving data publishing: A comprehensive survey. *IEEE Access : Practical Innovations, Open Solutions*, *9*, 8512–8545. doi:10.1109/ACCESS.2020.3045700

Makkar, K., Kumar, P., Poriye, M., & Aggarwal, S. (2022). A Comparative Study of Supervised and Unsupervised Machine Learning Algorithms on Consumer Reviews. *2022 IEEE World Conference on Applied Intelligence and Computing (AIC),* (pp. 598-603). IEEE. 10.1109/AIC55036.2022.9848880

Mallick, J., Kant, Y., & Bharath, B. D. (2013). Estimation of land surface temperature over Delhi using Landsat-7 ETM+. *Photonirvachak (Dehra Dun)*, *36*(2), 131–140.

Manfre, L. A., Hirata, E., Silva, J. B., Shinohara, E. J., Giannotti, M. A., Larocca, A. P. C., & Quintanilha, J. A. (2012). An Analysis of Geospatial Technologies for Risk and Natural Disaster Management. *ISPRS International Journal of Geo-Information*, *1*(2), 166–185. doi:10.3390/ijgi1020166

Manimaran, A., Goundar, S., Chandramohan, D., & Arulkumar, N. (2023). Application Areas, Benefits, and Research Challenges of Converging Blockchain and Machine Learning Techniques. *Integrating Blockchain and Artificial Intelligence for Industry 4.0 Innovations*, 1-16.

Manuel, A., Delgado, J., Parra Jounou, I., Ausín, T., Casacuberta, D., Cruz, M., Guersenzvaig, A., Moyano, C., Rodríguez-Arias, D., Rueda, J., & Puyol, Á. (2023). Ethical assessments and mitigation strategies for biases in AI-systems used during the COVID-19 pandemic. *Big Data & Society*, *10*(1), 10. doi:10.1177/20539517231179199

Ma, Q., Duan, Y., & Yao, Z. (2023). Meta-analysis of the impact of geospatial technologies on learning outcomes. *Education and Information Technologies*, *28*(12), 15739–15764. doi:10.1007/s10639-023-11712-w

Maria, C. (2014). GUDIII: Computational homology. *Computational Geometry*, *47*(3), 373–378.

Martin Werner, Y.-Y. C. (2023). Handbook of Big Geospatial Data. In *Y.-Y. C. Martin Werner, Handbook of Big Geospatial Data* (pp. 60–90). Springer.

McClain, B. P. (2022). *Python for Geospatial Data Analysis*. Theory, Tools, and Practice.

McClain, B. P. (2023). *Advanced Python for Geospatial Data Analysis*. O'Reilly Media.

McInerney, D., & Kempeneers, P. (2015). Open source geospatial tools: Applications in Earth observation. In *Open Source Geospatial Tools*. Applications in Earth Observation., doi:10.1007/978-3-319-01824-9_20

Meeds, E., Hendriks, R., Al Faraby, S., Bruntink, M., & Welling, M. (2015). MLitB: Machine learning in the browser. *PeerJ Computer Science, 2015*(7). doi:10.7717/peerj-cs.11

Meenakshi Kandpal, S. B. (2024). Geospatial Data Analytics using Machine Learning. *International Journal of Geo-information* .

Mei, R., Yan, H., He, Y., Wang, Q., Zhu, S., & Wen, W. (2022). *Considerations on Evaluation of Practical Cloud Data Protection*. CNCERT. doi:10.1007/978-981-19-8285-9_4

Mete, M. O. (2023). *GEOSPATIAL BIG DATA ANALYTICS FOR SUSTAINABLE SMART CITIES*, *XLVIII*(July), 141–146.

Mishra, V., Mukherjee, S., Kumar, R., & Stone, D. A. (2014). *Heat wave exposure in India in current climate and their expected amplification under climate change*.

Mittelstadt, B. D., Allo, P., Taddeo, M., Wachter, S., & Floridi, L. (2016). The ethics of algorithms: Mapping the debate. *Big Data & Society*, *3*(2), 2053951716679679. doi:10.1177/2053951716679679

Mohan, P., Neelakandan, S., Mardani, A., Maurya, S., Arulkumar, N., & Thangaraj, K. (2023). Eagle strategy arithmetic optimisation algorithm with optimal deep convolutional forest based fintech application for hyper-automation. *Enterprise Information Systems*, *17*(10), 2188123. do i:10.1080/17517575.2023.2188123

Mohapatra, M., Srivastava, A. K., Balachandran, S., & Geetha, B. (2017). Inter-annual variation and tendencies of tropical cyclones over the North Indian Ocean. *Mausam (New Delhi)*, *68*(2), 263–272.

Montesinos López, O. A., Montesinos López, A., & Crossa, J. (2022). Fundamentals of Artificial Neural Networks and Deep Learning. In *Multivariate Statistical Machine Learning Methods for Genomic Prediction*. Springer. doi:10.1007/978-3-030-89010-0_10

Muñoz-Díaz, D., & Rodrigo, F. S. (2004). Spatio-temporal patterns of seasonal rainfall in Spain (1912-2000) using cluster and principal component analysis: Relationship to the North Atlantic oscillation. *Annales Geophysicae*, *22*(5), 1471–1489. doi:10.5194/angeo-22-1435-2004

Nanda, A., Kumari, A., Gupta, S., & Bhattacharya, S. (2018). Topological analysis of Indian monsoon rainfall data using Mapper. In *International Conference on Pattern Recognition and Machine Intelligence* (pp. 268-275). Springer, Cham.

Ndossi, M. I., & Avdan, U. (2016). Application of open source coding technologies in the production of Land Surface Temperature (LST) maps from Landsat: A PyQGIS plugin. *Remote Sensing (Basel)*, *8*(5), 413. doi:10.3390/rs8050413

Nguyen, A. Q., Nguyen, H. T., Tran, V. C., Pham, H. X., & Pestana, J. (2021). A Visual Real-time Fire Detection using Single Shot MultiBox Detector for UAV-based Fire Surveillance. In *ICCE 2020 - 2020 IEEE 8th International Conference on Communications and Electronics* (pp. 338–343). IEEE. 10.1109/ICCE48956.2021.9352080

Nowotarski, C. J., & Jensen, C. (2019). Topology and topological data analysis for Big Data: Observations and a conceptual treatise. *Big Data and Cognitive Computing*, *3*(3), 33.

Of, S. A. (2020). Python libraries matplotlib, seaborn and pandas for visualization geospatial datasets generated by QGIS Polina LEMENKOVA 1, *LXIV*(1).

Olasz, A., Nguyen Thai, B., & Kristóf, D.A. Olasz a. (2016). A NEW INITIATIVE FOR TILING, STITCHING AND PROCESSING GEOSPATIAL BIG DATA IN DISTRIBUTED COMPUTING ENVIRONMENTS A. *ISPRS Annals of the Photogrammetry, Remote Sensing and Spatial Information Sciences*, *III*(July), 12–19. doi:10.5194/isprsannals-III-4-111-2016

Oluoch, I. (2024). Crossing Boundaries: The Ethics of AI and Geographic Information Technologies. *ISPRS International Journal of Geo-Information*, *13*(3), 87. doi:10.3390/ijgi13030087

Onsrud, H. J., Johnson, J. P., & Lopez, X. (1994). Protecting personal privacy in using geographic information systems. *Photogrammetric Engineering and Remote Sensing*, *60*(9), 1083–1095.

Ouchra, H., Belangour, A., & Erraissi, A. (2023). An overview of GeoSpatial Artificial Intelligence technologies for city planning and development. *2023 Fifth International Conference on Electrical, Computer and Communication Technologies (ICECCT)*, Erode, India. 10.1109/ICECCT56650.2023.10179796

Ouchra, H., Belangour, A., & Erraissi, A. (2023). Machine Learning Algorithms for Satellite Image Classification Using Google Earth Engine and Landsat Satellite Data: Morocco Case Study. *IEEE Access : Practical Innovations, Open Solutions*, *11*, 71127–71142. doi:10.1109/ACCESS.2023.3293828

Oviedo, F., Ferres, J. L., Buonassisi, T., & Butler, K. T. (2022). Interpretable and Explainable Machine Learning for Materials Science and Chemistry. *Accounts of Materials Research*, *3*(6), 597–607. doi:10.1021/accountsmr.1c00244

Ozmen, E. (2021). Event density visualization with python folium library for public complaint management: An application with 311 calls. *Research & Reviews in Science and Mathematics-I.*

Padmanabhan, A., Ziao, X., Vandewalle, R. C., Baig, F., Michel, A., Li, Z., & Wang, S. (2021). CyberGIS-compute for enabling computationally intensive geospatial research. *Proceedings of the 3rd ACM SIGSPATIAL International Workshop on APIs and Libraries for Geospatial Data Science, SpatialAPI 2021 - In Conjunction with the 29th International Conference on Advances in Geographic Information Systems, ACM SIGSPATIAL 2021*. ACM. 10.1145/3486189.3490017

Paluszek, M., Thomas, S., & Ham, E. (2022). MATLAB Machine Learning Toolboxes. In Practical MATLAB Deep Learning. doi:10.1007/978-1-4842-7912-0_2

Pang, B., Nijkamp, E., & Wu, Y. N. (2020). Deep Learning With TensorFlow: A Review. In Journal of Educational and Behavioral Statistics, 45(2). doi:10.3102/1076998619872761

Patroumpas, K., Giannopoulos, G., & Athanasiou, S. (2014). Towards geospatial semantic data management: strengths, weaknesses, and challenges ahead. In *Proceedings of the 22nd ACM SIGSPATIAL International Conference on Advances in Geographic Information Systems* (pp. 301-310). ACM. 10.1145/2666310.2666410

Pebesma, E. (2024). *Applied Geospatial data Science with python.*

Pei, T., Xu, J., Liu, Y., Huang, X., Zhang, L., Dong, W., Qin, C., Song, C., Gong, J., & Zhou, C. (2021). GIScience and remote sensing in natural resource and environmental research: Status quo and future perspectives. *Geography and Sustainability*, *2*(3), 207–215. doi:10.1016/j.geosus.2021.08.004

Pereira, G. W., Valente, D. S. M., de Queiroz, D. M., Coelho, A. L. de F., Costa, M. M., & Grift, T. (2022). Smart-Map: An Open-Source QGIS Plugin for Digital Mapping Using Machine Learning Techniques and Ordinary Kriging. *Agronomy (Basel)*, *12*(6), 1350. doi:10.3390/agronomy12061350

Pierdicca, R. (2021). *GeoAI:artificial intelligence for interpretation and processing of complex geomatic data*. EGU General Assembly.

Pinte, D., Jones, E., Kern, R., & Berkes, P. (2012). Python for geophysical data processing and visualization. *74th EAGE Conference and Exhibition Incorporating SPE EUROPEC 2012*. IEEE. 10.3997/2214-4609.20149884

Pohanka, T., Pechanec, V., & Hejlova, V. (2016). Python Web Server for Sensor Data Visualization. *International Multidisciplinary Scientific GeoConference SGEM*, *1*, 803–810. doi:10.5593/SGEM2016/B21/S08.101

Ponnusamy, J. & Venugopal, K. (2003). GIS Data Dissemination and Intellectual Property Rights. *VII, 336.*

Portele, C. (2022). Encoding of Geographic Information. In *Springer Handbook of Geographic Information* (pp. 95–106). Springer International Publishing. doi:10.1007/978-3-030-53125-6_4

Poux, F., & Billen, R. (2019). A smart point cloud infrastructure for intelligent environments. In *Laser Scanning* (pp. 127–149). CRC Press. doi:10.1201/9781351018869-9

Priyashani, N., Kankanamge, N., & Yigitcanlar, T. (2023). *Multisource Open Geospatial Big Data Fusion : Application of the Method to Demarcate Urban Agglomeration Footprints.*

Psaila, G., & Fosci, P. (2021). J-CO: A platform-independent framework for managing geo-referenced JSON data sets. *Electronics (Basel)*, *10*(5), 621. doi:10.3390/electronics10050621

Rai, P. K., Mishra, V. N., & Singh, P. (Eds.). (2022). *Geospatial technology for landscape and environmental management: sustainable assessment and planning*. Springer. doi:10.1007/978-981-16-7373-3

Rajamani, S. K., & Iyer, R. S. (2023). Use of Python Modules in Ecological Research. In *Perspectives on the Transition Toward Green and Climate Neutral Economies in Asia* (pp. 182–206). IGI Global. doi:10.4018/978-1-6684-8613-9.ch011

Rajeevan, M., Bhate, J., Kale, J. D., & Lal, B. (2010). High resolution daily gridded rainfall data for the Indian region: Analysis of break and active monsoon spells. *Current Science*, 296–306.

Ramalho, L. (2019). *Fluent Phyton, Advance Praise for Head First Python*. Orielly.

Ratna, S. B., Behera, S., Ratnam, J. V., Takahashi, K., & Yamagata, T. (2016). An index for tropical temperate troughs over southern Asia. *Climate Dynamics*, *46*(9-10), 3173–3192.

Rauch, S. (2022). *Analysing Long Term Spatial Mobility Patterns of Individuals and Large Groups Using 3D-GIS: A Sport Geographic Approach*. Tijdschrift voor Economische en Sociale Geografie.

Read, J., Reutemann, P., Pfahringer, B., & Holmes, G. (2016). MEKA: A multi-label/multi-target extension to WEKA. *Journal of Machine Learning Research*, 17.

Reddy, E., Cakici, B., & Ballestero, A. (2019). Beyond mystery: Putting algorithmic accountability in context. *Big Data & Society*, *6*(1), 2053951719826856. doi:10.1177/2053951719826856

Reece, A. S., & Hulse, G. K. (2020). Canadian cannabis consumption and patterns of congenital anomalies: An ecological geospatial analysis. *Journal of Addiction Medicine*, *14*(5), e195–e210. doi:10.1097/ADM.0000000000000638 PMID:32187114

Reichstein, M., Camps-Valls, G., Stevens, B., Jung, M., Denzler, J., Carvalhais, N., & Prabhat. (2019). Deep learning and process understanding for data-driven Earth system science. *Nature*, *566*(7743), 195–204. doi:10.1038/s41586-019-0912-1 PMID:30760912

Reus, N. M. De, Kerbusch, P. J. M., & Schadd, M. P. D. (n.d.). *Geospatial analysis for Machine Learning in Tactical Decision Support*, 1–18.

Rey, S. J., Anselin, L., Li, X., Pahle, R., Laura, J., Li, W., & Koschinsky, J. (2015). Open geospatial analytics with PySAL. *ISPRS International Journal of Geo-Information*, *4*(2), 815–836. doi:10.3390/ijgi4020815

Ricker, B., Schuurman, N., & Kessler, F. (2015). Implications of smartphone usage on privacy and spatial cognition: Academic literature and public perceptions. *GeoJournal*, *80*(5), 637–652. doi:10.1007/s10708-014-9568-4

Roscher, R., Bohn, B., Duarte, M. F., & Garcke, J. (2020). Explainable Machine Learning for Scientific Insights and Discoveries. *IEEE Access : Practical Innovations, Open Solutions*, *8*, 42200–42216. doi:10.1109/ACCESS.2020.2976199

Roy, A., Fouché, E., Morales, R. R., & Möhler, G. (2019). *In-Database Geospatial Analytics using Python.*

Roy, S., Saha, M., Dhar, B., Pandit, S., & Nasrin, R. (2021). Geospatial analysis of COVID-19 lockdown effects on air quality in the South and Southeast Asian region. *The Science of the Total Environment, 756*, 144009. doi:10.1016/j.scitotenv.2020.144009 PMID:33250248

Rozemberczki, B., Scherer, P., He, Y., Panagopoulos, G., Riedel, A., Astefanoaei, M., Kiss, O., Beres, F., López, G., Collignon, N., & Sarkar, R. (2021). PyTorch Geometric Temporal: Spatiotemporal Signal Processing with Neural Machine Learning Models. *International Conference on Information and Knowledge Management, Proceedings*. ACM. 10.1145/3459637.3482014

Rubinstein, I., & Hartzog, W. (2016). Anonymization and Risk (August 17, 2015). *91 Washington Law Review.*

Saibi, H., Bersi, M., Mia, M. B., Saadi, N. M., Al Bloushi, K. M. S., & Avakian, R. W. (2018). Applications of Remote Sensing in Geoscience. In Recent Advances and Applications in Remote Sensing. IEEE. doi:10.5772/intechopen.75995

Sami, K. (2009). *Georeferencing System for Abu Dhabi Spatial Data.*

Santosuosso, U., Cini, A., & Papini, A. (2020). Tracing outliers in the dataset of Drosophila suzukii records with the Isolation Forest method. *Journal of Big Data, 7*(1), 14. doi:10.1186/s40537-020-00288-8

Saputra, H. Y., & Radam, I. F. (2023). International Journal of Transportation Accessibility model of BRT stop locations using Geographically Weighted regression (GWR): A case study in Banjarmasin, Indonesia. *International Journal of Transportation Science and Technology, 12*(3), 779–792. doi:10.1016/j.ijtst.2022.07.002

Sarang, P. (2023). Centroid-Based Clustering. In *Thinking Data Science. The Springer Series in Applied Machine Learning*. Springer.

Sarkar, T. (2022). How to use python data science packages more productively. In *Productive and Efficient Data Science with Python: With Modularizing, Memory profiles, and Parallel/GPU Processing* (pp. 47–84). Apress. doi:10.1007/978-1-4842-8121-5_3

Schafer, A. (2023). *Geospatial Technology, GeoAI and Spatial Data Science applied to Engineering.*

Shabanpour, N., Razavi-Termeh, S. V., Sadeghi-Niaraki, A., Choi, S.-M., & Abuhmed, T. (2022). Integration of machine learning algorithms and GIS-based approaches to cutaneous leishmaniasis prevalence risk mapping. *International Journal of Applied Earth Observation and Geoinformation, 112*. doi:10.1016/j.jag.2022.102854

Shaham, S., Rafieian, S., Ding, M., Shirvanimoghaddam, M., & Lin, Z. (2019). *On the Importance of Location Privacy for Users of Location Based Applications*. ArXiv, abs/1911.01633.

Sharma, K., Chakraborty, A., & Nair, A. (2019). Cyclone risk analysis and cyclone adaptation measures for South Asia. In *Regional Climate Hazards and Adaptations in South Asia* (pp. 19–62). Elsevier.

Shokri, R., Stronati, M., Song, C., & Shmatikov, V. (2017, May). Membership inference attacks against machine learning models. In 2017 IEEE symposium on security and privacy (SP) (pp. 3-18). IEEE. doi:10.1109/SP.2017.41

Silma, D. (2024). Applied Geospatial Data Science with Deep learning.

Singh, G., Mémoli, F., & Carlsson, G. E. (2007). Topological methods for the analysis of high dimensional data sets and 3d object recognition. *PBG@ Eurographics*, 2, 091-100.

Soliman, A., & Terstriep, J. (2019). Keras Spatial Extending deep learning frameworks for preprocessing and on-the-fly augmentation of geospatial data. *Proceedings of the 3rd ACM SIGSPATIAL International Workshop on AI for Geographic Knowledge Discovery, GeoAI 2019.* ACM. 10.1145/3356471.3365240

Solymosi, R., Buil-Gil, D., Ceccato, V., Kim, E., & Jansson, U. (2023). Privacy challenges in geodata and open data. *Area*, *55*(4), 456–464. doi:10.1111/area.12888

Srivastava, N., & Saxena, N. (2023). Applications of Artificial Intelligence and Machine Learning in Geospatial Data. *Advances in Geospatial Technologies*, (May), 196–219. doi:10.4018/978-1-6684-7319-1.ch010

Stewart, A. J., Robinson, C., Corley, I. A., & Ferres, J. M. L. (n.d.). *TorchGeo : Deep Learning With Geospatial Data. The 30th International Conference on Advances in Geographic Information Systems (SIGSPATIAL' 22).* Association for Computing Machinery. 10.1145/3557915.3560953

Sui, D., & Goodchild, M. (2011). The convergence of GIS and social media: Challenges for GIScience. *International Journal of Geographical Information Science*, *25*(11), 1737–1748. doi:10.1080/13658816.2011.604636

Sunitha, G., Abouali, A., Galety, M. G., & Sriharsha, A. V. (2023). Dynamic Programming with Python. In *Advanced Applications of Python Data Structures and Algorithms* (pp. 102–122). IGI Global. doi:10.4018/978-1-6684-7100-5.ch005

Sunitha, G., Sriharsha, A. V., Yalgashev, O., & Mamatov, I. (2023). Interactive Visualization with Plotly Express. In *Advanced Applications of Python Data Structures and Algorithms* (pp. 182–206). IGI Global. doi:10.4018/978-1-6684-7100-5.ch009

Tamiminia, H., Salehi, B., Mahdianpari, M., Quackenbush, L., Adeli, S., & Brisco, B. (2020). Google Earth Engine for geo-big data applications : A meta-analysis and systematic review ISPRS Journal of Photogrammetry and Remote Sensing Google Earth Engine for geo-big data applications : A meta-analysis and systematic review. *ISPRS Journal of Photogrammetry and Remote Sensing*, *164*(May), 152–170. doi:10.1016/j.isprsjprs.2020.04.001

Taye, M. M. (2023). Theoretical Understanding of Convolutional Neural Network: Concepts, Architectures, Applications, Future Directions. *Computation (Basel, Switzerland), 11*(3), 52. doi:10.3390/computation11030052

Terstriep, J. (2023). *Keras Spatial.* https://pypi.org/project/keras-spatial/#description

Theobald, D. M. (2001). Topology revisited: Representing spatial relations. *International Journal of Geographical Information Science, 15*(8), 689–705. doi:10.1080/13658810110074519

Toms, S. (2022). *ArcPy and ArcGIS – Geospatial Analysis with Python.* Packt Publishing Ltd.

Toms, S., Crickard, P., & Van Rees, E. (2018). *Mastering geospatial analysis with Python: explore GIS processing and learn to work with GeoDjango, CARTOframes and MapboxGL-Jupyter.* Packt Publishing Ltd.

Ueckermann, M. P., Bieszczad, J., Entekhabi, D., Shapiro, M. L., Callendar, D. R., Sullivan, D., & Milloy, J. (2020). PODPAC: Open-source Python software for enabling harmonized, plug-and-play processing of disparate earth observation data sets and seamless transition onto the serverless cloud by earth scientists. *Earth Science Informatics, 13*(4), 1507–1521. doi:10.1007/s12145-020-00506-0

Urquhart, N. (2022). Linking to Real-world Data Sources and Services. In *Nature Inspired Optimisation for Delivery Problems: From Theory to the Real World* (pp. 191–207). Springer International Publishing. doi:10.1007/978-3-030-98108-2_9

Usama, M., Qadir, J., Raza, A., Arif, H., Yau, K. A., Elkhatib, Y., Hussain, A., & Al-Fuqaha, A. (2019). Unsupervised Machine Learning for Networking: Techniques, Applications and Research Challenges. *IEEE Access : Practical Innovations, Open Solutions, 7*, 65579–65615. doi:10.1109/ACCESS.2019.2916648

Valentini, G., & Masulli, F. (2002). NEURObjects: An object-oriented library for neural network development. In *Neurocomputing, 48,* 1–4. doi:10.1016/S0925-2312(01)00632-4

van der Kruk, E. S., Crompvoets, J., & van Loenen, B. (2018). Privacy issues in an urban mobility project: A case study of the CityBug app. *Applied Geography (Sevenoaks, England).*

Vatsavai, R. R., & Bhaduri, B. (2011). A hybrid classification scheme for mining multisource geospatial data. *GeoInformatica, 15*(1), 29–47. doi:10.1007/s10707-010-0113-4

Verma, S., Gupta, S., Shin, H., Panigrahi, A., Goswami, S., Pardeshi, S., Exe, N., Dutta, U., Joshi, T. R., & Bhojwani, N. (2022). GeoEngine: A Platform for Production-Ready Geospatial Research. *Proceedings of the IEEE Computer Society Conference on Computer Vision and Pattern Recognition, 2022-June*. IEEE. 10.1109/CVPR52688.2022.02073

Voisard, A. B. D. (2024). geospatial data modeling using Deep learning approach. IEEE.

Vopham, T., Hart, J. E., Laden, F., & Chiang, Y. Y. (2018). Emerging trends in geospatial artificial intelligence (geoAI): Potential applications for environmental epidemiology. In Environmental Health: A Global Access Science Source, 17(1). doi:10.1186/s12940-018-0386-x

Waghela, H., Patel, S., Sudesan, P., Raorane, S., & Borgalli, R. (2022). *Land Use Land Cover Classification using Machine Learning. 2022 International Conference on Automation, Computing and Renewable Systems (ICACRS)*, Pudukkottai, India. 10.1109/ICACRS55517.2022.10029176

Wang, Q. (2022). Support Vector Machine Algorithm in Machine Learning. *2022 IEEE International Conference on Artificial Intelligence and Computer Applications (ICAICA)*, Dalian, China. 10.1109/ICAICA54878.2022.9844516

Weber, A. M., Gupta, R., Abdalla, S., Cislaghi, B., Meausoone, V., & Darmstadt, G. L. (2021). Gender-related data missingness, imbalance and bias in global health surveys. *BMJ Global Health*, *6*(11), 6. doi:10.1136/bmjgh-2021-007405 PMID:34836912

Westra, E. (2010). *Python Geospatial Development.*

Westra, E. (20123). Python Geospatial Develoment. In Python Geospatial Develoment (pp. 160-180).

Westra, E. (n.d.). *Python Geospatial Development.*

Westra, E. (2015). *Python Geospatial Analysis Essentials*. Packt Publishing Ltd.

Whigham, P. A., McKay, R. I., & Davis, J. R. (1992). Machine induction of geospatial knowledge. Lecture Notes in Computer Science (Including Subseries Lecture Notes in Artificial Intelligence and Lecture Notes in Bioinformatics), 639 LNCS. doi:10.1007/3-540-55966-3_24

Wu, Q. (2020). geemap: A Python package for interactive mapping with Google Earth Engine. *Journal of Open Source Software*, *5*(51), 2305. doi:10.21105/joss.02305

Wu, Q. (2021a). Interactive mapping and geospatial analysis with Leafmap and Jupyter. *Proceedings of the 3rd ACM SIGSPATIAL International Workshop on APIs and Libraries for Geospatial Data Science, SpatialAPI 2021 - In Conjunction with the 29th International Conference on Advances in Geographic Information Systems, ACM SIGSPATIAL 2021*. ACM. 10.1145/3486189.3490015

Wu, Q. (2021b). Leafmap: A Python package for interactive mapping and geospatial analysis with minimal coding in a Jupyter environment. *Journal of Open Source Software*, *6*(63), 3414. doi:10.21105/joss.03414

Xie, Y., Eftelioglu, E., Ali, R. Y., Tang, X., Li, Y., Doshi, R., & Shekhar, S. (2017). Transdisciplinary foundations of geospatial data science. *ISPRS International Journal of Geo-Information*, *6*(12), 395. doi:10.3390/ijgi6120395

Xu, H., & Zhang, C. (2023). Development and applications of GIS-based spatial analysis in environmental geochemistry in the big data era. *Environmental Geochemistry and Health*, *45*(4), 1079–1090. doi:10.1007/s10653-021-01183-8 PMID:35066745

Yadav, A., Saraswat, S., & Faujdar, N. (2022). Geological Information Extraction from Satellite Imagery Using Machine Learning. *2022 10th International Conference on Reliability, Infocom Technologies and Optimization (Trends and Future Directions), ICRITO 2022*. IEEE. 10.1109/ICRITO56286.2022.9964623

Yadav, P., & Sharma, S. C. (2023). *Q-Learning Based Optimized Localization in WSN*. 2023 6th International Conference on Information Systems and Computer Networks (ISCON), Mathura, India. 10.1109/ISCON57294.2023.10112130

Yang, X., Peng, K., An, L., Huang, P., & Feng, P. (2022). *GMBLOF: A Machine Learning Algorithm of Novelty Detection Based on Local Outlier Factor*. 2022 5th International Conference on Pattern Recognition and Artificial Intelligence (PRAI), Chengdu, China. 10.1109/PRAI55851.2022.9904135

Yan, X., Liu, S., Liu, W., & Dai, Q. (2022). An improved coverage-oriented retrieval algorithm for large-area remote sensing data. *International Journal of Digital Earth*, *15*(1), 606–625. doi:10.1080/17538947.2022.2030816

Zeng, C., Zhang, F., & Luo, M. (2022). A deep neural network-based decision support system for intelligent geospatial data analysis in intelligent agriculture system. *Soft Computing*, *26*(20), 10813–10826. doi:10.1007/s00500-022-07018-7

Zhalehdoost, A., & Taleai, M. (2022). A Review of the Application of Machine Learning and Geospatial Analysis Methods in Air Pollution Prediction. *Pollution*, *8*(3), 904–933. doi:10.22059/POLL.2022.336044.1300

Zhang, G. (2022). *Mitigating Spatial Bias in Volunteered Geographic Information for Spatial Modeling and Prediction.*

Zhang, R., Zhang, J., Wang, Q., & Zhang, H. (2023). *DOIDS : An Intrusion Detection Scheme Based on DBSCAN for.*

Zhang, C., Wang, K., Yue, Y., Qi, X., & Zhang, M. (2023). Assessing Regional Ecosystem Conditions Using Geospatial Techniques—A Review. *Sensors (Basel)*, *23*(8), 4101. doi:10.3390/s23084101 PMID:37112442

Zhang, J., Wang, Y., Li, S., & Shi, S. (2020). An architecture for IoT-enabled smart transportation security system: A geospatial approach. *IEEE Internet of Things Journal*, *8*(8), 6205–6213. doi:10.1109/JIOT.2020.3041386

Zhang, Z., & Sejdić, E. (2019). Radiological images and machine learning: Trends, perspectives, and prospects. In *Computers in Biology and Medicine* (Vol. 108, pp. 354–370). Elsevier Ltd. doi:10.1016/j.compbiomed.2019.02.017

Zhao, W., Wang, M., & Pham, V. T. (2023). *Unmanned Aerial Vehicle and Geospatial Analysis in Smart Irrigation and Crop Monitoring on IoT Platform*. Research Gate.

Zhao, Q., Shi, Y., Liu, Q., & Fränti, P. (2015). A grid-growing clustering algorithm for Geospatial data. *Pattern Recognition Letters*, *53*, 77–84. doi:10.1016/j.patrec.2014.09.017

Zouhar, F., & Senner, I. (2020). Web-based visualization of big geospatial vector data. In Geospatial Technologies for Local and Regional Development: Proceedings of the 22nd AGILE Conference on Geographic Information Science 22 (pp. 59-74). Springer International Publishing. 10.1007/978-3-030-14745-7_4

About the Contributors

Mohammad Gouse Galety has been working in CS since 1999. He is now a Professor in the CS department of Samarkand International University of Technology, Samarkand, Uzbekistan. He holds two Ph. D.s in web mining and image processing of CS and two master's degrees in computers. The first master's degree is in computer applications, and the second is in computer science. His research interests span computer and information science, mainly web mining, IoT, and AI. He (co) authorizes over 50 journal papers and international conference proceedings indexed by Springer and Scopus, four patents, and five books. He is a Fellow of the IEEE and ACM. Since 1999, he has served in different national and international organizations like Sree Vidyanikethan Degree College, India; Emeralds Degree College, Tirupati, India; Brindavan College of Engineering, India; Kuwait Educational Center, Kuwait, Ambo University, Ethiopia; Debre Berhan University, Ethiopia, Lebanese French University, Erbil, Iraq, CUE, Erbil, Iraq. He teaches undergraduate and postgraduate students several courses on computer science and information technology/science engineering.

Arul Kumar Natarajan currently serves as an Assistant Professor in the Department of Computer Science at the Samarkand International University of Technology in Uzbekistan. He earned his Doctor of Philosophy degree from Bharathidasan University, India 2017. Concurrently, he is engaged in postdoctoral research in Generative AI for Cybersecurity at the Singapore Institute of Technology, Singapore. Throughout his 14-year teaching career, Dr. Arul has held esteemed positions at various institutions, including Christ University, Bishop Heber College in India, and Debre Berhan University in Ethiopia. Dr. Arul has made significant contributions to academia, specializing in cybersecurity and artificial intelligence, evidenced by his portfolio of scholarly works. He has authored 32 international publications indexed in Scopus and 03 international publications and has delivered 34 conference presentations. Additionally, he has edited and published two books with IGI Global Publisher, USA. In addition to his academic pursuits, Dr. Arul is a prolific innovator, holding 17 patents in India and 01 granted patent in the United Kingdom across diverse domains, including communication and computer science. He demonstrates proficiency in Networking and Cybersecurity, having completed the CCNA Routing and Switching Exam from CISCO and the Networking Fundamentals exam from Microsoft. He maintains a sincere interest in GenAI for Cybersecurity.

Tesfaye Fufa Gedefa who holds a Bachelor's Degree in Information Technology from Jima University (2012) and a Master of Science in Software Engineering from Adama Science and Technology University, transitioned from lecturing at Debre Berhan University to his current role as an Associate Researcher II at the Space Science and Geospatial Institute. His research journey encompasses diverse fields such as artificial intelligence, network security, cloud computing, big data, image processing, data science, and disaster management. With a keen interest in extending his expertise, Tesfaye plans to delve into new sensor development and calibration, big data analytics, data assimilation, and integration, alongside image fusion techniques for urban climate monitoring, disaster risk reduction, and food security. He is committed to applying space technology across multidisciplinary domains, leveraging Python programming to amplify the impact of his research endeavors. Through international collaborations and a dedication to advancing knowledge, Tesfaye aims to make meaningful contributions at the intersection of technology and societal challenges.

Tsegaye Demsis Lemma is a current Ph.D. candidate in Remote Sensing at the Department of Remote Sensing Research and Development, Entoto Observatory and Research Center, Ethiopian Space Science and Technology Institute (ESSTI), affiliated with Addis Ababa University. He previously served as an associate researcher and satellite control engineer in the satellite research, development, and operation directorate of ESSTI. He is extremely motivated and friendly, passionate about aerospace and remote sensing. His current work deals with Earth Observation for drought and flood forecasting in the Awash Basin, Ethiopia. Mr. Tsegaye has good personal skills in social, individual, and teamwork. He participates in satellite-driven products for disaster risk reduction, urban climate, and agriculture. He is a hard worker who struggles with challenges caused by new ideas. He performs well in a given task and has very good academic achievements. He plans to extend his research work (new sensor development and calibration, Big data, data assimilation, and integration(optical and SAR) and image fusion for vegetation mapping, urban climate monitoring, change detection and disaster risk reduction, and food security).

***

Rohini A currently serves as an Assistant Professor in the School of Computer Science and Engineering at Presidency University in Bengaluru, bringing over 15 years of comprehensive professional experience to her role. Her extensive teaching background spans multiple engineering institutions, including Sri Venkateswara College of Engineering, Chennai; Vels Institute of Science, Technology & Advanced Studies (VISTAS), Chennai; SRM Easwari Engineering College, Chennai; and Alpha College of Engineering, Bengaluru. In addition to her academic contributions, Ms. Rohini has gained valuable industry exposure as a Technical Trainer at CERPSOFT Solutions Pvt. Ltd., Bengaluru, and as a Development Engineer at Bharat Earth Movers Limited, Bengaluru, under the Government of India. She holds a Bachelor's degree(B.E) in Computer Science and Engineering from Golden Valley Institute of Technology (VTU, Karnataka) and a Master's degree(M.E) in Computer and Communication from Sri Venkateswara College of Engineering (Anna University, Chennai). Currently pursuing her PhD at Vels Institute of Science, Technology & Advanced Studies (VISTAS), Chennai, her research focus lies in Medical Image Processing, particularly in the domains of fMRI, AI, and ML. Ms. Rohini is recognized as a subject matter expert in various subjects such as Compiler Design, Operating Systems, Theory of Computation, Computing and Programming, Data Structures, and C Programming. She actively engages in sharing her expertise as a resource person, conducting technical sessions to provide training to students and faculty across different colleges. Her professional certifications include being an L&T Infotech Certified Trainer, Infosys Certified Trainer in "Data Analytics," and an IBM Certified Database Associate in DB2. Furthermore, she has received training from IIT Madras CSIE. She holds lifetime professional memberships with ISTE and IAENG. As the Chairperson of the IEEE Women chapter at Presidency University, she continues to contribute significantly to the academic and professional development of her community.

Munir Ahmad, Ph.D. in Computer Science, brings over 24 years of invaluable expertise in the realm of spatial data development, management, processing, visualization, and quality assurance. His unwavering commitment to open data, crowdsourced data, volunteered geographic information, and spatial data infrastructure has solidified him as a seasoned professional and a trusted trainer in cutting-edge spatial technologies. With a profound passion for research, Munir has authored more than 30 publications in his field, culminating in the award of his Ph.D. in Computer Science from Preston University Pakistan in 2022. His dedication to propelling the industry forward and sharing his extensive knowledge defines his mission. Connect with Munir to delve into the world of Spatial Data, GIS, and GeoTech. #SpatialData #GIS #GeoTech

Tsion Ayalew Dessalegn is a Senior Lecturer at the Department of Software Engineering, Bule Hora University. She holds her MSc. in Software Engineering from Bule Hora University, Ethiopia. She received her BSc. in Software Engineering from Haramaya University, Ethiopia. Her research interests are Software Engineering, Machine Learning, Artificial Intelligence, and Distributed cloud computing.

Ganesh Davanam received his B.Tech degree in Information Technology from JNT University, Hyderabad in 2006 and M.Tech degree in Computer Science and Engineering from Acharya Nagarjuna University in 2010. He Received his Ph.D from Koneru Lakshmaiah Educational Foundation, Guntur in 2021. During the period 2006-07 he worked as Assistant Professor in Information Technology department at AITS, Rajampet, India. Since 2007, he is Currently working as Associate Professor in CSE Department at Sree Vidyanikethan Engineering College, Tirupati, India. He has 15 years of teaching Experience in Engineering Colleges. He has Published 02 SCI, 10 ESCI and 15 Scopus papers in International journals. He has presented 35 Papers in Various Reputed National and International Conferences. He has submitted 02 patents and published 01 Copyright. He has authored 03 Book Chapters and Organized Various FDP's and International Conferences. He also published 02 Text books in the area of Cognitive Networks and Deep Learning. His current research interests are cognitive Networks, Artificial Intelligence and Machine Learning. He is a member of ISTE, ACM and CSI.

Sakthivel E currently holds the position of Assistant Professor in the School of Computer Science and Engineering at Presidency University in Bengaluru, boasting an impressive 16 years of comprehensive professional experience. His rich teaching background encompasses several esteemed engineering institutions, including Sri Venkateswara College of Engineering, Chennai; Nehru Institute of Engineering and Technology, Coimbatore; and PRIST University, Tanjavur. Beyond his academic endeavors, Mr. Sakthivel has accumulated valuable industry expertise, having served as a Senior Software Developer at NATELSOFT, Chennai, and as a Lead Engineer in Development at FIS Global Business India Pvt. Ltd., Chennai. Presently, he is actively involved in providing technical support for software development and delivering technical training at CERP Software Solutions Pvt. Ltd., Bengaluru—a prominent End-to-End Academic Software Solution provider. Mr. Sakthivel earned his Bachelor's degree (B.Tech) in Information Technology from SCAD College of Engineering and Technology (Anna University, Chennai) and his Master's degree (M.E) in Computer and Engineering from Thiruvalluvar College of Engineering and Technology (Anna University, Chennai). His research focus centers on Network Security, AI, and ML. Acknowledged as a subject matter expert, Mr. Sakthivel specializes in various subjects, including C Programming, JAVA Programming, Networks, and Web Programming. Actively engaged in knowledge dissemination, he serves as a resource person, conducting technical sessions to provide training to industry professionals from various companies, as well as students and faculty across diverse colleges. His professional certifications include recognition as an Infosys certified trainer, Wipro Mission 10x certified Professional, Oracle Certified Professional, and Cisco Certified Network Instructor (Level 1 to Level 4). Demonstrating his commitment to professional development, he holds lifetime memberships with the ISTE.

Nisha F is an accomplished professional with a strong educational background. She obtained her Bachelor's degree in Computer Science and Engineering from Bharathiyar University, Coimbatore, and pursued her Masters in Software Engineering from Sri Ramakrishna Engineering College, Coimbatore. Currently, Nisha holds the position of Assistant Professor in the Computer Science and Engineering Department at CMR University's School of Engineering and Technology, located at the Lakeside Campus in Bengaluru, Karnataka. Nisha's expertise lies in various areas, including web designing, web programming, data science, artificial intelligence, machine learning, and natural language processing. With her profound knowledge and skills in these domains, she actively contributes to the field of computer science education and research. Nisha is also a proud member of the International Association of Engineers (IANG), demonstrating her commitment to professional growth and collaboration within her field.

Shahid Iqbal Rai, an esteemed international scholar from Pakistan, is currently pursuing a Ph.D. in Data Science at the University of Rome Tor Vergata, Italy, specializing in Natural Language Processing (NLP) with a keen focus on Large Language Models (LLM). Armed with a Master of Science in Computer Science, Shahid's prior research honed in on "English to Urdu: optimizing sequence learning in neural machine translation." His academic journey features a significant role as a researcher at COMSATS University Islamabad, where his NLP contributions resulted in a published research article at an "IEEE" international conference. Since 2019, Shahid has been actively contributing to academia as a researcher and lecturer at The University of Lahore. His dedication to the realms of Artificial Intelligence and deep neural networks has positioned him as an innovator, leading to the development of groundbreaking techniques. Notably, he has mentored multiple students in the same domain, exemplifying his commitment to knowledge dissemination. This snapshot of Shahid's academic trajectory underscores his commitment to advancing research in NLP and AI, showcasing a remarkable blend of theoretical expertise and hands-on guidance.

M. Sujithra, M.C.A, M. Phil, Ph.D., is a distinguished academician and researcher currently serving as an Associate Professor in the Department of Computing - Data Science at Coimbatore Institute of Technology, Coimbatore, Tamil Nadu, India. With a career spanning 20 years, Dr. Sujithra has established herself as a prominent figure in the field of Data Science, Machine Learning, Information Security, Cloud Computing, and Network Security. Proficient in conducting innovative research, publishing scholarly articles, and delivering engaging lectures. Skilled in curriculum development, student mentorship, and academic leadership. Committed to pushing the boundaries of knowledge and driving impactful outcomes, I am eager to contribute to pioneering projects, promote interdisciplinary collaboration, and address societal challenges through research and education, fostering an environment that embraces innovation and exploration.

Islom Mamatov is a teaching at University and he is a software engineer. During last years of experience several local and international papers are published.

Froilan Mobo is a Doctor of Public Administration graduate from the Urdaneta City University Class of 2016 and a graduate of the 2nd Doctorate Degree (Ph.D.) in Development Education program at the Central Luzon State University, Nueva Ecija, Philippines, Class of 2022. After obtaining his Doctorate Degree, he was promoted to the position of Associate Professor V in the Philippine Merchant Marine Academy (PMMA), and this allowed him to work with different international research institutions, such as the Director / Research Consultant of the IKSAD Research Institute, Turkey; At present, he is in the process of finishing his 3rd Masters's degree leading to Social Studies Education at Bicol University.

Md Rokibul Hasan is a business analyst with a knack for diving into the exciting world of artificial intelligence (AI) and machine learning (ML). He pursued his master's degree in Managerial Economics at the University of Wroclaw. After that, he jetted off to the USA to grab an MBA with a focus on Business Analytics from Gannon University. He also got some impressive certifications like PMP, CSM. He firmly believes that emerging technologies hold the key to unlocking untapped potential and driving growth like never before.

Thangam S., BE, ME, Ph.D., Dr. Thangam S. currently serves as an Assistant Professor (SG) at the Amrita School of Computing, Bengaluru.,Amrita Vishwa Vidyapeetham, India. Her area of interest in research includes Service Oriented Architecture, Networks, IoT, Artificial Intelligence, Data Structures and Cloud Computing. She completed her Ph. D. in Computer Science and Engineering from Anna University, Chennai. India. She has 26 years of teaching experience. She has published her research works in 25 international journals and conferences. She is a member of ISTE..

Assefa Senbato Genale is a Senior Lecturer at the Department of Computer Science, Bule Hora University. Received his MSc in software engineering from Addis Ababa Science and Technology University, Ethiopia in 2019. He holds his BSc. in Computer Science from Arba Minch University, Ethiopia. He is currently working as Associate Dean of Informatics College. He has more than 10 published research papers and reviewed more than 10 papers at Springer Journal as a reviewer. Completed two university Projects and consulted both undergraduate and postgraduate students on their final project and thesis respectively. His research interests are Big data Analytics, Software Engineering, Artificial Intelligence, Machine Learning, and Deep Learning.

K. G. Suma is working as an Associate Professor in the Department of Computer Science and Engineering, VIT-AP, India. She has received her Ph.D. degree from Anna University Chennai, India in 2017. Her research areas include Image Processing, Medical Imaging, Machine Learning and Soft Computing. She carried several administrative roles in academics. She has won Young Women Engineer from IET in the year of 2020. She has published several Scopus and SCI indexed Journal publications, Book chapters, and attended national and international conferences and won prizes. She has 2 National/ International patents.

Gurram Sunitha is a Professor at Department of AI & ML, School of Computing, Mohan Babu University, Tirupati, A.P., India. She received her Ph.D. degree in Computer Science and Engineering from S.V.University, Tirupati. She has 23 years of experience in academia. Her research interests include Data Mining, Spatio-Temporal Analytics, Machine Learning and Artificial Intelligence. She has 13 patents and 6 books (International & National, Edited & Authored) published to her credit. She has published around 70+ research papers in various International Refereed Journals (SCIE/Scopus/Web of Science) and International Conferences indexed in Springer, IEEEXplore etc. She has been serving as reviewer for several reputed journals; She has been acting as Technical Program/Organizing/ Advisory/ Reviewer Committee Member and has been serving on the program committees and co-chaired various International Conferences across the world.

Chinthapatla Pranay Varna is pursuing his B.Tech. degree in Computer Science and Engineering at Sree Vidyanikethan Engineering College, Tirupati, India. His research interests include Machine Learning, Artificial Intelligence, Deep learning and Predictive analysis. These areas of interest indicate a strong passion for understanding and developing advanced algorithms and models to solve complex problems, make data-driven decisions, and uncover patterns and insights from large datasets

Robin Rohit Vincent holds a Bachelor's degree in Electrical and Electronics Engineering from M.S University (1998), a Master's degree in Computer Science and Engineering from Madurai Kamaraj University (2000), and a Ph.D. in Computer Science from Alagappa University (2003). As of 2021, he's an Associate Professor at Presidency University, Bangalore, specializing in Computer Science, Computer Engineering, and Information Science. With 23 years of experience in higher education, he's taught at five universities in India and abroad, and served as the Head of Educational Technology Centre at the University of Technology and Applied Sciences - Ibra in Oman. Dr. Vincent is also a CISCO certified instructor and an accomplished researcher, guiding Ph.D. students, reviewing journals, publishing in national and international journals, and holding patents. He's a member of various professional organizations including the International Association of Engineers (IANG), the Computer Society of India, All India Association for Educational Research (AIAER), and The Institution of Engineers (AMIE).

Desalegn Aweke Wako is a senior lecturer, and he is currently working as an assistant professor at the College of Informatics, Bule Hora University. His areas of research interests include artificial intelligence, data science, big data analytics, machine learning, and computer vision.

Index

G

I

K

L

M

P

Q

R

S

T

U

V

www.ingramcontent.com/pod-product-compliance
Lightning Source LLC
LaVergne TN
LVHW081344300926
849161LV00008B/2059